Plant Anatomy and Morphology:
Structure, Function and Development

Plant Anatomy and Morphology: Structure, Function and Development

Editor: Luke Fitzgerald

Callisto Reference
www.callistoreference.com

Callisto Reference,
118-35 Queens Blvd., Suite 400,
Forest Hills, NY 11375, USA

Visit us on the World Wide Web at:
www.callistoreference.com

ISBN: 978-1-64116-258-6 (Hardback)

Printed in China.

Cataloging-in-Publication Data

Plant anatomy and morphology : structure, function and development / edited by Luke Fitzgerald.
p. cm.
Includes bibliographical references and index.
ISBN 978-1-64116-258-6
1. Plant anatomy. 2. Plant morphology. 3. Plants--Development. 4. Plants.
5. Botany. I. Fitzgerald, Luke.
QK641 .P53 2020
581.4--dc23

Table of Contents

Preface ... IX

Chapter 1 **Micromorphological and Anatomical Features of Four Species of *Elytrigia* Desv. (Poaceae)** ... 1
Lin Meng and Peichun Mao

Chapter 2 **Morphological Variability of Evergreen Oaks (*Quercus*) in Turkey** ... 11
Yilmaz Aykut, Uslu Emel and Babaç M. Tekin

Chapter 3 **Pollen Morphology of *Verbascum* L. (Scrophulariaceae) in Northern and Central Iraq** ... 20
Muazaz Al-Hadeethy, Athiya Al-Mashhadani, Talib Al-Khesraji, Sahapat Barusrux, Hazim Al-Jewari, Piyada Theerakulpisut and Pimwadee Pornpongrungrueng

Chapter 4 **Leaf Micromorphological Features of the Genus *Cordia* L. (Boraginaceae)** ... 27
Wafaa Amer, Rim Hamdy and Riham Mahdy

Chapter 5 **Palynological Study of some Iranian Species of *Scabiosa* L. (Caprifoliaceae)** ... 40
Ebadi-Nahari Mostafa and Nikzat-Siahkolaee Sedigheh

Chapter 6 **Stomatal and Trichome Diversity in *Senna* Mill** ... 48
Ayesa Begum, Md. Oliur Rahman and Momtaz Begum

Chapter 7 **Foliar Epidermal, Stem and Petiole Anatomy of Meghalayan *Dioscorea* L. (Dioscoreaceae) and its Systematic Implication** ... 57
Nilofer Sheikh and Yogendra Kumar

Chapter 8 **Morphological and Anatomical Studies of the Newly Recorded Genus *Plagiopteron* Griff. (Celastraceae)** ... 68
Vu Tien Chinh, Ritesh Kumar Choudhary, Tran The Bach, Bui Hong Quang, Do Van Hai, Sangmi Eum and Joongku Lee

Chapter 9 **Pollen Morphology and Numerical Analysis of *Tamarix* L. (Tamaricaceae) in Egypt and its Systematic Implication** ... 74
Ahmed Elkordy and Ahmed Faried

Chapter 10 **Pollen Morphology of Iranian *Dracocephalum* L. (Lamiaceae) and its Taxonomic Significance** ... 88
Maryam Naderifar, Ali Sonboli and Abbas Gholipour

Chapter 11 **Achene Micromorphology of Seven Taxa of *Achillea* L. (Asteraceae) from Turkey** **100**
Tulay Aytas Akcin and Adnan Akcin

Chapter 12 **Palynological Features of Eleven *Aethionema* Taxa from Turkey and their Systematic Implications** **107**
Mehmet Cengiz Karaismailoglu

Chapter 13 **Systematic Implications of Achene Characteristics in Genera *Centaurea* L., *Cyanus* Mill., *Psephellus* Cass. and *Rhaponticoides* Vaill. (Asteraceae)** **115**
Mehmet Bona

Chapter 14 **Morphometrics of the Tribe *Merremieae* Austin (Convolvulacae) from India** **127**
S. A. Deshmukh and Vinod B. Shimpale

Chapter 15 **Morphometric Analysis of the Genus *Monochoria* (Pontederiaceae) in Sri Lanka** **135**
Savithri Sarojani Udage and Deepthi Yakandawala

Chapter 16 **Pollen Morphology of the Tribe Phyllodoceae (Ericoideae, Ericaceae) and its Taxonomic Significance** **145**
A. K. M. Golam Sarwar and Hideki Takahashi

Chapter 17 **Pollen Morphology of *Teucrium* L. (Lamiaceae, Ajugoideae)** **154**
Ream I. Marzouk, Salama M. El-Darier and Abdel Baset M. Askar

Chapter 18 **Morphological, Anatomical and Cytological Investigations on Three Taxa of *Centaurea* L. (Asteraceae)** **162**
Neslihan Taşar, Gülden Doğan, Yaşar Kiran, M. Oliur Rahman and Uğur Çakilcioğlu

Chapter 19 **Morphological and Anatomical Studies of the Newly Recorded *Rhus Chinensis* Mill. (Anacardiaceae)** **174**
Özgür Eminagaoglu and Melahat Ozcan

Chapter 20 ***Musa* × *Parahaekkinenii* (Musaceae): A New Artificial Interspecific Hybrid** **182**
Komban Parameswaran Smisha and Mamiyil Sabu

Chapter 21 **Mericarp Morphology of the Tribe Selineae (Apiaceae, Apioideae) and its Taxonomic Implications** **189**
Changyoung Lee, Jinki Kim, Ashwini M. Darshetkar, Ritesh Kumar Choudhary, Sang-Hong Park, Joongku Lee and Sangho Choi

Chapter 22 **Floral Polymorphism in *Polygonum Bistorta* L.** **201**
Ming-Lin Chen, Xiao-Yu Wang and Aubrey L. Funke

Chapter 23 **A Comparative Study of Achene Morphology in Korean Polygonaceae**.......... **212**
Min-Jung Kong, Jun-Ho Song, Beom-Cheol An, Sung-Won Son, Gang-Uk Suh, Mi-Jin Chung and Suk-Pyo Hong

Permissions

List of Contributors

Index

Preface

The main aim of this book is to educate learners and enhance their research focus by presenting diverse topics covering this vast field. This is an advanced book which compiles significant studies by distinguished experts in the area of analysis. This book addresses successive solutions to the challenges arising in the area of application, along with it; the book provides scope for future developments.

Plant anatomy is the study of the internal structure of plants. It often involves sectioning of tissues and microscopy, to study plants at the cellular level. Plant anatomy is divided into structural categories such as root anatomy, stem anatomy, wood anatomy, leaf anatomy, fruit/ seed anatomy and flower anatomy. The study of the external structure and physical form of plants is known as plant morphology. It is useful in the visual identification of plants. Plant morphology studies the reproductive and vegetative structures of plants. It examines the pattern of development along with the process by which structures originate and mature when a plant grows. This book includes some of the vital pieces of work being conducted across the world, on various topics related to plant anatomy and morphology. It strives to provide a fair idea about these disciplines and to help develop a better understanding of the latest advances within these fields. The extensive content of this book provides the readers with a thorough understanding of the subject.

It was a great honour to edit this book, though there were challenges, as it involved a lot of communication and networking between me and the editorial team. However, the end result was this all-inclusive book covering diverse themes in the field.

Finally, it is important to acknowledge the efforts of the contributors for their excellent chapters, through which a wide variety of issues have been addressed. I would also like to thank my colleagues for their valuable feedback during the making of this book.

Editor

MICROMORPHOLOGICAL AND ANATOMICAL FEATURES OF FOUR SPECIES OF *ELYTRIGIA* DESV. (POACEAE)

LIN MENG[1] AND PEICHUN MAO

Beijing Research and Development Center for Grass and Environment, Beijing Academy of Agriculture and Forestry Sciences, Beijing 100097, China

Keywords: Elytrigia Desv.; Micromorphology; Anatomy; SEM.

Abstract

The micromorphological and anatomical characters of *Elytrigia caespitosa* (K. Koch) Nevski, *E. intermedia* (Host) Nevski × *E. elongata* (Host) Nevski, *E. intermedia* (Host) Nevski and *E. repens* (L.) Desv. *ex* Nevski have been studied using Scanning Electron Microscope (SEM) to determine interspecific variation. The results show that the root transverse section consists of epidermis, cortex and stele. Two rings of vascular bundles and a central pith cavity appear in stem morphology. The leaves of *E. caespitosa* have either single or twin, horseshoe-shaped short cells born along the costal zone of the upper epidermis, which lack prickle hairs and contain spherical or oblique-shaped papillae. In *E. intermedia,* the parallel subsidiary cells are distributed on the upper epidermis, and there are no short cells in the leaves. Dome-shaped subsidiary cells appear on the upper epidermis of *E. intermedia* × *E. elongata* and *E. repens*, but *E. intermedia* × *E. elongata* showes spot-shaped papillae, and its bulliform cells sank into the "hinge cells". *E. repens* has no papillae, and its bulliform cells are not sunken into the mesophyll. Therefore, the differences in micromorphological characters on the upper epidermis of the leaf could be useful in classifying and determining phylogenetic relationships among the species.

Introduction

Plant morphological feature is largely controlled by the genes of a species, but it can also be influenced by the environment (Sattler and Rutishauser, 1997; Liu, 2006). Therefore, the micromorphological and anatomical characters of root, stem and leaf can reflect the relationship between habitat and phylogenetics of plants (Liu, 2006). *Elytrigia* Desv. is a perennial rhizomatous grass of the family Poaceae. There are about 50 species of *Elytrigia* throughout the world (Chen and Jia, 2000; Lv *et al.,* 2007), and many of them are ecologically and economically important. *Elytrigia intermedia* is valued for its high quality forage, *E. repens* for stabilizing slopes and sandy soil (Chen and Jia, 2000), and *E. intermedia* and *E. elongata* for breeding distant hybrids of wheat (Lv *et al.,* 2007; Webb and Alrneida, 1990). Several authors have shown that the leaf epidermal morphology of the grass family has taxonomic significance because of the fine morphological structure (Chen *et al.,* 1993; Cai and Guo, 1995). The anatomy of roots, stems and leaves of *E. elongata* and *E. intermedia* were examined and analyzed using Scanning Electron Microscope (SEM) showing that the two species have significant differences in leaf epidermal micromorphology (Shi *et al.,* 2009), e.g. three to four rows of papillae are distributed along the costal zone of *E. elongata* leaves, but three to four rows of prickle hairs are distributed along the costal zone of *E. intermedia*. Therefore, the present study aims to contribute to the micromorphological and anatomical features of the roots, stems and leaves of the four species, *E. caespitosa,*

[1]Corresponding author: E-mail: menglin9599@sina.com

E. intermedia × *E. elongata, E. intermedia* and *E. repens,* using Scanning Electron Microscopy (SEM), and to evaluate the differences of the micromorphological characters for systematic purposes.

Materials and Methods

Seeds of *Elytrigia caespitosa* (K. Koch) Nevski, *E. intermedia* (Host) Nevski × *E. elongata* (Host) Nevski, *E. intermedia* (Host) Nevski and *E. repens* (L.) Desv. *ex* Nevski were collected from the National Plant Germplasm System (NGPS) of USA in 2007; and planted at the experimental sites of the National Experiment Station of Precision Agriculture (NESPA), Xiao Tangshan, about 55 km far from Beijing (Lat. 39°34' N, Long. 116°28' E) in 2008. Twenty seedlings of each species were planted in 80 cm × 80 cm rows. Five healthy individuals for each of the four species were selected at the heading stage in May, 2009. Samples were collected by cutting 5 mm segments from the middle sections of the second functional leaves, mature roots, and between the stem stipe of each plant. There were two repetitions totaling 40 specimens. All specimens were pre-fixed for about 3-4 h in 3% glutaraldehyde, and then fixed for more than 12 h in 1% osmic acid (H_2OSO_4). The stationary liquid was formulated using a pH 7.2 phosphate buffer solution (PBS). All specimens were cleaned ultrasonically 3-5 times using PBS, then dehydrated in a 30%, 50%, 70%, 85%, 95% and 100% alcohol solution for 15 min each step by step, and laid in isoamyl acetate. The CO_2 critical point was obtained using a HCP-2 Critical Point Dryer (Hitachi Co. Ltd., Japan), and coated with gold by IB-5 Vacuum Ion Sputter (Eiko Engineering Co. Ltd., Japan).

Mature leaves, roots and stems are examined with S-570 Scanning Electron Microscope (Hitachi Co. Ltd., Japan), and are analyzed with WD-5 Online Photo Management System for SEM (Analysis and Examination Center, Wuhan University of China). The characteristics of the epidermal micromorphology of the upper leaves, including the length of long cells, the shape, density and distribution of short cells and papillae, the density and distribution of prickle hairs and the stomata cell parameters, and the leaf transverse section parameters, such as leaf thickness and characteristics of bulliform cells are thoroughly examined and analyzed. Root anatomical features, including the number of vascular bundles, diameter of stele, thickness of cortex and diameter of metaxylem, and the stem anatomical features including the density of vascular bundles, thickness of cuticle and diameter of inner cycle vessel, are also examined and analyzed. Each index has 10 data points, measured using Photoshop, and means are calculated. Statistical analyses are conducted using SPSS, 13.0 version. The standard leaf micro-morphological terminology is mainly adopted from Chen *et al.* (1993), Cai and Guo (1995), Cai (2000), Kocsis *et al.* (2004) and Torre (2004), whereas terminology for the morphology of roots, stems and leaf transverse sections are adopted from Liu (2006).

Results and Discussion

Root micromorphological characters:

Generally, three parts including the epidermis, cortex and stele are found in the root transverse section of four *Elytrigia* species. The epidermis is made up of a layer of tightly packed cells with many epidermal hairs (root hairs). Under the epidermis is exodermis with larger 2-3 layered cortex cells, which has thicker cell walls and no intercellular spaces. The endodermis is a single layer of small, tightly packed cells, each of which has five thick sides and are the horse-shoe type with the passage cell. There are 2-3 layers of cells adjoining the endodermis to the thick cell wall and the radialized sequence, and the cell volume becomes larger from inner to outer layers. The stele includes the pericycle, vascular bundles and marrow. The pericycle is a layer of parenchymatous cells adjoining the endodermis. The exarch vascular bundle is polyarch xylem, and the centre of the stele contains the marrow filled with parenchyma (Fig.1). The comparative result of root anatomical features of the four *Elytrigia* species is presented in Table 1.

Table 1. Comparison of root anatomical features of four *Elytrigia* species.

Species	No. of vascular bundles	Diameter of stele (μm)	Thickness of cortex (μm)	Diameter of metaxylem (μm)
E. caespitosa	13	466.57 ± 5.36 a	102.25 ± 2.56 a	36.13 ± 4.41 a
E. intermedia × *E. elongata*	10	442.43 ± 15.81 a	59.52 ± 7.17 c	31.03 ± 6.38 b
E. intermedia	9	456.97 ± 14.09 a	73.76 ± 11.53 b	35.83 ± 2.48 a
E. repens	5	323.69 ± 17.77 b	91.90 ± 3.92 ab	29.54 ± 3.61 b

Different small letters in the same column indicate significant differences at $P<0.05$.

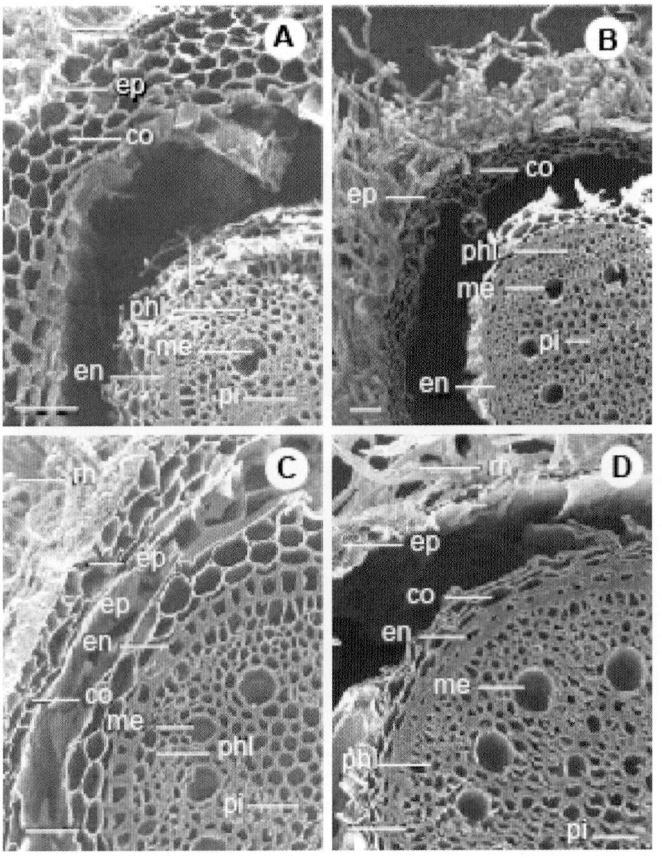

Fig. 1. Root transverse sections of four *Elytrigia* species. A) *E. repens*, B) *E. intermedia*, C) *E. intermedia* × *E. elongata* and D) *E. caespitosa*. rh: root hair; ep: epidenmis; co: cortex; phl: phloem; en: endodermis; me: metaxylem; pi: pith (A, B and D×200, C×150, scale bars = 50 μm).

There are approximately 13 vascular bundles in *E. caespitosa*, which is significantly different than *E. repens* ($P<0.05$), that has only 5 vascular bundles. *E. intermedia* × *E. elongata,* and *E. intermedia* have 10 and 9 vascular bundles, respectively, and do not differ significantly ($P>0.05$). However, the thickness of the cortex of *E. caespitosa* (about 102.25 μm) is 1.72 times than that of

E. intermedia × *E. elongata*. There are no significant differences among stele diameters of *E. caespitosa*, *E. intermedia* × *E. elongata* and *E. intermedia* ($P>0.05$), but diameters are significantly larger than those of *E. repens* ($P<0.05$), which is only 323.69 μm.

Table 2. Comparison of stem anatomical features of four *Elytrigia* species.

Species	Density of vascular bundles (number mm^{-2})	Thickness of cuticle (μm)	Diameter of inner cycle vessel (μm)
E. caespitosa	9 ± 0.21 ab	4.01 ± 0.09 a	34.47 ± 0.80 b
E. intermedia × *E. elongata*	8 ± 0.23 b	4.89 ± 0.14 a	35.08 ± 1.01 b
E. intermedia	8 ± 0.14 b	5.22 ± 0.09 a	51.67 ± 0.89 a
E. repens	10 ± 0.35 a	4.26 ± 0.15 a	31.15 ± 1.08 b

Different small letters in the same column indicate significant differences at $P<0.05$.

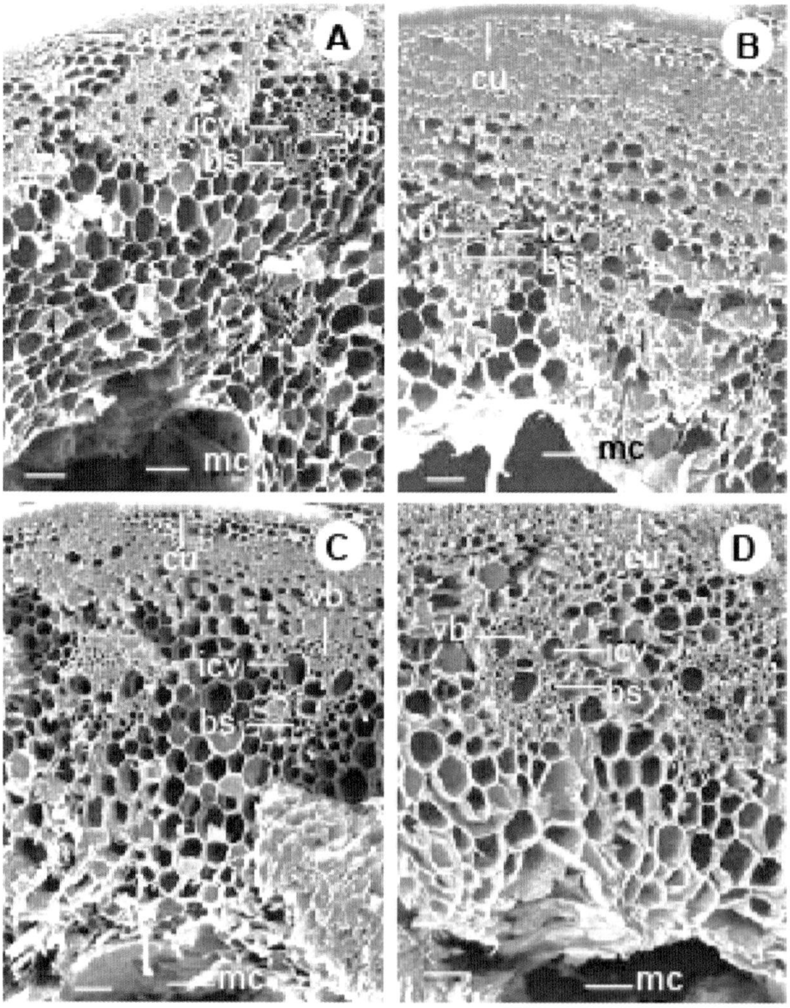

Fig. 2. Stem transverse sections of four *Elytrigia* species. A) *E. repens*, B) *E. intermedia*, C) *E. intermedia* × *E. elongata* and D) *E. caespitosa*. vb: vascular bundle; bs: bundle sheath; icv: inner cycle vessel; cu: cuticle; mc: medullary cavity (×100, scale bars = 50 μm).

Stem micromorphological characters:

Stem transverse sections of the four *Elytrigia* species consist of four parts: the epidermis, ground tissue, vascular bundles and the medullary cavity (Fig. 2). Cuticle thickness in the epidermis is only about 4.01-5.22 μm. Under the epidermis are two to three layers of sclerenchyma tissue (fiber) with thick cell walls, which are composed of parenchymatous cells. The collateral vascular bundle with sheath and no cambium are arranged in two rings. The outer ring contains small vascular bundles, which are embedded under the fiber and form the mechanical tissue, and parenchymatous cells are presented among the vascular bundles. The inner ring is composed of larger vascular bundles, and these are distributed among the parenchyma between the medullary cavity and the outer mechanical tissue (Fig. 2).

The cuticle thickness of four species do not differ significantly ($P>0.05$), but the density of the vascular bundles in *E. repens* differs from *E. intermedia* × *E. elongata* and *E. intermedia* ($P<0.05$). The diameter of the inner cycle vessel in *E. intermedia* is wider than that of *E. caespitosa*, *E. intermedia* × *E. elongata* and *E. repens*, but there are no significant differences in the inner cycle vessel diameters among the latter three species (Table 2).

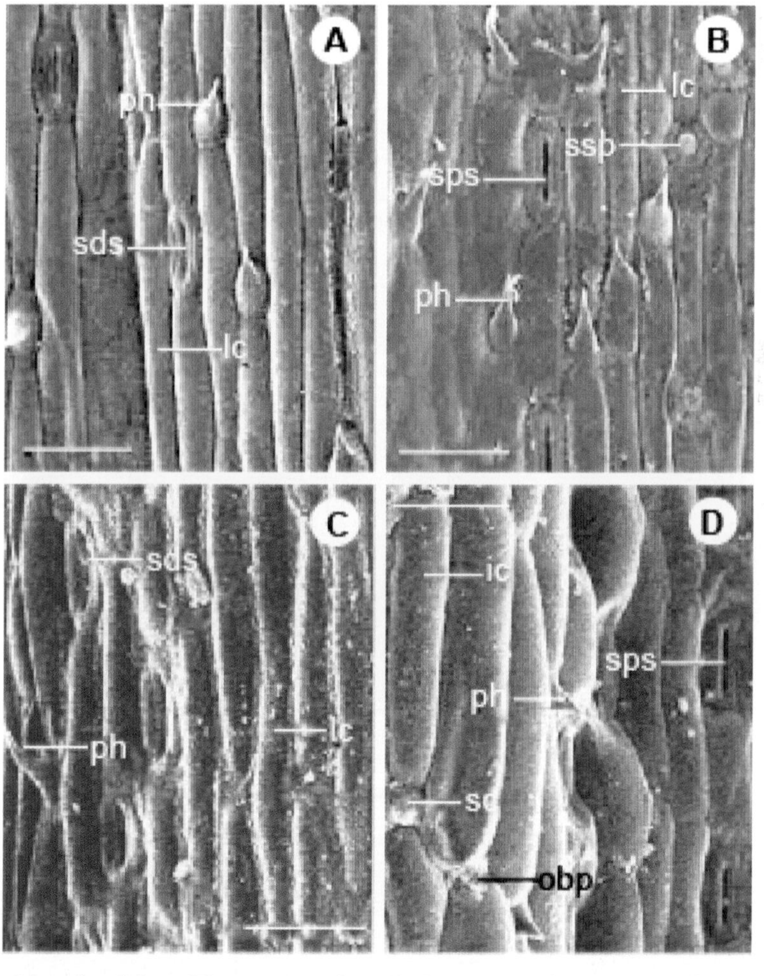

Fig. 3. Upper leaf epidermis of four *Elytrigia* species. A) *E. repens*, B) *E. intermedia*, C) *E. intermedia* × *E. elongata* and D) *E. caespitosa*. lc: long cell; sc: short cell; ph: prickle hair; obp: oblique papillae; ssp: spot-shaped papillae; sds: stomata with dome-shaped subsidiary cell; sps: stomata with parallel subsidiary cell (×200, scale bars = 50 μm).

Table 3. Comparison of upper leaf epidermal micromorphology of four *Elytrigia* species.

Features	Species			
	E. caespitosa	*E. intermedia* × *E. elongata*	*E. intermedia*	*E. repens*
Long cells				
Length (μm)	128.89 ± 2.98 d	183.89 ± 5.31 b	160.00 ± 2.77 c	233.33 ± 8.08 a
Cell wall	Thin	Thick	Relatively thick	Thin
Density (number mm^{-2})	78 ± 41	0^{d}	0	0
Short cells				
Length (μm)	31.11 ± 4.81	0	0	0
Shape and distribution	Saddle-shaped, single or twin born in costal zone, absent in intercostal zone	0	0	0
Stomata cells				
Distribution	Absent in costal zone, 3 rows born in intercostal zone	2 rows born in each sides of the rib in costal zone, separated from prickle hair and absent in intercostal zone	1-2 rows on each side of the rib in costal zone, alternated with prickle hairs and absent in intercostal zone	1-2 rows along each side of the rib in costal zone, absent in intercostal zone
Density (number mm^{-2})	84 ± 1.94 b	96 ± 2.77 a	66±1.14 c	74 ± 2.56 c
Length (μm)	53.33 ± 1.23 ab	49.44 ± 1.43 b	52.78 ± 0.91 ab	57.78 ± 2.00 a
Width (μm)	25.00 ± 0.58 a	16.67 ± 0.48 c	19.44 ± 0.34 b	17.22 ± 0.60 bc
Subsidiary cells	Parallel	Dome-shaped	Parallel	Dome-shaped
Papillae				
Density (number mm^{-2})	132 ± 3.05 a	18 ± 0.52 c	99 ± 1.71 b	0
Shape and distribution	Spherical or oblique papillae distributed along the sides of costal rib alternating with long cells	Spot-shaped papillae scattered in costal zone and absent in intercostal zone	Spot-shaped papillae with 4-6 cluster born in costal zone and absent in intercostal zone	0
Prickle hair				
Density (number mm^{-2})	0	27 ± 0.78 c	130 ± 2.25 a	73 ± 2.53 b
Distribution	0	2-3 rows of prickle hair in costal zone, or 1 row born in each side of rib, and absent in intercostal zone	1-2 rows in each side of rib, and absent in intercostal zone	3-5 rows in costal zone, and absent in intercostal zone

Different small letters in the same row indicate significant differences at $P<0.05$. [d]represents that the indices are absent in this species. 0 indicates absent.

Leaf micromorphological characters:

The micromorphology and anatomy of the leaves in the four *Elytrigia* species comprise three layers: epidermis, mesophyll and leaf veins. The epidermis is composed of long cells, subsidiary cells and stomatal and bulliform cells. Quadrate long cells with thin or thick cell walls are arranged in parallel along the length of the costal and intercostal zones, and many of them show slight wave bending pattern (Fig. 3). The cells of *E. repens* are the longest (233.33 μm), whereas those of *E. caespitosa* are the shortest (128.89 μm). The short cells only appear in the epidermis of *E. caespitosa* at a density of about 78 mm^{-2}. They are saddle shaped, single or twin, and born along the costal zone, but absent in the intercostal zone. Three rows of stomatal cells are present in the intercostal zone of the upper epidermis of *E. caespitosa*, and 1-2 rows appear along each side of the ribs in the costal zone for *E. intermedia* × *E. elongata*, *E. intermedia* and *E. repens*. The density, length and width of the stomatal cells of the four species are expressed differently, and the subsidiary cells of *E. caespitosa* and *E. intermedia* are in parallel, but *E. intermedia* × *E. elongata* and *E. repens* are dome-shaped. The papillae of *E. caespitosa* are spherical or oblique in shape and are distributed along the sides of the costal rib alternating with the long cells, at a density of about 132 mm^{-2}. *E. intermedia* × *E. elongata* and *E. intermedia* show spot-shaped papillae scattered throughout the costal zone. The density of papillae in *E. intermedia* × *E. elongata* is 18 mm^{-2} but papillae are absent in *E. repens*. The density of prickle hairs in *E. intermedia* is 130 mm^{-2}, which is significantly different than that of *E. repens* (73 mm^{-2}) and *E. intermedia* × *E. elongata* (27 mm^{-2}) ($P<0.05$) (Table 3).

The transverse section of leaves consists of epidermis, vascular bundles, mesophyll cells and fiber, i.e., sclerenchyma tissue (Fig. 4). Bulliform cells are present in the intercostal zones of all four species. However, those of *E. intermedia* × *E. elongata* are sunken into the mesophyll and form “hinge cells”, whereas those of the other species do not show this pattern. The ratio of leaf thickness of intercostal zones to thickness of bulliform cells of the four species is approximately 17.8-28.8%, and that of *E. intermedia* × *E. elongata* and *E. repens* are significantly different ($P<0.05$) compared with *E. caespitosa* and *E. intermedia*. The average costal and intercostal thicknesses of *E. caespitosa* and *E. intermedia* × *E. elongata* are greater than those of *E. intermedia* and *E. repens* (Table 4) and the veins are hump shaped. There are two layers of vascular bundles in the veins: smaller, thick cell walls in the inner layer, and larger, thin cell walls in the outer layer. Fiber is distributed along each side of the vascular bundles. The mesophyll tissue is made up of parenchymatous cells with large intercellular spaces, and there is no cellular differentiation between the palisade and spongy tissues (Fig. 4).

Table 4. Comparison of leaf transverse section of four *Elytrigia* species.

Species	Leaf thickness			Bulliform cell		
	Costal zones (μm)	Intercostal zones (μm)	Average (μm)	Characters	Thickness (μm)	Ratio of bulliform cell (%)[d]
E. caespitosa	250.33 ± 5.78 b	166.48 ± 3.84 a	208.41 ± 4.81 a	Not sunken	29.67 ± 0.69 b	17.8 ± 0.41 b
E. intermedia × *E. elongata*	280.04 ± 8.08 a	154.63 ± 4.46 a	217.33 ± 6.27 a	Sunken into mesophyll and form 'hinge cell'	41.04 ± 1.18 a	26.5 ± 0.76 a
E. intermedia	186.74 ± 3.23 c	122.67 ± 2.12 b	154.70 ± 2.68 b	Not sunken	23.63 ± 0.41 c	19.3 ± 0.33 b
E. repens	197.37 ± 6.84 c	100.41 ± 3.48 c	126.89 ± 4.40 c	Not sunken	28.93 ± 1.00 b	28.8 ± 1.00 a

Different small letters in the same column indicate significant differences at $P<0.05$.
[d] means that the ratio of bulliform cell thickness to leaf thickness.

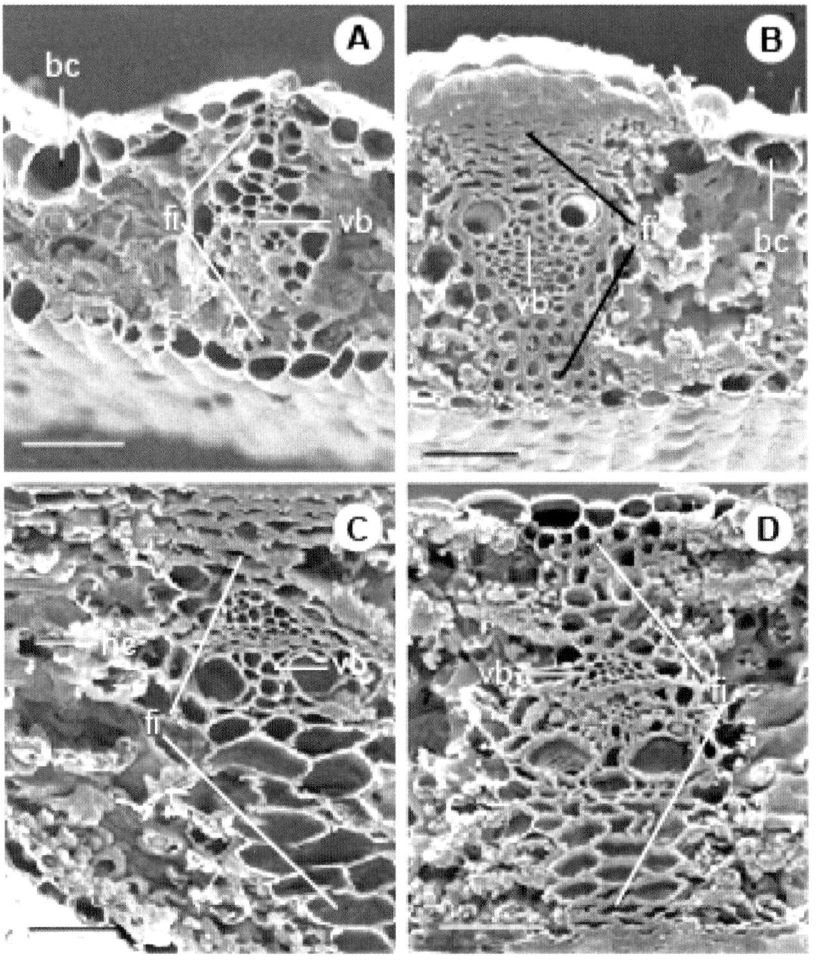

Fig. 4. Leaf transverse sections of four *Elytrigia* species. A) *E. repens*, B) *E. intermedia*, C) *E. intermedia* × *E. elongata* and D) *E. caespitosa*. bc: bulliform cell; hc: hinge cell; fi: fibre; vb: vascular bundle (×300, scale bars = 50 μm).

Leaf micromorphological and anatomical characters are useful tool for plant identification and bears taxonomic significance (Cai and Guo, 1995; Shi *et al.,* 2009; Cai and Zhang, 2006; Kahraman *et al.,* 2010). The present investigation of the leaf anatomy of four *Elytrigia* species agree with those in previous studies of *E. intermedia* and *E. elongata* (Shi *et al.*, 2009).

Based on leaf micromorphological and anatomical characters of four *Elytrigia* species a dichotomous key is presented below:

1.	Single or twin short cells born along the costal zone, no prickle hairs on the upper epidermis.	*E. caespitosa*
–	No short cells, but prickle hairs on the upper epidermis.	2
2.	Subsidiary cells on the upper epidermis are parallel.	*E. intermedia*
–	Subsidiary cells on the upper epidermis are dome-shaped.	3
3.	Long cells thick-walled; papillae spot-shaped; bulliform cells sunken into mesophyll forming "hinge cell".	*E. intermedia* × *E. elongata*
–	Long cells thin-walled; papillae absent; bulliform cells not sunken into mesophyll.	*E. repens*

Bulliform cells are unique characteristics of the leaves of endemic xerophytic graminaceous plants (Cai, 2000). These cells are also known as motor cells, which can cause the leaf blades to curve or spread through the leaf during periods of water loss and absorption. Usually, the bulliform cells appear in groups along the interveins in the upper epidermis of grasses. Wang and Wang (1989) found that the bulliform cells sunk into the mesophyll and formed "hinge cells" on the upper epidermis of *Bouteloua breviseta*, increasing leaf curvature. However, Qiang *et al.* (2007) in *Carex orbicularis* and Guo *et al.* (2007) in *Blysmus sinocompressus* showed that bulliform cells with a little and smaller morphology appeared above the main vein of the leaves. In our study bulliform cells appear in all four *Elytrigia* species, but they are only embedded in the mesophyll into the "hinge cell" in *E. intermedia* × *E. elongata*. The parallel or dome-shaped subsidiary cells in the leaves of grasses exhibit plant adaptation to drought and cold (Cai and Guo, 1995). Therefore, having parallel subsidiary cells in *E. caespitosa* and *E. intermedia* and dome-shaped subsidiary cells in *E. intermedia* × *E. elongata,* and *E. repens* confirm that these species are capable to resist heavy drought and cold.

Acknowledgements

This work was supported by the Natural Science Foundation of China (No. 30571321; No. 31272489), Key Projects in the National Science & Technology Pillar Program in the Eleventh Five-year Plan Period (No. 2008BADB3B05) and the Natural Science Foundation of Beijing (No. 6082009). We are grateful to Mr G. D. Shi for helping in examining the specimens.

References

Cai, L.B. 2000. Leaf epidermical characters of some species of *Roegneria* and their taxonomic significance. Bull. Bot. Res. **20**(4): 372-378.

Cai, L.B. and Guo, Y.K. 1995. Studies on constituent cells of leaf epidermis, systematics and phylogenetic path of the family Poaceae. Acta Botanica Boreali-Occidentalia Sinica **15**(4): 323-335.

Cai, L.B. and Zhang, T.L. 2006. Genetic relationship between *Leymus* and its related taxa in terms of the anatomical characteristics of their leaves. Acta Botanica Boreali-Occidentalia Sinica **26**(3): 537-543.

Chen, M.J. and Jia, S.X. 2000. Forage Plants in China. China Agriculture Press, Beijing, China, pp. 125-128.

Chen, S.L., Jin, Y.X. and Wu, Z.J. 1993. Leaf epidermal characteristic atlas of the family Poaceae. Jiangsu Science and Technology Press, Nanjing, China, pp. 59-63.

Guo, M., Qiang, K.B. and Zhang, X.Q. 2007. Study on morphological anatomy of leaf in *Blysmus sinocompressus*. J. Gansu Agri. Univ. **42**(1): 82-87.

Kahraman, A., Celep, F. and Dogen, M. 2010. Anatomy, trichome morphology and palynology of *Salvia chrysophylla* Stapf (Lamiaceae). S. Afri. J. Bot. **76**(2): 187-195.

Kocsis, M., Darok, J. and Borhid, A. 2004. Comparative leaf anatomy and morphology of some neotropical *Rondeletia* (Rubiaceae) species. Plant Syst. Evol. **248**: 205-218.

Liu, M. 2006. Morphology and Anatomy of Seed Plant (3rd edition). Science Press, Beijing, China, pp.121-268.

Lv, W.D., Xu, P.B. and Pu, X. 2007. Summary of the situation for applying genetic resources from *Elytrigia* in *Triticum aestivum* breeding. Acta Prataculturae Sinica **16**(6): 36-140.

Qiang, K.B., Guo, M. and Zhang, X.Q. 2007. Studies on the morphology and anatomy of leaves of *Carex orbicularis*. Acta Prataculturae Sinica **16**(2): 76-83.

Sattler, R. and Rutishauser, R. 1997. The fundamental relevance of morphology and morphogenesis to plant research. Annals Bot. **80**(5): 571-582.

Shi, G.D., Mao, P.C., Zhang, G.F., Meng, L. and Zhang, D.G 2009. SEM observation on anatomical structure of *Elytrigia elongata* (Host) Nevski and *E. intermedia* (Host) Nevski. Prataculturae Science **26**(8): 52-56.

Torre, S. 2004. Morphology and Anatomy: Leaves. *In:* Roberts, A.V. (Ed.), Encyclopedia of Rose Science. Academic Press, London. Pp. 497-504.

Wang, X.L. and Wang, J. 1989. Plant Morphological Structure and Environment. Lanzhou University Press, Lanzhou, China, pp. 65-148.

Wang, Y.J. 2005. Bulliform cell of grass family. Biology Education **30**(11): 7-9.

Webb, M.E. and Alrneida, M.T. 1990. Micromorphology of the leaf epidermis in the taxa of the *Agropyron - Elymus* complex (Poaceae). Bot. J. Linn. Soc. **103**: 153-158.

2

MORPHOLOGICAL VARIABILITY OF EVERGREEN OAKS (*QUERCUS*) IN TURKEY

YILMAZ AYKUT*, USLU EMEL[1] AND BABAÇ M. TEKIN[1]

Department of Molecular Biology and Genetics, Faculty of Science and Arts, Uşak University, 64200 Uşak, Turkey

Keywords: *Ilex;* Morphometric, UPGMA; Turkey.

Abstract

The genus *Quercus* L. has a problematic taxonomy because of widespread hybridization among them. Evergreen *Quercus* contain three species in section *Ilex* Loudon namely, *Q. ilex* L., *Q. coccifera* L. and *Q. aucheri* Jaub. *et* Spach in Turkey. Here, two species, *Q. coccifera* and *Q. aucheri* are usually confused with each other. However, *Q. coccifera* and *Q. calliprinos* are accepted as different species but this subject is still controversial. Morphometric leaf and fruit variations of *Q. ilex*, *Q. coccifera* and *Q. aucheri* in 26 populations were measured for 25 characters. Variations within and among populations of species were detected by cluster analysis and principal component analysis. This study shows that populations of *Q. coccifera* from the south region of Turkey form a second group within *Q. coccifera*. Secondly, *Q. coccifera* show more similarity to *Q. aucheri* than *Q. ilex*, and finally there are two groups within *Q. coccifera*, which may be evaluated as *Q. coccifera* and *Q. calliprinos*.

Introduction

The genus *Quercus* L., popularly known as oaks shows highest morphological variations among species and populations (Hokanson *et al.*, 1993; Kremer and Petit, 1993), especially its leaf characters are the most valuable in the classification and delimitation of species (Borazan and Babaç, 2003). The major reason for the phenotypic diversification of oaks is the high frequency of hybridization among species (Borazan andBabaç, 2003; Jensen, 1995).

Leaves are good indicators of putative hybridization and oaks can be easily identified by their leaves. The leaves of hybrid species have typically asymmetric shapes and are irregular (Jensen, 1995). Because of common interspecific hybridization in the genus *Quercus*, individuals that exhibit intermediate morphological characters can be seen widely. Sometimes, it is not possible to identify oak species due to high morphological variation. In this case, acorns are secondary important materials in oaks for classification and determination of hybridization (Jensen, 1988).

In Turkey three evergreen *Quercus* species exist, viz. *Q. aucheri, Q. coccifera* and *Q. calliprinos. Q. aucheri* is not very widely distributed, it only exists in the south western Anatolia region of Turkey and in the Greek islands like Rhodos. *Q. coccifera* is confused with *Q. calliprinos* Webb. (Toumı and Lumaret, 2001) because of small or medium shrub formation and acorn shape and evergreen nature. These two species may be evaluated within *Q. coccifera* as *Q. coccifera* subsp. *coccifera* and *Q. coccifera* subsp. *calliprinos* (Webb) Holmboe as two different species. This subject is still controversial (Salvatore and Paola, 1976; Toumı and Lumaret, 2001). *Q. ilex* the last member of *Ilex* section has two morphological types. These are the rotundifolia type containing small and round leaves and the ilex type containing big pointed leaves. The

*Corresponding author: Email:aykutyilmaza@gmail.com
[1]Department of Biology, Faculty of Science and Arts, Abant Izzet Baysal University, Bolu, Turkey.

rotundifolia morphotype exists in North Africa and the interior region of Spain (Tutın *et al.*, 1964). The ilex morphotype exists along the Atlantic coast of France. These two morphotypes are two different species (Tutın *et al.*, 1964) or two subspecies (Saenz De Rıvas, 1967) or only two varieties (Maıre, 1961). Additionally, the presence of intermediate forms for these two morphotypes is reported in the Mediterranean region of France and in the north and east coasts of Spain. The aims of the present study were firstly to examine the morphological relationships among the populations sampled and their potential hybrids in Turkey, secondly to designate the status of confused two species, *Q. coccifera* and *Q. calliprinos*, and finally to evaluate and compare results provided from the leaf and acorn character.

Materials and Methods

The populations sampled are located in the 17 provinces that include the regions of North West, West, South and South West of Turkey. A total of 26 populations belonging to three species of *Ilex* section, namely *Quercus coccifera*, *Q. ilex* and *Q. aucheri* were collected. While 16 populations were designated to reveal variations within *Q. coccifera* (Table 1 and Fig. 1). *Q. aucheri* and *Q. ilex* were sampled in 5 populations owing to their distributions in a restricted region.

Table 1. Study populations with population number, location, coordinates and altitude (C = *Q. coccifera*, A = *Q. aucheri*, I = *Q. ilex*) *Q. coccifera* (C), *Q. aucheri* (A) *and Q. ilex* (I).

Q. coccifera (C), *Q. aucheri* (A) and *Q. ilex* (I)		Coordinates		Altitude (m)
Pop. No.	Location	N	E	
C$_1$	İzmir-Balıkesir border area, Altınova barrage road	39^{0}12.903	026^{0}49.302	70
C$_2$	İzmir-between Dikili-Çandarlı, 20 km. to Çandarlı	39^{0}01.253	026^{0}55.505	40
C$_3$	Manisa-between Kırkağaç-Akhisar, 1-2 km. after Çandarlı	39^{0}05.800	027^{0}40.257	190
C$_4$	Çanakkale-Ezine-Bozcaada pier	39^{0}47.950	026^{0}12.115	50
C$_5$	Gökçeada-between Gökçeada-Dereköy	40^{0}09.689	025^{0}49.586	60
C$_6$	Mersin-5-10 km. after Seratvul	36^{0}50.997	033^{0}18.402	1400
C$_7$	Karaman-between Mut-Ermenek, 45 km. before Ermenek	36^{0}37.276	032^{0}55.182	1300
C$_8$	Antalya-between Korkuteli-Bucak, 25 km. before Bucak	37^{0}15.582	030^{0}19.362	920
C$_9$	Aydın-Eski Çine, Ovacık village	37^{0}32.889	028^{0}05.310	300
C$_{10}$	Aydın-Söke, between Bağarası-Akçakaya village	37°40.350	027°31.347	40
C$_{11}$	Muğla-between Muğla-Kale, 59 km. before Kale	37^{0}08.142	028^{0}32.157	800
C$_{12}$	Denizli- between Kale-Tavas, 1-2 km. before Tavas	37°33.069	029°03.150	940
C$_{13}$	Uşak-between Sivaslı-Uşak, 12 km. after Sivaslı	38°34.259	029°36.303	825
C$_{14}$	Gaziantep- between Yavuzeli-Araban	37^{0}22.975	037^{0}33.292	740
C$_{15}$	Kahramanmaraş- between k.maraş- göksun	37^{0}43.514	036^{0}40.038	1075
C$_{16}$	Hatay-between Kırıkhan-Hassa	36^{0}36.554	036^{0}23.591	350
A$_1$	Antalya-between Kemer-Kumluca	36^{0}25.429	030^{0}25.447	530
A$_2$	Aydın-Çine,Across from the cemetery Kuruköy	37^{0}33.558	028^{0}04.047	180
A$_3$	Aydın-Priene-Söke	37^{0}44.967	029^{0}16.369	90
A$_4$	İzmir-Selçuk-Zeytinköy	37^{0}59.569	027^{0}17.226	65
A$_5$	Muğla-between Milas-Bodrum, Dörttepe village	37^{0}11.242	027^{0}37.142	8
I$_1$	Zonguldak-Alaplı, Sabırlı village	41^{0}08.901	031^{0}23.147	180
I$_2$	Zonguldak-between Alaplı-Düzce	41°08.443	031°20.596	4
I$_3$	Düzce- between Yığılca-Alaplı	41^{0}09.136	031^{0}23.627	60
I$_4$	İstanbul-between Anatolian Fortrees-Kavacık	41^{0}04.220	029^{0}05.085	65
I$_5$	Gökçeada-between Gökçeada-Dereköy	40^{0}09.689	025^{0}49.586	60

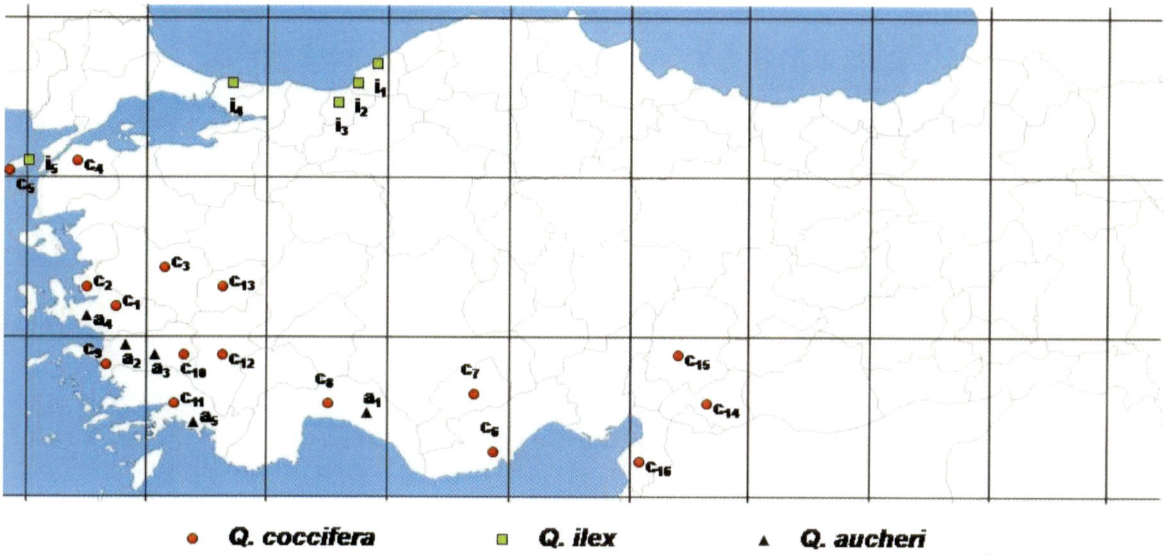

Fig. 1. Distribution of studied populations of *Q. coccifera*, *Q. ilex* and *Q. aucheri* in Turkey.

Leaf and fruit samples for identification and statistical analyses of each population were collected from 260 trees. In total, 2600 leaf and fruit materials were measured.All leaf samples were collected at the same height and location after leaf growth had stopped to avoid seasonal and positional variations as reported by Blue and Jensen (1988). Fruit samples were selected from mature acorns and cupules were also used as fruit characters.

Ten characters of leaves (Table 2 and Fig. 2) ad 15 characters from fruits (Table 3 and Fig. 3) were used. Most of the leaf characters were adopted from different sources (Bruschı *et al.*, 2000; Kremer *et al.*, 2002; Borazan and Babaç, 2003; Bruschı *et al.,* 2003; Gonzalez-Rodrıguez *et al.,* 2004; Ponton *et al.*, 2004; Boratynskı *et al.*, 2008).

Table 2. The leaf characters used in the morphological analysis.

LL	:	Lamina length
PL	:	Petiole length
MWL	:	Maximal width of lamina
MW	:	Middle width of lamina
DTW	:	The distance between the widest point and the leaf tip
DBW	:	The distance between the widest point and the leaf base
TLL	:	Total leaf length (LL+LP)
P%	:	Petiole length (PL) x 100/total leaf length (TLL)
MW%	:	Middle width of lamina (MW) x 100/total leaf length (TLL)
MWL%	:	Maximal width of lamina (MWL) x 100/total leaf length (TLL)

Table 3. The fruit characters used in the morphological analysis.

Nut characters	Cupsule characters	Stalk characters	Index characters
1. Nut length	4. Cupule outer diameter	9. Stalk length	11. Cupule length/Nut length
2. Nut diameter	5. Cupulethickness	10. Stalk thickness	12. Cupule depth/Cupule length
3. Nut scar diameter	6. Cupule scale length (maximum)		13. Nut diameter/Nut length
	7. Cupule depth		14. Cupule thickness/Cupule outer diameter
	8. Cupule length		15. Acorn mass

The fruit characters were also selected from literature (Nikoliç and Orloviç, 2002; Tılkı and Alptekin, 2005). Arithmetic means of all trees were calculated for each character. Then, means of the populations were calculated for each characters. Principal Component Analysis (PCA) and Cluster Analysis (CA) using Statistical version 8.0 were carried out for the analysis of variations in leaf and fruit samples.

Fig. 2. Morphological leaf characters

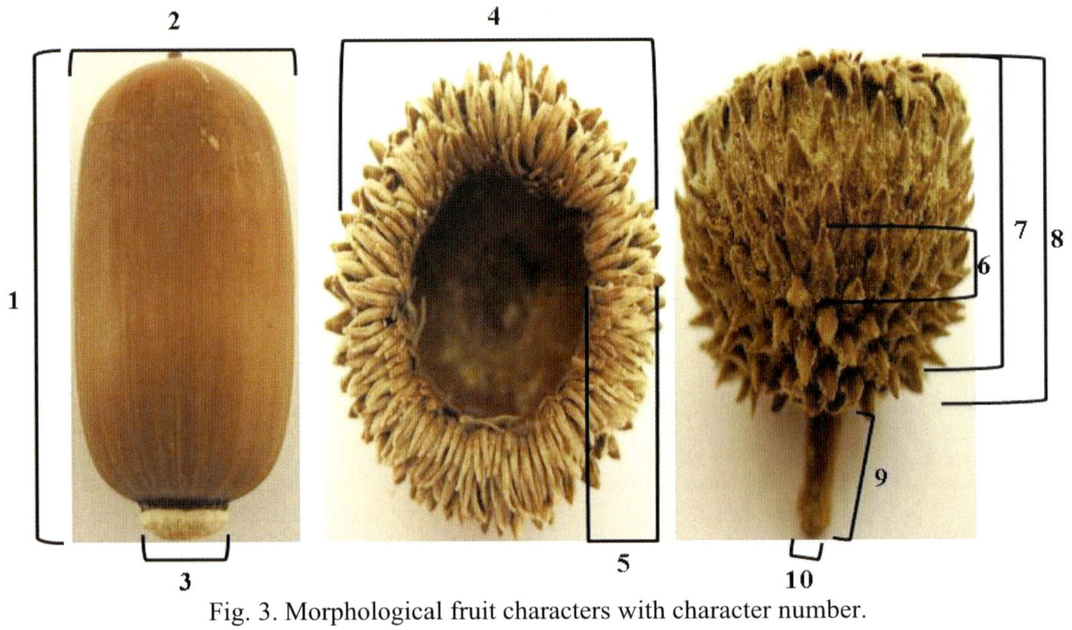

Fig. 3. Morphological fruit characters with character number.

Results

The UPGMA cluster analysis performed on the populations for leaf characters recognized two main groups one contains all populations of *Q. ilex* and another group consisted of 21 populations belonging to *Q. coccifera* and *Q. aucheri*. (Fig. 4) When the first group having the populations of *Q. ilex* is evaluated, it can be stated that populations of *Q. ilex* tend to form more morphologically discrete group than *Q. coccifera* and *Q. aucheri* populations and geographically close populations show more similarity like $İ_1$ and $İ_2$ populations (Fig. 4). Differences in geographical distribution are effective on species diversity. The biggest difference in the populations of *Q. ilex* is observed in $İ_5$ population and the locality of this population is an island in Aegean sea.

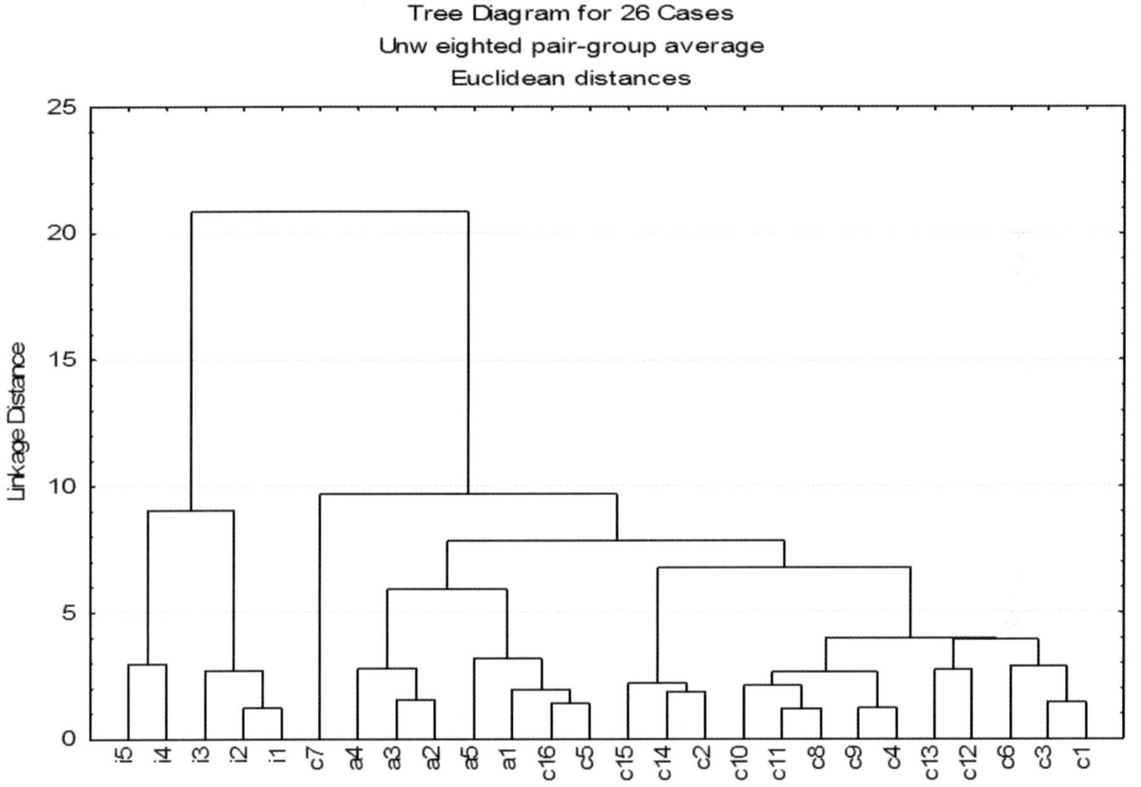

Fig. 4. Phenogram resulting from Cluster Analysis with UPGMA for the leaf materials.

The largest variation in second main group comprising of 21 populations is observed in C_7 population of *Q. coccifera* that occurs at high altitude (Table 1 and Fig. 4). Except C_7 population, the second main group is divided into two sub-groups, one consists of all populations of *Q. aucheri* and only C_{16} and C_5 populations of *Q. coccifera*, while other sub-group consists of complete populations of *Q. coccifera* (Fig. 4). The highest variation within this sub-group are observed in C_{14}, C_{15} and C_2 populations. Populations of *Q. aucheri* show the differences from populations of *Q. coccifera* within the second main group but this difference is not clear as in *Q. ilex*.

PCA results (Fig. 5) show the high similarity with CA (Fig. 4) results. PCA analyses clearly separate the *Q. ilex* populations from others. Similarly, two main groups are observed from PCA analysis. While one of these groups consists of populations belonging to *Q. ilex*, other two species are evaluated in the second group. C_{14}, C_{15} and C_{16} populations show the most differences within *Q. coccifera*. Results revealed that fruit characters, *Q. ilex*is is separated from the other two species (Figs 6 & 7). CA results show clearly the presence of two main groups. The first main group consists of populations of *Q. ilex* and the second main group consists of the populations of *Q. coccifera* and *Q. aucheri*. This result shows the high similarity with the results of leaves.

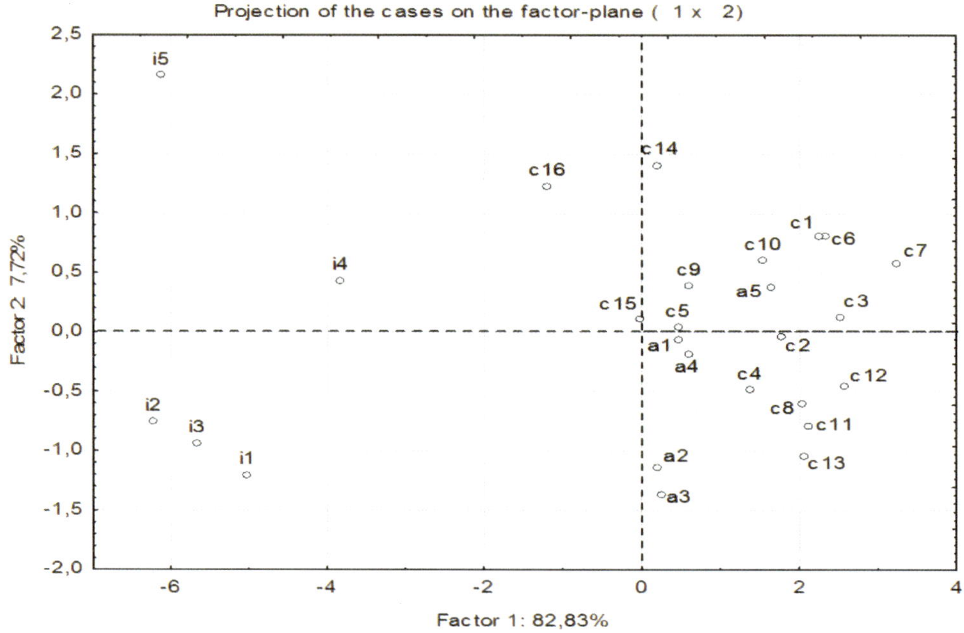

Fig. 5. Resulting projection of Principal Component Analysis for the leaf materials.

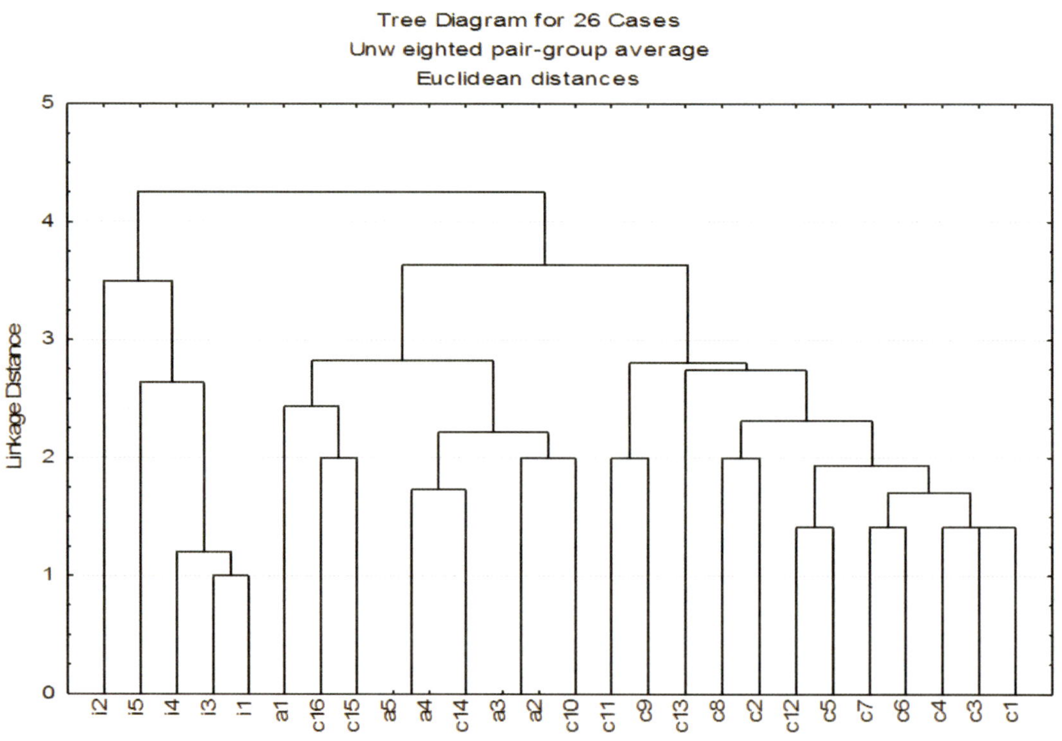

Fig. 6. Phenogram resulting from Cluster Analysis with UPGMA for the fruit materials.

Populations C_{10}, C_{14}, C_{15} and C_{16} form a discrete group with *Q. aucheri* in the CA graph (Fig. 6). However, remaining populations of *Q. coccifera* form other group (Fig. 6). Similar results showing differences among the species are observed in PCA (Fig. 7).

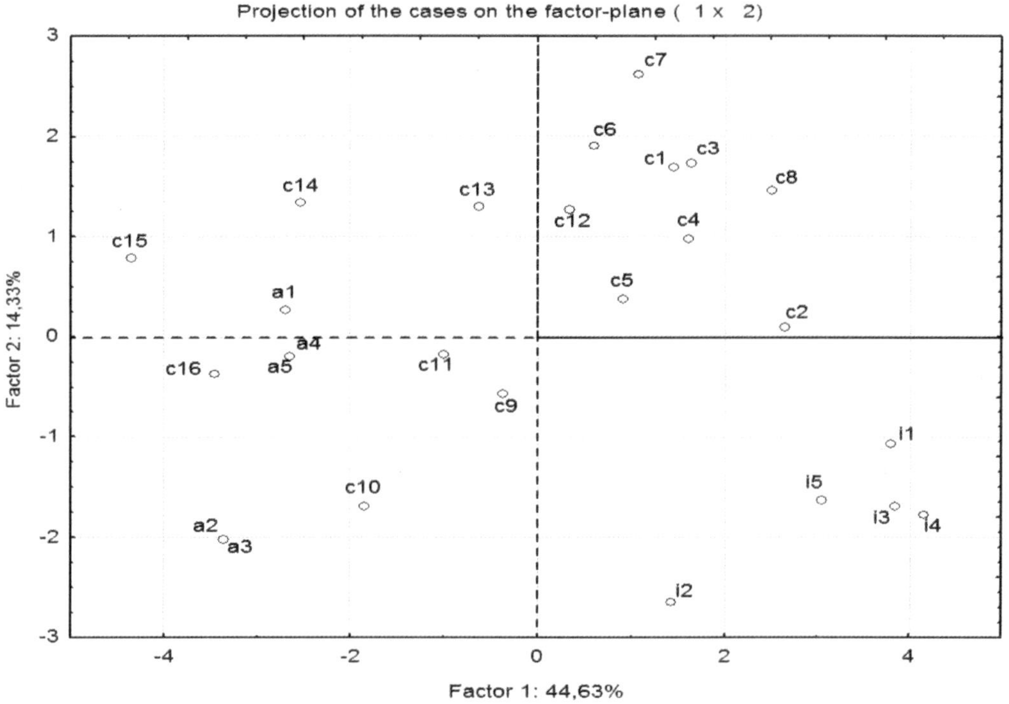

Fig. 7. Resulting projection of Principal Component Analysis for the fruit materials.

Discussion

The present study revealed that PCA and CA analyses could be used to solve taxonomic problems and to understand the relations among three species belonging to *Ilex* section of the genus *Quercus*. Leaf materials were generally used for the comparison of the oaks (Borazan and Babaç, 2003; Bruschı *et al.,* 2003; Ponton *et al.*, 2004; Gonzalez-Rodrıguez and Oyama, 2005; Franjıc *et al.,* 2006; Boratynski *et al.*, 2008) but here, for the first time the fruit materials together with the leaf materials were examined in detail.

The results of morphometric studies provided the satisfactory findings for phenetic groupings of taxa in *Ilex* section. The most significant differences were found on the *Q. ilex* populations. This species was separated from the remaining species on the basis both leaf and fruit. However, *Q. coccifera* populations were grouped next to *Q. aucheri* populations. On the other hand, the results of *Q. coccifera* and *Q. aucheri* are not clearly separated from each other. Especially, due to similar leaf and fruit characters in both taxa, they showed introgression with each other in both CA and PCA plots (Figs 4-7). However, these results draw attention to the presence of a group away from *Q. coccifera*. The first group consists of the populations sampled from North West, West and South West regions of Turkey. The populations sampled from the south region of Turkey such as C_{14}, C_{15} and C_{16} was included into the second group showing the similarity to the populations of *Q. aucheri*. Result from the leaf and fruit studies supported these groupings. Similar results are also observed by Salvatore and Paola (1976), Toumı (1995),ToumıandLumaret (2001)and Yılmaz *et al.(*2013).

Geographically separation of *Q. coccifera* suggests that there are variation within this species. The restricted group of *Q.coccifera* located only in the south region of Turkey is geographically closer to Syria, Israel and Palestine. In Palestine there are two subspecies *Q. calliprinos* Webb. viz. *Q. calliprinos* sub sp. *coccifera* and *Q. calliprinos* sub sp. *calliprinos* (Zohary, 1966).

Our results suggested that the two groups showing geographical differences within *Q. coccifera* may be quite possibly strengthen the existence of two species as *Q. coccifera* and *Q. calliprinos* (Yılmaz *et al.,* 2013).

Acknowledgements

The authors would like to thank Abant İzzet Baysal University Directorate of Scientific Research Projects (BAP) for providing financial support.

References

Boratynskı, A.,Marcysıak, K., Lewandowska, A., Jasınska, A., Iszkulo, G. And Burczyk, J. 2008. Differences in leaf morphology between *Quercus petraea* and *Q. robur* adult and young individuals. Silva Fenn. **42:** 115–124.

Borazan, A. and Babaç, M.T. 2003. Morphometric leaf variation in oaks (*Quercus*) of Bolu,Turkey. Ann. Bot. Fenn. **40:** 233–242.

Bruschı, P.,Vendramın, G.G., Busottı, F. And Grossonı, P. 2000. Morphological and molecular differentiation between *Quercus petraea* and *Quercus pubescens* (Fagaceae) in Northern and Central Italy. Ann. Bot. **85:** 325–333.

Bruschı, P., Vendramın, G.G., Busottı, F. and Grossonı, P. 2003. Morphological and molecular diversity among Italian populations of *Quercus petraea* (Fagaceae). Ann. Bot. **91:** 707–716.

Gonzalez-Rodrıguez, A., Arıas, D.M.,Valencıa, S. and Oyama, K. 2004. Morphological and RAPD analysis of hybridization between *Quercus affinis* and *Q. laurina* (Fagaceae), two Mexican red oaks. Am. J. Bot. **91:** 401–409.

Gonzalez-Rodrıguez, A. and Oyama, K. 2005. Leaf morphometric variation in *Quercus affinis* and *Q. laurina* (Fagaceae), two hybridizing Mexican red oaks. Bot. J. Linn. Soc .**147**: 427–435.

Hokanson, S.C., Isebrands, J.G., Jensen, R.J. and Hancock, J.F. 1993. Isozyme variation in oaks of the Apostle Islands in Wisconsin: Genetic structure and levels of inbreeding in *Quercus rubra* and *Quercus ellipsoidalis* (Fagaceae). Am. J. Bot. **80**: 1349–1357.

Jensen, R.J. 1988. Assesing patterns of morphological variation of *Quercus* spp. in mixed-oak communities. Am. Midl. Nat. **120**: 120–135.

Jensen, R.J. 1995. Using leaf shape to identify taxa in a mixed-oak community in land between the lakes, Kentucky. Proc. Sixth Symposium Nat. Hist. lower Tennessee and Cumberland River Valleys 177-188. Center Field Biol., Austin Peay State Univ., Clarksville.

Kremer, A., Dupouey, J.L., Deans, J.D., Cottrell, J., Csaıkl, U., Fınkeldey, R., Espınel, S., Jensen, J., Kleınschmıt, J., Van Dam, B., Ducousso, A., Forrest, I., De Heredıa, U.L., Lowe, A.J., Tutkova, M., Munro, R.C., Steınhoff, S. and Badeau, V. 2002. Leaf morphological differentiation between *Quercus robur* and *Quercus petraea* in stable across western European mixed oak stands. Ann. Forest Sci. **59**: 777–787.

Kremer, A. and Petıt, R.J. 1993. Gene diversity in natural populations of oak species. Ann. Forest Sci. **50**: 186–202.

Maıre, R. 1961. Flore de l'Afrique du Nord, Vol. 7. Ed Lechevallier, Paris.

Nıkolıc, N.P. and Orlovıc, S.S. 2002. Genotypic variability of morphological characteristics of English oak (*Quercus robur* L.) acorn. Proceeding for Natural Sciences **102**: 53–58.

Ponton, S., Dupouey, J.L. and Dreyer, E. 2004. Leaf morphology as species indicator in seedlings of *Quercus robur* L. and *Q. petraea* (Matt.) Liebl.: modulation by irradiance and growth flush. Ann. Forest Sci .**61**: 73–80.

Saenz De Rıvas, C. 1967. Estudios sobre *Quercus ilex* L., *Quercus rotundifolia* Lamk. Anales Del Instituto Botanico A. J. Cavanilles **25**: 243–262.

Salvatore, G. and Paola, G. 1976. ''*Quercus calliprinos*'' Webb e ''*Quercus coccifera*''L.:Ricerche sull'anatomia fogliare e valutazioni tassonomiche e corologiche. Gıornale Botanico Italliano **110**: 89–115.

Tılkı, F. and Alptekın, C.U. 2005. Variation in acorn characteristics in three provenances of *Quercus aucheri* Jaub. et Spach and provenance, temperature and storage effects on acorn germination. Seed Sci. Technol. **33**: 441–447.

Toumı, L. 1995. Etude de la structure genetique et introgressions eventuelles chez les chenes sclerophylles mediterraneens a l'aide de marquers alloenzymatiques. Ph. D. Thesis, University of Aix-Marsielle III, Marsielle.

Toumı, L. and Lumaret, R. 2001. Allozyme characterization of four Mediteranean evergreen oak species. Biochem. Syst. Ecol. **29**: 799–817.

Tutın, T.G., Heywood, V.H., Burges, N.A., Moore, D.M., Valentıne, D.H., Walters, S.M. and Webb, D.A. 1964. Flora Europaea. Cambridge University Press, London.

Yılmaz, A., Uslu, E. and Babaç M.T. 2013. Molecular diversity among Turkish oaks (*Quercus*) using random amplified polymorphic DNA (RAPD) analyses. Afr. J. Biotechnol.**12**: 6358-6365.

Zohary, M. 1966. Flora Palaestina. Jerusalem Academic Press. Israel.

3

POLLEN MORPHOLOGY OF *VERBASCUM* L. (SCROPHULARIACEAE) IN NORTHERN AND CENTRAL IRAQ

MUAZAZ AL-HADEETHY, ATHIYA AL-MASHHADANI, TALIB AL-KHESRAJI[1], SAHAPAT BARUSRUX[2], HAZIM AL-JEWARI, PIYADA THEERAKULPISUT[3] AND PIMWADEE PORNPONGRUNGRUENG[3,4]

Department of Biology, College of Education-Ibn Al-Haitham, University of Baghdad, Baghdad, Iraq

Keywords: Pollen morphology; SEM; Taxonomy; *Verbascum*; Iraq.

Abstract

The present study highlights the taxonomic values of the pollen features of 20 species of *Verbascum* L. in Iraq. The pollen grains were acetolysed and observed under Light and Electron microscopes. Our results showed that the pollen grains of these species are usually radially symmetrical, isopolar, tricolporate and display reticulate sculpturing. The variations were found in shape, size, and exine thickness. Most of the species are subprolate, prolate spheroidal, and prolate, whereas oblate spheroidal pollens were found only in *V. thapsus*. The pollen size among the species studied ranges from 16.2-32.5 µm. The majority of species are medium sized except *V. agrimoniifolium* and *V. oreophilum* that are small sized, and four species i.e. *V. palmyrense*, *V. sinuatum*, *V. songaricum*, and *V. thapsus* having both small and medium sized pollen grains. Moreover, the result showed that the exine thickness ranges from 0.87-4.75 µm, but this value can overlap between some species. Therefore, only the shape of pollen grains could be used to classify these species into groups.

Introduction

The genus *Verbascum* L. (Scrophulariaceae) is mostly herbaceous with few shrubs and climbers, comprises about 350 species, and it is distributed all over the world (Zohary, 1974). *Verbascum* is widely spread in the Arabian Peninsula, especially in Saudi Arabia, Oman and Yemen (Huber-Morath, 1978; Ghadanfar, 1992; Wood, 1997). In Iraq, *Verbascum* is represented by 26 species (AL-Rawi, 1964; AL-Bermani, 1981).

The taxonomy of *Verbascum* species in Iraq is still uncertain, especially that as related to the geographical distribution of the species, the morphological variation and the condition of herbarium samples (AL-Bermani, 1981). Therefore, study of different characteristics would help in resolving the relationships and delimitation of these species. Pollen morphology is proved to be a useful character for improving the accuracy in the identification and classification of several plant groups (Erdtman, 1952; Rajbhandary *et al.*, 2012; Özler *et al.*, 2013; Sawar and Takahashi, 2013). Despite pollen morphology of *Verbascum* species was studied by several authors (Erdtman, 1952; Filippini *et al.*, 1990; Perveen, 1993; Lobin and Pormbski, 1994; Bukhari and Alfarhan, 2006; Kheiri *et al.*, 2006; Pehlivan *et al.*, 2008), the pollens of *Verbascum* found in Iraq has never been investigated. Therefore, this study was designed to investigate constant and diagnostic pollen morphological characteristics to be used as a tool for improving the accuracy in the identification and classification of *Verbascum,* and further authentication of the taxa.

[1]Department of Biology, College of Education, University of Tikrit, Tikrit, Iraq

[2]Centre for Research and Development of Medical Diagnostic Laboratories (CMDL), Faculty of Associated Medical Sciences, Khon Kaen University, Khon Kaen, Thailand

[3]Applied Taxonomic Research Center, Department of Biology, Faculty of Science, Khon Kaen University, Khon Kaen, Thailand

[4]Corresponding author: Email: ppimwa@kku.ac.th

Materials and Methods

Pollen sample of 20 species of *Verbascum* L. grown in the north and central areas of Iraq, were collected between April 2012 and May 2013 (Table 1). The voucher specimens have been deposited in the herbarium of College of Education, Ibn Al-Haitham, University of Baghdad, Iraq. The pollen was prepared using the acetolysis method as described by Erdtman (1960). The measurements and observations were carried out using both Light microscope (LM) and Scanning Electron microscope (SEM). For LM analysis, the acetolysed pollens were placed in a small vial, and 5-7 drops of silicone oil were added, and then the samples were mounted on a glass slide sealed with paraffin and observed under light microscope. The measurements of polar axis (P), equatorial diameter (E) and exine thickness of pollen were done using 10 reading samples for each specimen. For SEM investigation, the acetolysed pollen was suspended in 100% ethanol. Then the suspension was dried on an aluminum stub, coated with gold and observed using LEO 1450VP Electron Microscopy. The terminology and pollen size classes follow Walker and Doyle (1975) and Punt *et al.* (1994).

Table 1. List of *Verbascum* L. species in Northern and Central of Iraq employed in the present study.

Species	Locations	Date of collections	Collection numbers
1. *Verbascum agrimoniifolium* (C. Koch) Hub. - Mor.	Sulaymaniyah Province	14 Jun 2012	M. AL-Hadeethy 166A
2. *V. alceoides* Boiss. & Hausskn. *ex* Boiss.	Sulaymaniyah Province	6 May 2012	M. AL-Hadeethy 235
3. *V. alepense* Benth.	Nineveh Province	11 May 2013	M. AL-Hadeethy 245
4. *V. andrusii* Post	Nineveh Province	27 Jun 2012	M. AL-Hadeethy 129
5. *V. assurense* (Bornm. & Hand. - Mazz.) Hub. - Mor.	Diyala Province	4 Apr 2012	M. AL-Hadeethy 172
6. *V. calvum* Boiss. & Kotschy *ex* Boiss.	Sulaymaniyah Province	14 Jun 2012	M. AL-Hadeethy 250
7. *V. carduchorum* Bornm.	Sulaymaniyah Province	21 Apr 2012	M. AL-Hadeethy 168
8. *V. cheiranthifolium* Boiss.	Sulaymaniyah Province	4 May 2012	M. AL-Hadeethy 225
9. *V. damascenum* Boiss.	Anbar Province	5 Apr 2013	M. AL-Hadeethy 152
10. *V. geminiflorum* Hochst.	Sulaymaniyah Province	6 May 2012	M. AL-Hadeethy 230
11. *V. laetum* Boiss. & Hausskn. *ex* Boiss.	Sulaymaniyah Province	20 Apr 2012	M. AL-Hadeethy 144
12. *V. macrocarpum* Boiss.	Sulaymaniyah Province	4 May 2012	M. AL-Hadeethy 200A
13. *V. oreophilum* C. Koch	Sulaymaniyah Province	5 May 2012	M. AL-Hadeethy 223
14. *V. palmyrense* Post	Nineveh Province	27 Jun 2012	M. AL-Hadeethy 220
15. *V. pseudodigitalis* Nábělek	Sulaymaniyah Province	20 Apr 2012	M. AL-Hadeethy 123
16. *V. sinaiticum* Benth.	Nineveh Province	27 Jun 2012	M. AL-Hadeethy 112
17. *V. sinuatum* L.	Kirkuk Province	7 Aug 2012	M. AL-Hadeethy 166B
18. *V. songaricum* Schrenk	Sulaymaniyah Province	4 May 2012	M. AL-Hadeethy 200B
19. *V. speciosum* Schrad.	Sulaymaniyah Province	21 Apr 2012	M. AL-Hadeethy 182
20. *V. thapsus* L.	Sulaymaniyah Province	20 Apr 2012	M. AL-Hadeethy 145

Results and Discussion

The results of the pollen morphology of *Verbascum* are summarized in Table 2. It was found that the pollen morphology of the species studied is uniform. The general characteristic is monad, radially symmetrical, isopolar, tricolporate and reticulate in exine sculpturing (Figs 1 & 2).

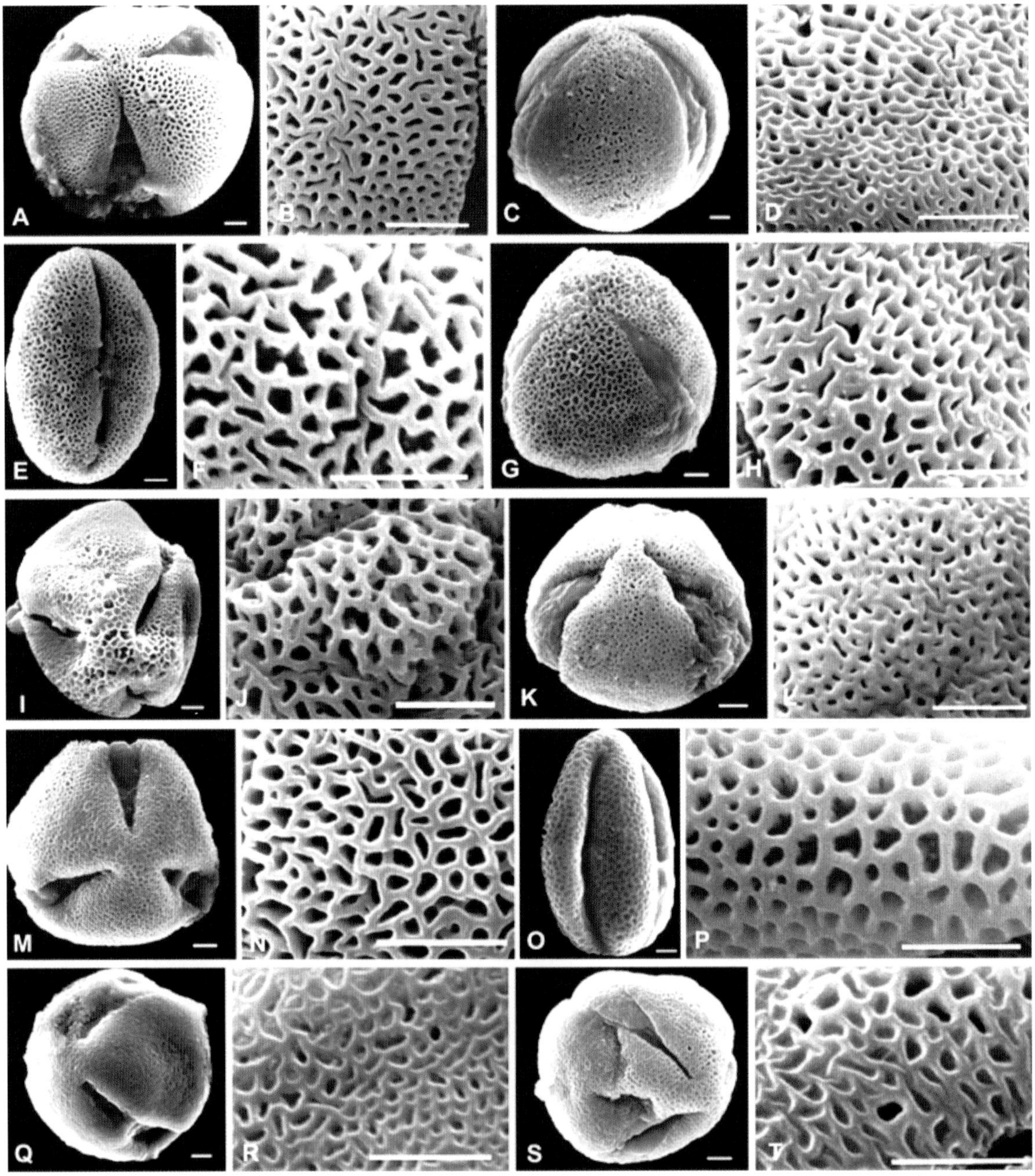

Fig. 1. SEM micrographs of the pollen grains of *Verbascum*. A, B. *V. agrimoniifolium*; C, D. *V. alceoides;* E, F. *V. alepense;* G, H. *V. andrusii*; I, J. *V. assurense;* K, L. *V. calvum;* M, N. *V. carduchorum*; O, P. *V. cheiranthifolium;* Q, R. *V. damascenum;* S, T. *V. geminiflorum*; A, G, I, K, M and Q. polar view; B, D, F, H, J, L, N, P, R and T. exine sculpturing; C, E, O, S. equatorial view (Scale bar = 2 μm).

Table 2. Pollen morphological characteristics of *Verbascum* in Northern and Central Iraq.

Species	Polar axis (µm)	Equatorial axis (µm)	P/E Ratio	Shape	Size	Exine thickness (µm)	Exine sculpturing
1. *V. agrimoniifolium*	16.5-20.5 (17.54±1.39)	15.5-21.0 (17.39±1.65)	1.01	Prolate spheroidal	small	0.87-1.05 (0.93±0.10)	Reticulate
2. *V. alceoides*	27.5-32.5 (28.05±1.78)	22.5-25.0 (24.13±1.13)	1.16	Subprolate	medium	2.75-3.00 (2.65±0.41)	Reticulate
3. *V. alepense*	25.0-27.5 (26.38±1.04)	21.3-22.5 (21.88±0.63)	1.21	Subprolate	medium	3.00-3.72 (3.22±0.39)	Reticulate
4. *V. andrusii*	26.25-27.5 (26.88±0.63)	18.8-20.0 (19.25±0.61)	1.40	Prolate	medium	1.25-2.5 0 (2.15±0.48)	Reticulate
5. *V. assurense*	27.5-32.5 (30.88±1.68)	25.0-27.5 (26.50±1.22)	1.17	Subprolate	medium	3.75-4.25 (4.18±0.44)	Reticulate
6. *V. calvum*	27.5-30.0 (29.00±1.09)	22.5-25.0 (24.13±1.13)	1.20	Subprolate	medium	1.25-2.5 0 (2.08±0.51)	Reticulate
7. *V. carduchorum*	25.0-28.8 (27.13±1.38)	22.5-25.0 (24.00±1.22)	1.13	Prolate spheroidal	medium	3.25-4.00 (3.13±0.76)	Reticulate
8. *V. cheiranthifolium*	26.3-27.5 (26.88±0.63)	20.0-22.5 (21.38±1.18)	1.26	Subprolate	medium	1.25-2.5 0 (2.00±0.58)	Reticulate
9. *V. damascenum*	27.5-30.0 (29.25±1.00)	20.0-22.5 (21.6±0.98)	1.35	Prolate	medium	3.75-4.75 (4.00±0.58)	Reticulate
10. *V. geminiflorum*	26.25-27.5 (26.50±0.93)	20.0-22.5 (21.13±1.04)	1.25	Subprolate	medium	3.00-4.75 (3.58±1.28)	Reticulate
11. *V. laetum*	30.0-32.5 (31.23±0.97)	23.8-25.0 (24.50±0.61)	1.27	Subprolate	medium	2.50-3.75 (3.05±0.46)	Reticulate
12. *V. macrocarpum*	27.5-28.8 (28.13±0.63)	20.0-22.5 (21.75±1.00)	1.29	Subprolate	medium	2.50-3.75 (3.23±0.52)	Reticulate
13. *V. oreophilum*	19.1-22.5 (20.71±0.61)	18.8-20.0 (19.25±0.61)	1.08	Prolate spheroidal	small	1.75-2.50 (1.90±0.53)	Reticulate
14. *V. palmyrense*	22.5-25.0 (24.00±1.09)	20.0-22.5 (21.25±1.12)	1.13	Prolate spheroidal	small-medium	1.25-2.5 0 (2.05±0.53)	Reticulate
15. *V. pseudodigitalis*	27.5-30.0 (28.75±0.97)	20.0-21.3 (20.88±0.57)	1.38	Prolate	medium	2.25-2.50 (2.53±0.21)	Reticulate
16. *V. sinaiticum*	27.5-30.0 (29.13±0.98)	20.0-21.3 (20.88±0.57)	1.40	Prolate	medium	1.50-2.5 0 (2.00±0.49)	Reticulate
17. *V. sinuatum*	19.6-25.2 (23.86±1.09)	20.6-23.2 (23.55±0.61)	1.01	Prolate spheroidal	small - medium	2.00-2.5 0 (2.30±0.22)	Reticulate
18. *V. songaricum*	23.8-26.3 (25.38±0.98)	20.0-21.3 (20.38±0.57)	1.25	Subprolate	small - medium	1.25-3.00 (2.13±0.75)	Reticulate
19. *V. speciosum*	25.0-27.5 (26.50±1.09)	20.0-22.5 (21.25±1.25)	1.25	Subprolate	medium	3.50-4.00 (3.70±0.29)	Reticulate
20. *V. thapsus*	22.5-25.0 (23.75±0.77)	24.0-27.0 (25.00±0.50)	0.95	Oblate-spheroidal	small - medium	1.24-2.16 (1.91±0.35)	Reticulate

In *Verbascum,* pollen grains vary in shape, size and exine thickness. The shape can be prolate spheroidal, subprolate, prolate and oblate-spheroidal. The pollens are small to medium-sized. Most of the species have medium-sized pollens (25.0-32.5 µm) except in *V. agrimoniifolium* and *V. oreophilum,* where small-sized pollens (16.5-24.0 µm) are observed. *Verbascum palmyrense*, *V. sinuatum*, *V. songaricum*, and *V. thapsus* have both small and medium sized pollen grains. The polar axis (P) ranged from 16.5-32.5 µm. The largest polar axis was recorded in *V. alceoides, V. assurense* and *V. laetum* with 32.5 µm long and the smallest was found in *V. agrimoniifolium* with

16.5 μm long. The equatorial diameter (E) ranged from 15.5-27.5 μm and the largest one was present in *V. assurense* with 27.5 μm, while the smallest was found in *V. agrimoniifolium*. In addition, the exine thickness of the pollen of species studied ranged from 0.87 to 4.75 μm.

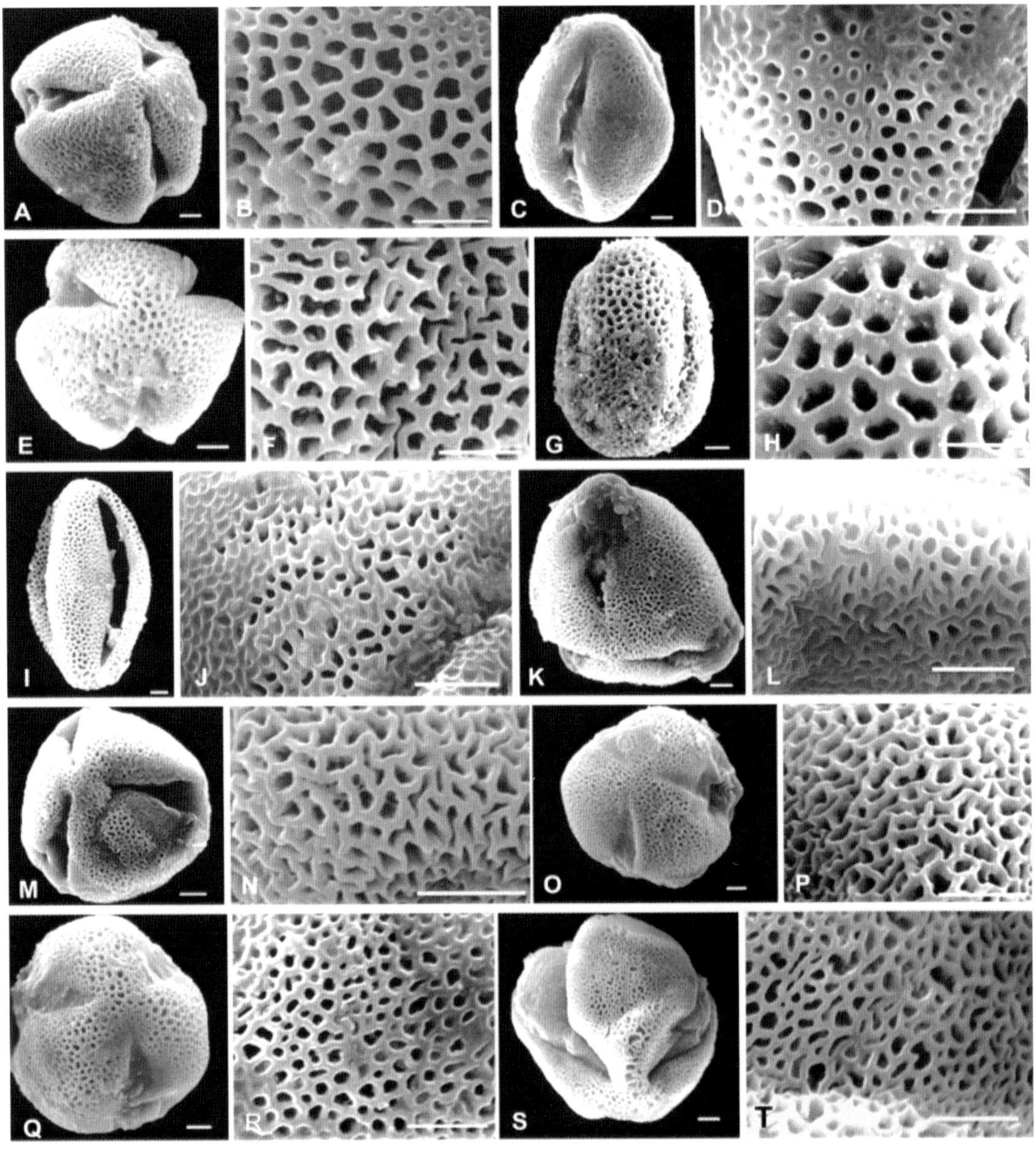

Fig. 2. SEM micrographs of the pollen grains of *Verbascum*. A, B. *V. laetum*; C, D. *V. macrocarpum*; E, F. *V. oreophilum;* G, H. *V. palmyrense;* I, J. *V. pseudodigitalis*; K, L. *V. sinaiticum;* M, N. *V. sinuatum*; O, P. *V. songaricum*; Q, R. *V. speciosum;* S, T. *V. Thapsus*; A, E, K, M, O, Q and S. polar view; B, D, F, H, J, L, N, P, R and T. exine sculpturing; C, G and I. equatorial view (Scale bar = 2 μm).

The overlapping of exine thickness is found among the taxa studied, thus this character is of little taxonomic value. Results obtained from the present study agree with previous studies on some species of *Verbascum* (Erdtman, 1952; Filippini *et al.*, 1990; Lobin and Pormbski, 1994;

Bokhri and Alfarhan, 2006; Asmat *et al.*, 2011). Several authors reported two types of exine sculpturing for *Verbascum* i.e. microreticulate and macroreticulate (Kheiri *et al.*, 2006; Pehlivan *et al.*, 2008; Karavelioğullari *et al.*, 2010). Pehlivan *et al.* (2008) concluded that the pollen morphological characteristics observed under the light microscope seems to have little value for taxonomic study but the sculpturing investigated under SEM provides a better result for dividing the *Verbascum* taxa into two groups with reticulate and coarsely reticulate. However, all taxa included in our study showed a uniform reticulate exine sculpturing pattern.

Our results showed that only the shape of pollen grains can be used for grouping the *Verbascum* into four groups viz. the Subprolate group including: *V. alceoides*, *V. alepense*, *V. assurense*, *V. calvum*, *V. cheiranthifolium*, *V. geminiflorum*, *V. laetum*, *V. macrocarpum*, *V. songaricum*, *V. speciosum,* the Prolate spheroidal group including: *V. agrimoniifolium*, *V. carduchorum*, *V. oreophilum*, *V. palmyrense*, and *V. sinuatum;* the Prolate group includes: *V. andrusii*, *V. damascenum*, *V. pseudodigitalis*, *V. sinaiticum* and the Oblate spheroidal with only one species *V. thapsus.* Even though the pollen characters alone cannot be used for species identification, they provide some information for the classification of these plants. Moreover, using this character in combination with other characters may help in clarifying some problematic taxa.

Acknowledgement

We gratefully acknowledge the staff of the Department of Biology, Faculty of Science, Khon Kaen University (KKU), Thailand, for their technical support and laboratory assistance. This study was supported by grants from the Scholarships & Cultural Affairs Directorate, Ministry of Higher Education Scientific Research, Iraq.

References

AL-Bermani, A.K. 1981. Systematic study of the genus *Verbascum* (Scrophulariaceae) as it occurs in Iraq. M.Sc. Thesis in Biology/ Botany, University of Baghdad, Baghdad, Iraq.

AL-Rawi, A. 1964. Wild plants of Iraq with their distribution. Tech. Bull. **14**. Dir. Gen. of Agr. Res. Proj. Iraq, Ministry of Agriculture, Government Press, pp. 1-323.

Asmat, T., Khan, M., Ahmed, M., Zafar, M., Manzoor, F., Munir, M., Akhtar, K., Bashir, S., Mukhtar, T., Ambreen, M. and Abbasi, S. 2011. Pollen morphology of selected species of Scrophulariaceae of district Dir Upper, Pakistan J. Med. Plants Res. **5**: 6423-6428.

Bukhari, N.A. and Alfarhan, A.H. 2006. A study of some characteristics of pollen types of the genus *Verbascum* in Saudi Arabia. Saudi J. Biol. Sci. **13**: 81-86.

Erdtman, G. 1952. Pollen Morphology and Plant Taxonomy: Angiosperms. Chronica Botanica Co., Massachusettes, USA, pp. 1-553.

Erdtman, G. 1960. The acetolysis method, a revised description. Svensk. Bot. Tidskr. **54**: 561-564.

Filippini, R., Cappelletti, E.M. and Caniato, R. 1990. Botanical identification of powdered plant drugs *Verbascum* flowers. Int. J. Crude Drug Res. **28**: 129-133.

Ghadanfar, S.A. 1992. An Annotated Catalogue of the Vascular Plants of Oman and their Vernacular Names. National Botanic Garden of Belgium, Meise, Belgium, pp. 1-153.

Huber-Morath, A. 1978. *Verbascum* L. *In:* Davis, P.H. (Ed.), Flora of Turkey and the East-Aegean Islands. Vol. **6**. Edinburgh University Press, Edinburgh, UK, pp. 461-603.

Karavelioğullari, F.A., Celik, S., Başer, B. and Yavru, A. 2010. *Verbascum ergin-hamzaoglui* (Scrophulariaceae), a new species from South Anatolia, Turkey. Turk. J. Bot. **35**: 275-283.

Kheiri, S., Khayami, M., Osaloo, S.K, and Mahmoodzadeh, A. 2006. Pollen morphology of some species of *Verbascum* (Scrophulariaceae) in Urmia. Pak. J. Biol. Sci. **9**: 434-436.

Lobin, W. and Pormbski, S. 1994. The genus *Verbascum* (Scrophulariaceae) on the Cape Verde Islands, W. Africa. Willdenowia **24**: 65-81.

Özler, H., Pehlivan, S., Celep, F., Doğan, M., Kahraman, A., Yavru, A., Başer, B. and Bagherpour, S. 2013. Pollen morphology of *Hymenosphace* and *Aethiopis* sections of the genus *Salvia* (Lamiaceae) in Turkey. Turk. J. Bot. **37**: 1070-1084.

Pehlivan, S., Başer, B. and Karavelioğullari, F.A. 2008. Pollen morphology of the genus *Verbascum* L. (Group A) in Turkey. Biodicon. **1**: 1-24.

Perveen, A. 1993. A preliminary study of the pollen flora of Karachi. Ph.D. Thesis. Department of Botany. University of Karachi, Karachi, Pakistan.

Punt, W., Blackmore, S., Nilsson, S. and Le Thomas, A. 1994. Glossary of Pollen and Spore Terminology. LPP Foundation, Utrecht, Netherland, pp. 1-246.

Rajbhandary, S., Hughes, M. and Shrestha, K.K. 2012. Pollen morphology of *Begonia* L. (Begoniaceae) in Nepal. Bangladesh J. Plant Taxon. **19**: 191-200.

Sarwar, A.K.M.G. and Takahashi, H. 2013. Pollen morphology of *Rhododendron* L. and related genera and its taxonomic significance. Bangladesh J. Plant Taxon. **20**: 185-199.

Walker, J.W. and Doyle, J.A. 1975. The bases of angiosperm phylogeny: palynology. Ann. Mo. Bot. Gard. **62**: 664-723.

Wood, J.R. 1997. A Handbook of the Yemen Flora. Royal Botanic Gardens, Kew, London, UK, pp. 1-434.

Zohary, M. 1974. Flora of Palestine. The Israel Academy of Sciences and Humanities. Jerusalem, Israel, pp. 1-519.

4

LEAF MICROMORPHOLOGICAL FEATURES OF THE GENUS *CORDIA* L. (BORAGINACEAE) FROM EGYPT

WAFAA AMER, RIM HAMDY[1] AND RIHAM MAHDY[2]

Department of Botany & Microbiology, Faculty of Science, Cairo University, Giza 12613, Egypt

Keywords: *Cordia*; Trichomes; Stomata; SEM; Egypt.

Abstract

Foliar epidermal characters of nine *Cordia* L. species found in Egypt are investigated. Glandular trichomes are observed in *C. boissieri* A. DC. and *C. cylindristachya* (Ruiz & Pav.) Roem. & Schult., while eglandular trichomes are predominant in all studied species. Different stomatal characters were examined including: shape, size, rims, stomatal level, aperture, cuticular deposition and wax ornamentation. The retrieved results showed a great variation among the studied species. The observed trichomes and stomata were used to construct two different taxonomic keys.

Introduction

Cordia L. is a large pantropical genus growing across the tropical and subtropical areas of America, Africa, Asia and Oceania (Al-Shehbaz, 1991) with 320 species of trees and shrubs. The genus is represented in the flora of Egypt by eight cultivated and one wild species.

Cordia myxa L. was recorded in ancient Egypt (Täckholm, 1961; Lucas, 1962; Darby *et al.*, 1977 and Manniche, 1989). Its fruit was eaten fresh and used to make wine. It may also have been raised for the production of birdlime (Van der veen, 2011). *Cordia sinensis* Lam. is the only relict species in its wild form; traced in moist ground in oases and Gebel Elba (Täckholm, 1974; Boulos, 2009). It was also raised in the Israel-Palestine area and in the Arabian Peninsula (Feinbrun-Dothan, 1978).

Taxonomic treatments of *Cordia* L. have long been recognized by many workers (Johnston, 1930; Kazmi, 1970; Taroda and Gibbs, 1986; Al Nowaihi *et al.,* 1987; Warfa, 1988; Verdcourt, 1991; El Mahi, 2012 and Mehrabian *et al.*, 2014). Metcalfe and Chalk (1950) described the general anatomical properties of the family including some details of the genus *Cordia* and underlined the taxonomic value of trichomes and stomata and their involvement in systematics.

The micromorphological characters of the epidermal system (stomata, trichomes, epidermal wax deposits) have been received much attention in the recent past years. According to Rejdal (1991); monographs are considered incomplete without micromorphology of the epidermis. In the context, however, little attention was given to the genus *Cordia*. Recently Ventrella and Marinho (2008) studied foliar micromorphological of *C. verbenacea* DC from Brazil, Sharma *et al.* (2013) made identification of *Cordia macleodii* Hook. through stomatal index. As the earlier studies confirmed the importance of epidermal micromorphology in taxonomy (Stockey and Taylor 1978 a, b; Stace, 1984; Palmer and Gerbeth-Jones, 1986 and 1988; Yoon 1994; Hassan, 2004; Özcan, 2002). It reveals that foliar epidermal characters are important in taxonomic identification and species relationships in *Cordia* species. Despite the previous studies, genus *Cordia* L. received very little attention in Egypt. Therefore, the present study was undertaken to evaluate the features

[1]Corresponding author: Email: rimhamdy@yahoo.com

[2]Timber Trees and Forestry Department, Horticulture Research Institute, Agriculture Research Center, Giza, Egypt.

of foliar trichomes and stomata in the genus *Cordia* L. growing in Egypt and to assess these characters for specific and interspecific delimitation. It was also aimed to construct two differential keys on the trichomes and stomatal characters to identify *Cordia* species from Egypt.

Materials and Methods

Nine species of *Cordia* used in this study are *C. africana* Lam., *C. boissieri* A. DC., *C. crenata* Delile, *C. cylindristachya* (Ruiz & Pav.) Roem. & Schult., *C. dentata* Poir., *C. dichotoma* G. Forst., *C. myxa* L., *C. sebestena* L. and *C. sinensis* Lam.

The localities of these species along with the voucher numbers are appended in Table 1. Fresh specimens collected and studied from different gardens as well as herbarium specimens deposited in Cairo University Herbarium (CAI), The Agriculture Research Centre, Flora of Phytotaxonomy Herbarium (CAIM) and Orman garden herbarium. Leaf material was examined by Scanning Electron Microscopy (SEM). Small pieces (c. 7 mm^2) of dried mature leaf material were fixed to SEM stubs using double-sided adhesive tape, coated with gold in a sputter coat (SPI-Module), examined and photographed with a Jeol (JSM-5500 LV) using high vacuum mode at the Electron Microscope Unit, Regional Center of Mycology and Biotechnology at Azhar University, Cairo (Egypt).

Terminology of trichomes used are according to Al-Shammary and Gornell (1994); Bigazzi *et al.* (1999) and Taia (2006), while for stomata terminology Wilkinson (1983 &1992) and El Hadidy (2004) were followed.

Table 1. List of species of *Cordia* L. along with voucher specimens used in the present study.

No.	Species	Voucher specimens
1	*Cordia africana* Lam.	Aswan: Aswan botanic garden, 20. 2. 1996, H. Rafaeel 29135 (CAIM) Giza: Mazhar botanic garden, Baragile, 24. 8. 2014, R. Mahdy s.n. (CAI)
2	*C. boissieri* A. DC.	Giza: Villa El Maghraby, 30. 4. 2004, T.Labib (Orman garden herbarium) Giza: Mazhar botanic garden, Baragile, 24. 8. 2014, R. Mahdy s.n. (CAI)
3	*C. crenata* Delile	Sudan: Jebel Dambobei, Karora, 25. 12. 1966, M. Kass, M. Mobarak& H. Omar 1166 (CAI)
4	*C. cylindristachya* (Ruiz & Pav.) Roem. & Schult.	Giza: Faculty of Agriculture, 5. 6. 1976, B. Diwan s.n. (Orman garden Herbarium)
5	*C. dentata* Poir.	Giza: Orman garden, 5. 6. 1933, Khattab g1349 (CAIM) Giza: Zoolgical garden, 22. 5. 2014, R. Mahdy s.n. (CAI)
6	*C. dichotoma* Forst.f.	Giza: Mazhar botanic garden, Baragile, 6. 5. 2011, M. El Said s.n. Mazhar Herbarium Giza: Mazhar botanic garden, Baragile, 24. 8. 2014, R. Mahdy s.n. (CAI)
7	*C. myxa* L.	Aswan: Aswan botanic garden, 12. 5. 1998, H. Rafaeel, 29366 (CAIM) Giza: Zoolgical garden, 22. 5. 2014, R. Mahdy s.n. (CAI)
8	*C. sebestena* L.	Giza: Orman garden, 28.7. 2007, Mostafa (Orman garden herbarium) Giza: Mazhar botanic garden, Baragile, 24. 8. 2014, R. Mahdy s.n. (CAI)
9	*C. sinensis* Lam.	Giza: Orman garden, 10.7.1933, J. Shabetaig1354 (CAIM) Giza: Zoolgical garden, 22. 5. 2014, R. Mahdy s.n. (CAI)

Results and Discussion

The genus *Cordia* offers different types of trichomes and stomata are summarized in Tables 2 and 3. Both glandular and eglandular trichomes are found in *Cordia* (Table 2). Glandular trichomes are observed only in *C. boissieri* (40-46 μm, Fig.1A) and *C. cylindristachya* (33-38 μm, Fig.1B); the earlier possess clavate head while the later posses globular head. Eglandular trichomes are evident in all species, however they vary in density and distribution; presence or absence of basal part (cell), with circular rim or pluricellular, shape of upper part (apical cell); with cylindrical, flattened or conical, their length and ornamentation.

The eglandular trichomes are simple unbranched; the majorities are medium-long sized upper part. Their length ranges from 108-833 μm. These eglandular trichomes may be without basal part as in *Cordia dichotoma* (Fig. 1C) or with basal part. The basal part either in circular rim as in *C. africana* (Fig. 1D), in one pluricellular (*C. myxa* Fig. 2F and *C. crenata* Fig. 2A, B), or in more than one row of cells as shown in *C. dentata* (Fig. 2G). The eglandular trichomes in the studied taxa appeared cylindrical in *C. africana* (Fig. 1D), or flattened in *C. dentata* (Fig.1F), rarely cone-like as in *C. crenata* (Fig. 2A). Trichome surface in the studied taxa appeared with variable surface namely: smooth (in *C. dichotoma*, Fig. 1C), granulate (in *C. africana*, Fig.1D), scaly (in *C. dentata*, Fig. 1F) or echinate (in *C. boissieri*, Fig. 1E).

Trichomes of the upper leaf surface of the studied taxa ranges from sparse (number of hairs / $1cm^2$, <300) in *Cordia africana, C. myxa, C. dichotoma* and *C. dentata* (Table 2). *Cordia sinensis* and *C. sebestena* possess moderate (≥300-600). The dense (≥600) appeared in *C. boissieri, C. cylindristachya* and *C. crenata*, while in the lower surface trichomes appeared dense in most of the studied species, sparse in *C. dentata* and very dense in *C. boissieri* (Table 2).

The following key was constructed on the basis of trichome features of the studied *Cordia* species as shown under SEM:

1	Eglandular and glandular present	2
-	Only eglandular trichomes present	3
2	Glandular with clavate head	*C. boissieri*
-	Glandular with globular head	*C. cylindristachya*
3	Basal part absent	*C. dichotoma*
-	Basal part present	4
4	Formed of circular rim	*C. africana*
-	Formed of pluricellular cells	5
5	Pluricellular base formed of more than two rows	*C. dentata*
-	Pluricellular base formed of one row	6
6	Distinct in cylindrical upper part	*C. crenata*
-	Not distinct in cylindrical upper part	7
7	Scaly upper part	*C. myxa*
-	Granulate upper part	8
8	Granules coarse	*C. sinensis*
-	Granules fine	*C. sebestena*

The epidermal and stomatal features of the studied *Cordia* species are summarized in Table 3. Stomata as seen under SEM showed high diversity in stomatal shape: round (L= 1-1.2W) in *C. africana, C. boissieri* and *C. cylindristachya* (Fig. 3A, B & Fig. 4C, D, E, F); narrowly elliptic

Table 2. Trichomes features of the studied *Cordia* species.

Cordia species	Trichome Density (average number/ cm²)		Glan-dular	Eglandular									
				With Basal part							Without Basal part		
				Circular Rim Base			Pluricellular Base						
	Upper epidermis (U)	Lower epidermis (L)		Hair Length/ µm	Shape	Ornamen-tation	Hair Length/µm	Shape	Number of Rows	Ornamen-tation	Hair Length/ µm	Shape	Ornamen-tation
C. africana	+	+++	-	186	Cylindrical (L)	Granulate	-	-	-	-	464-528	Flattened (L)	Scaly
											55	Conical (L)	Granulate
C. boissieri	+++	++++	+	82-126	Cylindrical (U/L)	Echinate	-	-	-	-	400-692	Flattened (U/L)	Smooth
C. crenata	+++	+++	-	-	-	-	208-280	Cylindrical (L)	One	Granulate	-	-	-
							76-96	Conical (U)					
C. cylindristachya	+++	+++	+	-	-	-	-	-	-	-	151-360	Flattened (L)	Smooth
							70-90	Conical (U)	One	Granulate	-	-	-
C. dentata	+	-	-	400-700	Flattened	Scaly	152-346	Cylindrical (U/L)	More than two	Granulate	152-346	Cylindrical (U/L)	Granulate
C. dichotoma	+	+++	-	-	-	-	-	-	-	-	210-371	Cylindrical	Granulate L
											486-552		Smooth U/L
C. myxa	+	+++	-	108-233	Cylindrical (L)	Granulate	331-446	Cylindrical (U/L)	One	Scaly	-	-	-
C. sebestena	++	+++	-	-	-	-	292-417	Cylindrical (U/L)	One	Granulate	-	-	-
C. sinensis	++	+++	-	-	-	-	117-167	Cylindrical (U/L)	One	Granulate	-	-	-

Hair density: average number of hairs: Sparse=+ (<300 hairs /cm²), Moderate= ++ (≥300-600 hairs /cm²), Dense=+++ (≥600 hairs /cm²), and very dense=++++ (hairs uncountable). Hair distribution: (U)= upper epidermis,(L)= lower epidermis and (U/L)=in both upper & lower epidermis.

Table 3. The epidermal and stomatal features of the studied *Cordia* species.

Cordia species	Epidermal Features		Stomata				Stomatal Aperture			Stomatal Rim	
	Cuticular Deposition	Wax Ornamentation	Level	Length (µm)	Width (µm)	Shape	Length (µm)	Width (µm)	Shape	Number	Width
C. africana	Rims + Striations	Absent	Raised	15-18-20	13-15-17	Round	8-*9*-9	1-*2*-3	Short & Wide	One	Wide
C.boissieri	Rims + Striations	Absent	Raised	15-17-18	16-17-18	Round	8-*9*-9	3-*4*-4	Short & Wide	Two	Wide
C. crenata	Striations	Granules & Particles	Leveled	23-26-28	15-16-17	Broadly elliptical	10-*14*-17	2-*3*-3	Long & Slit-Like	Two	Narrow
C. cylindristachya	Ridges	Particles	Raised	15-18-20	13-16-18	Round	9-*11*-13	3-*4*-5	Long & Slit-Like	One	Wide
C. dentata	Striations	Flakes & Granules	Leveled	18-20-22	12-14-15	Broadly elliptical	7-*8*-9	2-*3*-3	Short & Wide	Two	Wide
C. dichotoma	Ridges	Filaments & Granules	Leveled	26-29-32	11-16-20	Narrowly elliptical	15-*16*-17	3-*4*-4	Long & Wide	Two	Narrow
C. myxa	Ridges	Flakes & Particles	Leveled	25-29-33	15-16-17	Narrowly elliptical	15-*17*-18	2-*3*-3	Long & Wide	One	Wide
C.sebestena	Striations	Particles	Leveled	15-18-20	11-11-12	Broadly elliptical	6-*9*-11	1-*2*-2	Short & Narrow	One	Narrow
C. sinensis	Striations	Particles	Leveled	17-21-25	8-11-13	Narrowly elliptical	8-*10*-12	1-*2*-2	Long & Slit-Like	One	Narrow

Stomata shape: Round=(L= 1-1.2W), Narrowly elliptical=(L= 1.8-1.9W), Broadly elliptical=(L=1.4-1.6W). Aperture shape: Long: (aperture length average >10 µm), Short: (aperture length average ≤10 µm). Length and Width cited as average value in the middle figure.

(L= 1.8-1.9W) in *C. sinensis*, *C. myxa* and *C. dichotoma* (Fig. 3E, F & Fig. 5A,B,C,D); while broadly elliptic (L=1.4-1.6W) in *C. sebestena*, *C. crenata* and *C. dentata* (Fig. 3C,D, Fig. 4A, B & Fig. 5E,F).

The stomatal size ranged from 15-33 µm long and 8-20 µm wide. The largest in *C. myxa* and *C. dichotoma* while the smallest in *C. sebestena*. Some of the stomata show raised level as in *C. africana*, *C. boissieri* and *C. cylindristachya*, while other species are in the same level.

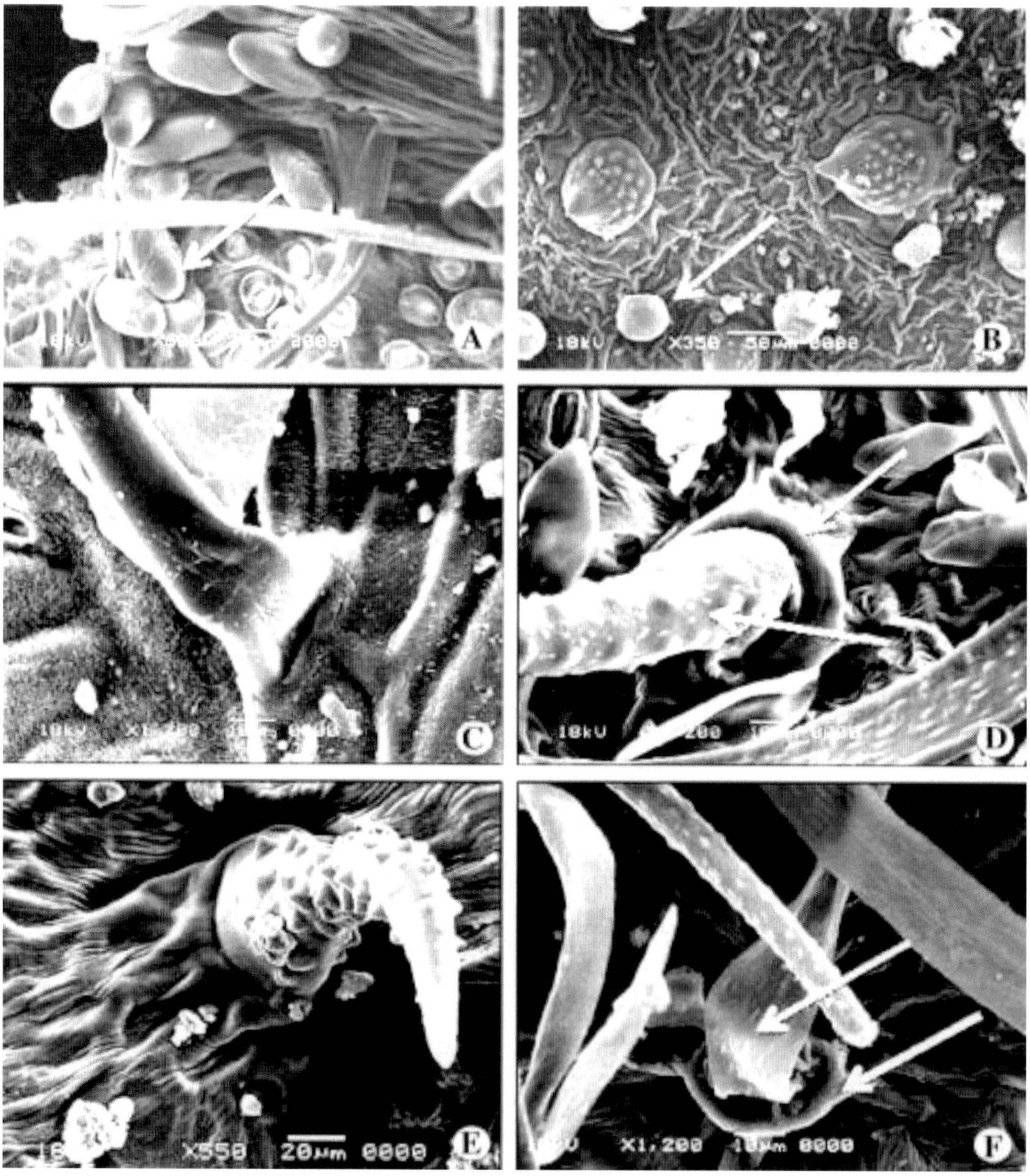

Fig. 1. SEM micrographs showing foliar trichomes in *Cordia* species. A-B Glandular trichomes: A. Clavate head of *C. boissieri*. and B. Globular head of *C. cylindristachya*. C-F Eglandular trichomes: C. Without basal part, cylindrical upper part of *C. dichotoma*, with smooth surface. D-E circular rim basal part: D. Cylindrical upper part of *C. africana*, with granulate surface, E. *C. boissieri*, with echinate surface. F. Flattened upper part of *C. dentata* with scaly surface.

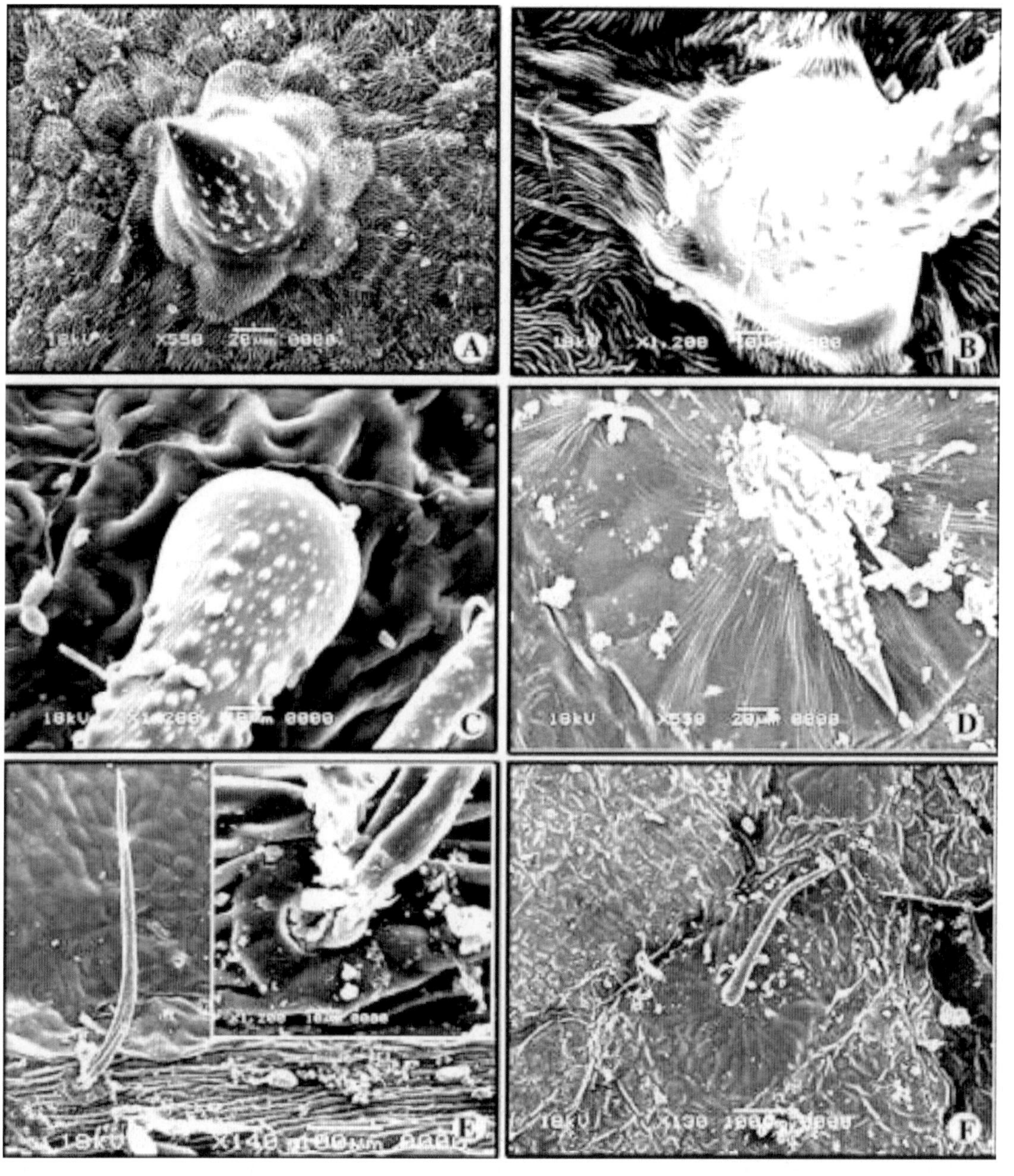

Fig. 2. SEM micrographs showing foliar trichomes in *Cordia* species. A-G eglandular trichomes with pluricellular basal part: A-F. In one row cells, A, B. *C. crenata*, A. conical upper part, B. cylindrical upper part, C. cylindrical upper part of *C. sebestena*, D. cylindrical upper part of *C. sinensis*, with granulate surface, E, F Cylindrical upper part of *C. myxa*, with scaly surface and G. pluricellular basal part in more than two rows of *C. dentata*.

The presence of stomatal rim is a diagnostic feature: single in *C. africana* (Fig. 3B), *C. sebestena* (Fig. 3D), *C. sinensis* (Fig. 3F), *C. cylindristachya* (Fig. 4F), *C. myxa* (Fig. 5B) and double in *C. crenata* (Fig. 4A), *C. boissieri* (Fig. 4D), *C. dichotoma* (Fig. 5D) and *C. dentata* (Fig. 5F).

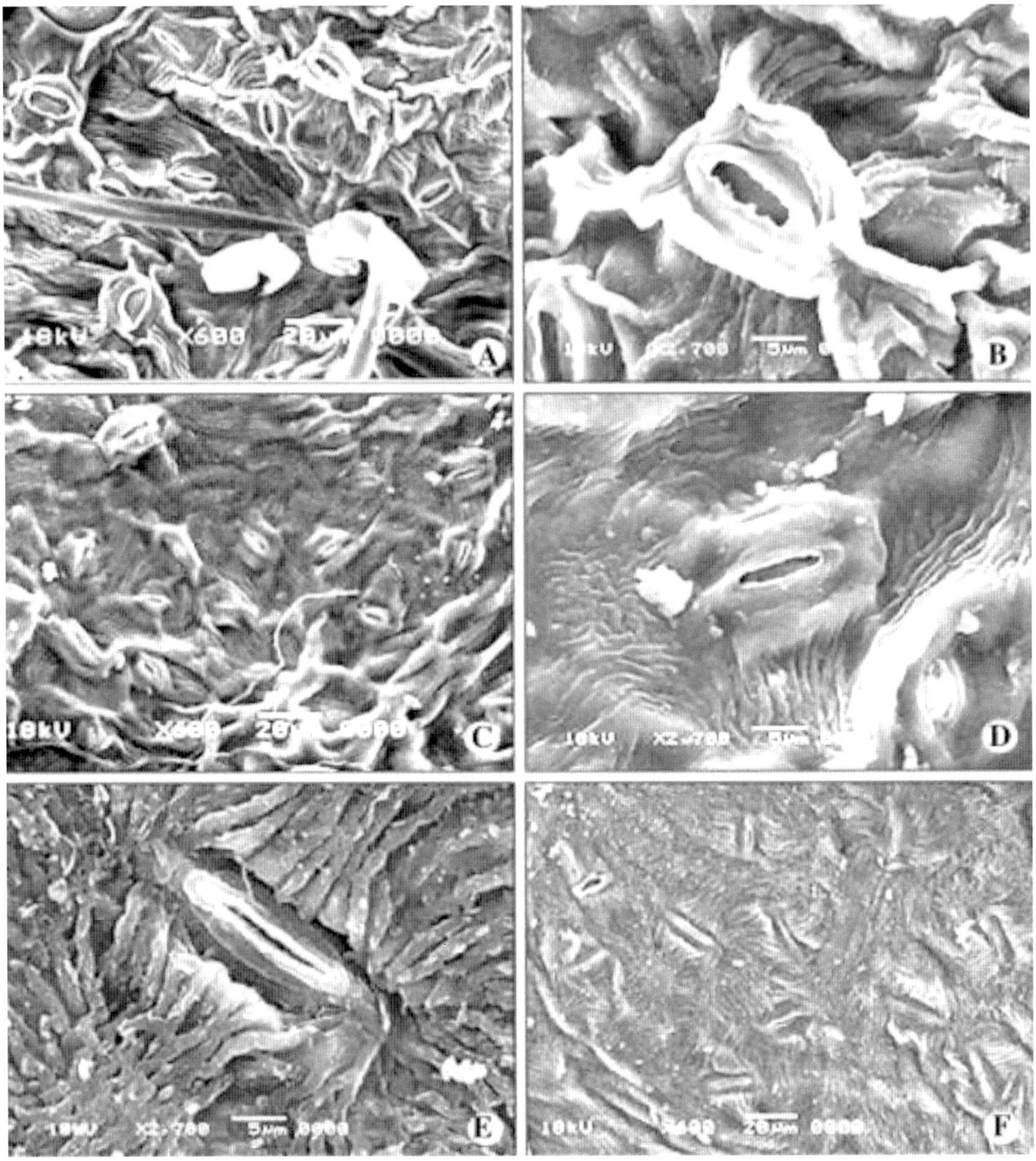

Fig. 3. SEM micrographs showing the stomatal pattern and magnified stomium in *Cordia* species. A, B: *C. africana*, C, D: *C. sebestena*, E, F: *C. sinensis*.

Stomatal opening (aperture) is variable among the studied taxa: long (average length >10 μm) in *C. dichotoma*, *C. myxa*, *C. crenata*, *C. cylindristachya* and *C. sinensis*; short (average length ≤10 μm) in *C. africana*, *C. boissieri*, *C. dentata* and *C. sebestena*.

Cuticular deposits are observed in the form of striations or ridges; striations forming lateral wings perpendicular to the long axis of the subsidiary cells are observed in *C. sinensis* (Fig. 3E, F). Sometimes these striations radiating extend across more than one epidermal cell as in *C. sebestena* (Fig. 3C,D) and *C. dentata* (Fig. 5E,F). Striations distinguished *C. crenata* (Fig. 4A, B) and *C. sinensis* (Fig. 3E, F) from the rest of the studied species. However in *C. africana* the stomata are enclosed by undulated rim. Also striations extend across more than one epidermal cells (Fig. 3A, B). These striations sometimes form short lateral wings perpendicular to the long axis of the subsidiary cells in *C. boissieri* (Fig. 4C, D). *Cordia cylindristachya* can be distinguished from the other studied species with the presence of stomata located on epidermal ridges forming lateral wings perpendicular to the long axis of the stomatal cells (Fig. 4E, F). These ridges are crescentic and mostly parallel (occasionally perpendicular) to the long axis of the subsidiary cells as observed in *C. myxa* (Fig. 5A, B) while appears crescentic perpendicular to the long axis of the subsidiary cells in *C. dichotoma* (Fig. 5 C, D).

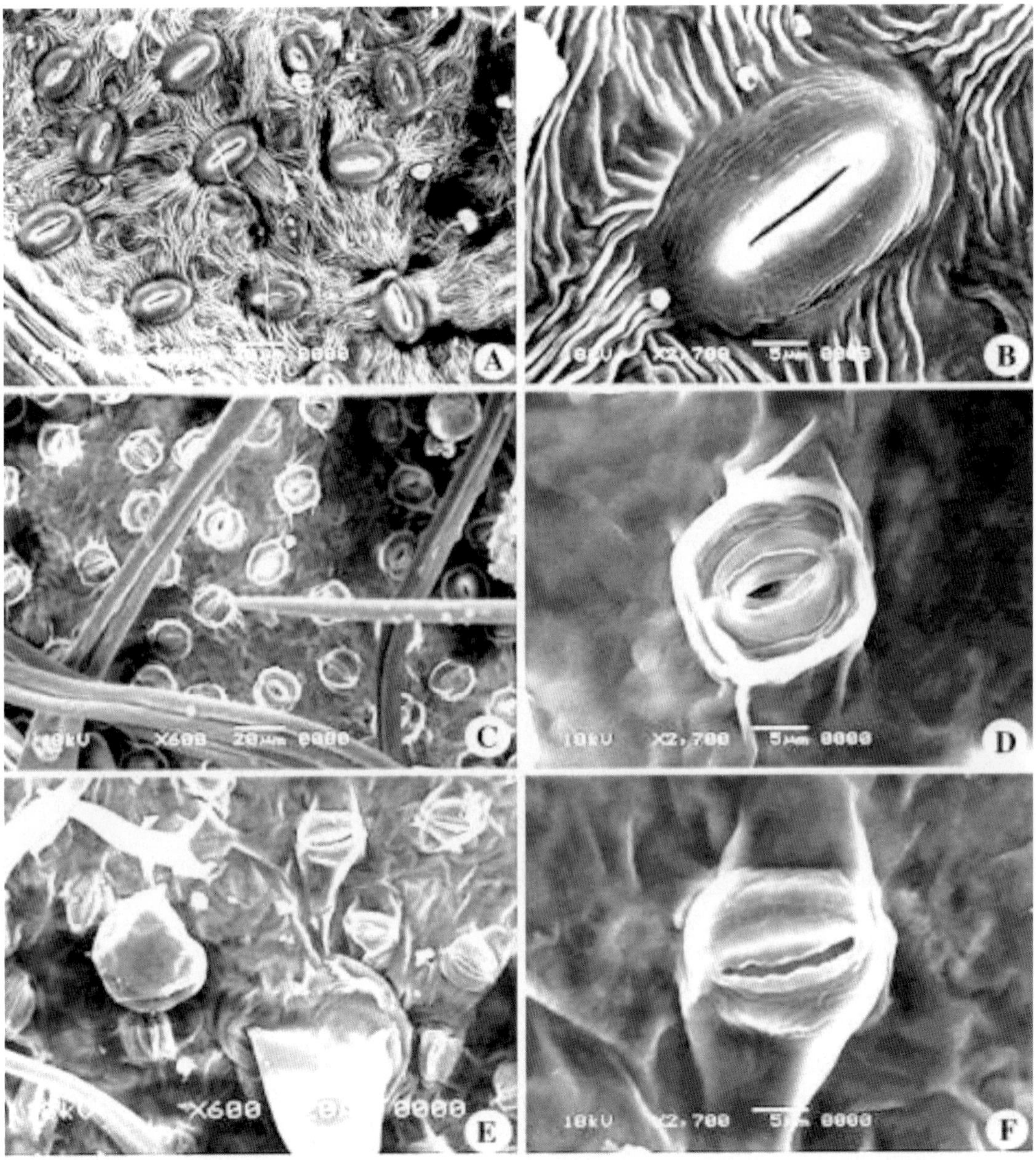

Fig. 4. SEM micrographs showing the stomatal pattern and magnified stomium in *Cordia* species. A, B: *C. crenata*, C, D: *C. boissieri*, E, F: *C. cylindristachya*.

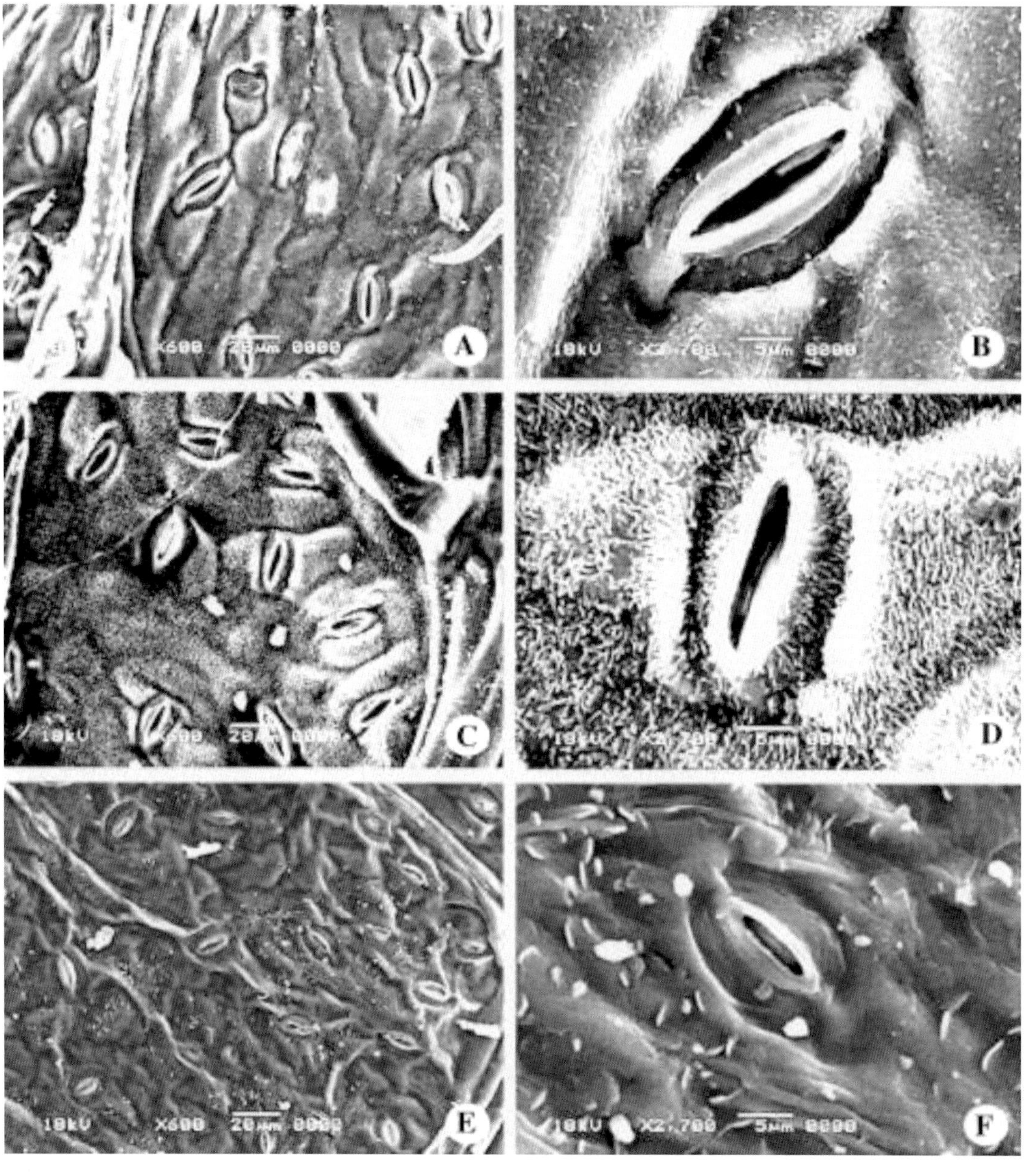

Fig. 5. SEM micrographs showing the stomatal pattern and magnified stomium in *Cordia* species. A, B: *C. myxa*, C, D: *C. dichotoma*, E, F: *C. dentata*.

The SEM of the epidermal system showed the presence of special wax structures (filaments, particles, flakes and granules, Table 3) in the studied taxa. Particles are observed in *C. cylindristachya*, *C. sebestena* and *C. sinensis*. These particles mixed with flakes in *C. myxa* (Fig. 5 B) or mixed with granules in *C. crenata* (Fig. 4B), while flakes and granules are observed in *C. dentata* (Fig. 5F) filaments and granules in *C. dichotoma* (Fig. 5D). The following key was constructed on the basis of the stomatal and epidermal features of the studied taxa as shown under SEM.

1	Obvious epidermal striations	2
-	Obscure epidermal striations	5
2	Stomata with cuticular rim	*C. africana*
-	Stomata without cuticular rim	3
3	Cuticular deposition with unconspicuous striations extending across more than one epidermal cell	*C. sebestena*
-	Cuticular deposition with conspicuous striations, not extending across more than one epidermal cell	4
4	Striations only perpendicular to the long axis of the subsidiary cells	*C. sinensis*
-	Striations radiating all over the stomata	*C. crenata*
5	Stomata round	6
-	Stomata elliptic	7
6	Stomata on ridge without stomatal rim	*C. cylindristachya*
-	Stomata not on ridge, with stomatal rim	*C. boissieri*
7	Stomata broadly elliptic	*C. dentata*
-	Stomata narrowly elliptic	8
8	Epidermis with dense waxy filaments	*C. dichotoma*
-	Epidermis not so	*C. myxa*

In our study *Cordia myxa* and *C. dichotoma* have been found closely related (Verdcourt, 1991) as attested by the following shared characters: lamina broadly ovate to suborbicular, inflorescence dichotomous branching, flowers salverform and white, stigma flattened and subfoliaceous with irregular or erose-denticulate margin, fruit yellowish orange. However the epidermal features do not support a close association between them, using micromorphological characters during this study to distinguish both species; the eglandular trichomes without basal part is observed in *C. dichotoma* (Fig. 1C) while basal part is pluricellular in *C. myxa* (Fig. 2F). Also wax ornamentation is formed of particles in addition to flakes in *C. myxa* (Fig. 5A, B) and filaments and granules in *C. dichotoma* (Fig. 5C, D).

Stomata in *Cordia* are usually confined to lower side (Metcalfe and Chalk, 1950). The results outlined in Table (3), indicate that stomatal rim and wax ornamentation is found to be significant characters; stomatal rim was double in *C. boissieri*, *C. crenata*, *C. dentata* and *C. dichotoma* and single in *C. africana*, *C. cylindristachya*, *C. myxa*, *C. sebestena* and *C. sinensis* (Figs.3, 4,5). Wax ornamentation is formed of particles and granules in *C. crenata* (Fig. 4A, B), *C. myxa* with particles and flakes (Fig. 5A, B), *C. dichotoma* with filament and granules (Fig. 5C, D) while *C. dentata* characterized flakes and granules (Fig. 5E, F).These results are in accordance with that of Akcin *et al.*, (2013), who claimed that, the stomatal rims and wax ornamentations are important characters for delimiting fourteen species of *Onosma* L.

Cordia cylindristachya was distinguished from other species, by stomata located on epidermal ridge (Fig. 4E, F) which was confirmed earlier by Metcalfe and Chalk (1950) on some *Cordia* species.

Finally, we conclude that the micromorphological characters of *Cordia* species such as presence of eglandular and glandular trichomes, presence or absence, shape and ornamentation of basal trichome part, stomatal rim and wax ornamentation possess diagnostic taxonomic value which can delimit *Cordia* species.

References

Akçin Ö.E., Şenel G. and Akçin, Y. 2013. Leaf epidermis morphology of some *Onosma* (Boraginaceae) species from Turkey. Turk. J. Bot. **37**: 55-640.

Al-Nowaihi, A.S., Khalifa, S.F. and Hamed, K. 1987. A contribution to the taxonomy of Boraginaceae. Phytologia **62**: 107-125.

Al-Shammary, K.I.A., and Gornall, R.J. 1994. Trichome anatomy of the Saxifragaceaes. l. from the southern hemisphere. Bot. J. Linn. Soc. **114**: 99-131.

Al-Shehbaz, I.A. 1991. The genera of Boraginaceae in the Southeastern United States. J. Arnold Arb. Suppl. **1**: 1-169.

Bigazzi, M., Selvi, F. and Fiorini, G. 1999. A reappraisal of the generic status of *Gastrocotyle*, *Hormuzakia* and *Phyllocara* (Boraginaceae) in the light of micromorphological and karyological evidence. Edinb. J. Bot. **56**(2): 229-251.

Boulos, L. 2009. Flora of Egypt Checklist, El Hadara Publishing, Cairo, Egypt, 410p.

Darby, W., Ghallioungui, P., Griffith, L. 1977. Food: The gift of Osiris. Vol. **1 & 2,** Academic press, London, New York, San Francisco.

Dasti, A.A., Bokhari, T.Z., Malik, S.A. and Akhtar, R. 2003. Epidermal morphology in some members of family Boraginaceae in Baluchistan. Asian J. Plant Sci. **2**(1): 42-47.

El Hadidy, A.M.H. 2004.Morphological studies on fruits and seeds of the genus *Lotus* L. in Egypt, Proceedings of First International Conference on Strategy of Egyptian Herbaria March 9-11, 2004; Giza, Egypt.129-159.

El Mahi, H.O.A. 2012. Taxonomic revision of the family Boraginaceae in Sudan. Ph.D. Thesis, Botany Department, Faculty of Science, University of El Khartoum.

Feinbrun-Dothan, N. 1978. Boraginaceae, Flora Palaestina, Vol. **3**, Israel Academy of Science and Humanities, Jerusalem.

Hassan, S.A. 2004. Stomatal diversity in certain species of Apocynaceae and its bearing on taxa delimitation. Taeckholmia **24** (1): 21-31.

Johnston, I.M. 1930. Studies in Boraginaceae. VIII. Contrib. Gray Herb. Harvard Uni. **92**: 1-89.

Johnston, I.M. 1935a. Studies in Boraginaceae. X. The Boraginaceae of Northeastern South America. J. Arnold Arb. **16**: 1-64.

Johnston, I.M. 1935b. 3. New or otherwise noteworthy species. J. Arnold Arb. **16**: 173-181.

Johnston, I.M. 1937. Studies in the Boraginaceae. XII - 1.Trigonotis in Southwestern China. 2. Novelties and Critical Notes. J. Arnold Arb. **18**: 10-13.

Johnston, I.M. 1940. Studies in Boraginaceae. XV. Notes on some Mexican and Central American species of *Cordia*. J. Arnold Arb. **21**: 336-355.

Johnston, I.M. 1948. Studies in the Boraginaceae. XVI - Species chiefly from Mexico and Western United States. J. Arnold Arb. **29**: 227-240.

Johnston, I.M. 1949a. Studies in the Boraginaceae. XVII. A. *Cordia* section *Varronia* in Mexico and Central America. J. Arnold Arb. **30**: 85-106.

Johnston, I.M. 1949b. Studies in the Boraginaceae. XVII. Boraginsaceae of the Southern West Indies. J. Arnold Arb. **30**(2): 111-138.

Johnston, I.M. 1951. Studies in the Boraginaceae. XX. Representatives of three subfamilies in Eastern Asia. J. Arnold Arb. **32**(2): 1-26.

Johnston, I.M. 1956. Studies in the Boraginaceae. XXVIII. New or otherwise interesting species from America and Asia. J. Arnold Arb. **37**: 288-306.

Kazmi, S.M.A. 1970. A revision of the Boraginaceae of West Pakistan and Kashmir. J. Arnold Arb. **51**: 133–184.

Lucas, A. 1962. Ancient Egyptian materials and industries. Fourth edition, Revised by J.R. Harris, E. Arnold, London.

Manniche, L. 1989. An ancient Egyptian herbal, British Museum Publications, Austin.

Mehrabian, A.R., Sheidai, M. and Mozaffarian, V. 2014. Micromorphology of leaf trichomes in *Onosma* (Boraginaceae) and their systematic relevance in Iran. Phytol. Balcan. **20** (1): 33- 48.

Metcalfe, C.R. and Chalk, L. 1950. Anatomy of Dicotyledons, Second edition, Clarendon Press, Oxford.

Özcan, T. 2002. SEM observation on petals and fruits of some Turkish endemic *Buplerum* L. (Umbelliferae) species. Bot. J. Linn. Soc. **138** (4): 441-449.

Palmer, P.G. and Gerbeth-Jones, S. 1986.A scanning electron microscope survey of the epidermis of East African grasses IV. Smithson. Contr. Bot. **49**: 1-120.

Palmer, P.G. and Gerbeth-Jones, S. 1988. A scanning electron microscope survey of the epidermis of East African grasses V and West African supplement. Smithson. Contr. Bot. **67**: 1-153.

Rejdali, moh.F.L.S. 1991. Leaf micromorphology and taxonomy of North African species of *Sideritis* L. (Lamiaceae). Bot. J. Linn. Soc. **107**(1): 67-77.

Sharma, A., Acharya, R.N., Shukla, V.J., Harisha, C.R. and Gupta, S.K. 2013. Identification of *Cordia macleodii* Hook. through DNA finger printing and stomatal index. Universal Journal of Pharmacy - UJP **2**(3): 98-101.

Stace, C.A. 1984. The taxonomic importance of the leaf surface. *In*: Heywood, V.H., Moore, D.F. (Eds), Current Concepts of Plant Taxonomy. Academic Press, London, pp. 31–39.

Stockey, R.A. and Taylor, T.N. 1978a. Scanning electron microscopy of epidermal patterns and cuticular structure in genus *Araucaria*. Scan. Electron Microsc. **2**: 223-228.

Stockey, R.A. and Taylor, T.N. 1978b. Cuticular features and epidermal patterns in the genus *Araucaria* de Jussieu. Bot. Gaz. **139**: 490-498.

Täckholm, V. 1961. Botanical identification of the plants found at the Monastery of Phoebammon. *In*: Bachatly, C. (Ed.) La Monastère de Phoebammon dans la Thébaïde. Cairo.

Täckholm, V. 1974. Student's flora of Egypt, second edition, Cairo University, Beirut.

Taia, W.K. 2006. Family Boraginaceae: Hair variations and their significance in the systematic of the Genera. Asian J. Plant Sci. **5**(3): 441-454.

Taroda, N.1984. Taxonomic studies on Brazilian species of *Cordia* L. (Boraginaceae). Ph.D. Thesis, University of Saint Andrews, Saint Andrews. pp. 1-232.

Taroda, N. and Gibbs, P. 1986. Studies of the genus *Cordia* L. (Boraginaceae) in Brazil. 1. A new infrageneric classification and conspectus. Rev. Bras. Bot. **9**: 31-42.

Van der Veen, M. 2011. Consumption, trade and innovation. Journal of African Archaeology Monograph series (JAAMs) **6**: 1-313.

Ventrella, M.C. and Marinho, C.R. 2008.Morphology and histochemistry of glandular trichomes of *Cordia verbenacea* DC. (Boraginaceae) leaves. Rev. Bras. Bot. **31**(3): 457-467.

Verdcourt, B. 1991. Boraginaceae. *In:* Polhili, R.M. (Ed.), Flora of Tropical East Africa. A. A. Balkema, Rotterdam, Brookfield, Netherlands, 124 p.

Warfa, A.M. 1988. *Cordia* (Boraginaceae) in NE Tropical Africa and Tropical Arabia. Acta Universitatis Upsaliensis - Comprehensive Summaries of Uppsala Dissertations from the Faculty of Science **174**: 1-80.

Wilkinson, H.P. 1983. Leaf anatomy of *Gluta* (L.) Ding Hou (Anacardiaceae). Bot. J. Linn. Soc. **86**: 375-403.

Wilkinson, H.P. 1992. Leaf anatomy of the Pittosporaceae R. Br. Bot. J. Linn. Soc. **110**: 1-59.

Yoon, C.Y. 1994. A taxonomic study on the genus *Angelica* L. in Korea and adjacent region. Ph.D. Thesis, Korea University, Korea.

5

PALYNOLOGICAL STUDY OF SOME IRANIAN SPECIES OF *SCABIOSA* L. (CAPRIFOLIACEAE)

EBADI-NAHARI MOSTAFA[1] AND NIKZAT-SIAHKOLAEE SEDIGHEH[2]

Department of Biology, Faculty of Science, Azarbaijan Shahid Madani University, Tabriz, Iran

Keywords: Caprifoliaceae; Pollen characters; *Scabiosa;* SEM; Taxonomy; UPGMA.

Abstract

The pollen morphology of six species of *Scabiosa* L. (Caprifoliaceae) from Iran has been examined by scanning electron microscopy (SEM). Pollen grains were tricolpate in *S. columbaria* and triporate in the rest studied species. Two types of exine ornamentation were revealed: spinulate and gemmate. The pollen shape in polar view varied from triangular to circular among investigated taxa. Statistical analysis showed that some quantitative morphological features such as polar axis (P), equatorial axis (E) and aperture diameter were main characters in identification of the taxa studied. These taxa were separated from each other using cluster analysis and placed within two clusters. Our result based on UPGMA analysis is in agreement with morphological classification and recent findings on taxonomic position of the *Scabiosa*.

Introduction

Scabiosa L. belonging Caprifoliaceae includes approximately 80 annual or perennial herbs, distributed in Europe and the Mediterranean Basin, southern Africa and eastern Asia (Reveal and Chase, 2011). Some of the species are known as a source of herbal medicine for the treatment of many human diseases (Bonet *et al.*, 2007). Few *Scabiosa* species are cultivated as ornamental plants.

Linnaeus (1753) distinguished three genera that include *Scabiosa* L., *Dipsacus* L. and *Knautia* L. Later on, *Scabiosa sensu* Linnaeus was segregated into different genera: *Pterocephalus* Adans, *Succisa* Haller, *Cephalaria* Schrad. *ex* Roem & Schult. After remaining species of *Scabiosa* in different sections, were raised to new genera: *Lomelosia* Raf, *Sixalis* Raf, *Pseudoscabiosa* Devesa, *Pterocephalidium* G. Lopez (Greuter and Raus, 1985; Adanson, 1763; Haller, 1768; Devesa, 1984a, b; Lopez-Gonzales, 1987). Rechinger and Lack (1991) did not accept the nomenclatural changes suggested by Greuter and Raus (1985). Jamzad (1993) maintained a traditional and broad concept of genus *Scabiosa* too.

Taxonomic problems and species complexity are very common in this genus. Hybridization is common and, as a result, the number of reported species (and subspecies) has widely varied (Bobrov, 1957; Grossman, 1975; Jasiewicz, 1976).Since taxonomic position of some species is ambiguous, finding various remarkable characters will be useful to determine the taxonomic status of species. Pollen morphology has been able to reposition several disputed genera and interpret problems related to the origin and evolution of many taxa (Nair, 1980) and to derive classification of angiosperms (Cronquist, 1981).

Pollen morphology of this family has been studied by various authors (Ting, 1949; Nowicke and Skvarla, 1979; Feng *et al.*, 2000). Erdtmann (1952) studied 35 genera of Dipsacaceae, and results showed the presence of two types of pollen apertures in this family: porate and colpate.

[1] Corresponding author , Email: ebadi2023@yahoo.com
[2] Faculty of Biological Sciences, Shahid Beheshti University, Tehran, Iran. Email: Nikzat.sedighe@gmail.com

Caputo and Cozzolino (1994) divided Dipsacaceae into two major clades based on morphological and palynological characters. Khalik (2010) studied various genera of Dipsacaceae and proved the taxonomic value of pollen characters. Pollen shape, size, exine ornamentation, number and features of apertures represent useful characters for distinguishing among species.

There are 22 species of *Scabiosa* recorded in the Flora Iranica (Rechinger, 1989) that species are divided into two subgenera (*Scabiosa* and *Asterocephalus*) and three sections including *Scabiosa*, *Asterocephalus* and *Olivierianae*. Greuter and Raus (1985) treated Iranian species of *Scabiosa* into two genera as *Lomelosia* (= *Scabiosa* sect. *Astrocephalus* and sect. *Olivierinae*) and *Scabiosa s. str.* (*Scabiosa s. l* sect. *scaboisa*).

The taxonomic grouping of the genera *Scabiosa* and *Lomelosia* is exactly unknown (Mayer and Ehrendorfer, 1999). For example on the basis of The plant list (2013), situation of many Iranian species is unresolved (*L. flavida*, *L. calocephala*, *L. bicolor*, *L. esfandiarii*, *L. prophyroneura*, *L. schimperiana*, *L. machrochaete*, *L. kermanensis* and *L. leucactis*) and some of them are accepted names as *Scabiosa* (*S. rotata*, *S. micrantha*, *S. argentea*). As well as, *L. caucasica* and *L. olivieri* accepted as genus *Lomelosia* and *S. persica* and *S. columbaria* accepted as genus *Scabiosa*.

As the circumscription of these groups are not obvious completely and it has not been stabilized yet, in this work only the genus *Scabiosa sensu* Rechinger and Lack (1991) has been subject of studies.

There are no reports on pollen morphology of the *Scabiosa* from Iran. The present study aims to survey the pollen morphology of six Iranian species of *Scabiosa* L. (belonging to three sections) as *S. columbaria* L., *S. micrantha* Desf., *S. persica* Boiss., *S. calocephala* Boiss., *S. olivieri* Coult. and *S. flavida* Boiss. & Hausskn. using scanning electron microscopy and evaluating its significance in taxonomy of the genus.

Materials and Methods

The plant samples for the study were collected from natural populations in different regions of Iran during spring and summer of 2013. Four to five individuals were collected randomly from each locality. Details of localities are given in Table 1. The voucher specimens were deposited in Herbarium of Shahid Beheshti University (HSBU).

Table 1. Localities and voucher numbers of the taxa studied.

Genus	Species	Locality	Voucher No.
Scabiosa L.	*S. columbaria* L.	Mazandaran, Siahbisheh	HSBU4004
	S. micrantha Desf.	North Khorasan, 45 km of Bojnurd	HSBU4005
	S. persica Boiss.	West Azarbaijan, Piranshahr	HSBU4000
	S. alocephala Bioss.	Tehran, SorkhHesar Park	HSBU4001
	S. olivieri Coult.	Tehran, Telo	HSBU4003
	S. flavida Boiss. & Hausskn	North Khorasan, 45 km East of Bojnurd	HSBU4002

For SEM, pollen samples were mounted on stubs using double-sided adhesive tape. Macro and microphotographs which showed general view of pollen surface were taken by Phenomprox SEM at an accelerating voltage 10.0 kV. For measurements "Image Tools ver. 3.00" software with high degree of accuracy and confidence was used. Some palynological characteristics, such as equatorial diameter (*E*), polar axis length (*P*), *P/E*, log *P/E*, exine ornamentation and pore dimensions, were described for each sample.

In order to show relationships of species, we performed a cluster analysis based on Euclidean Distances with un-weighted pair-group method with arithmetical mean (UPGMA) method by using the program PAST ver. 2.17c (Hammer *et al.*, 2001).

Results

Palynological characters were randomly measured by using minimum 20 pollen grains. The quantitative and qualitative palynological data of six investigated taxa were shown in Table 2.

Table 2. Evaluated pollen characters in the studied taxa.

	S. columbaria	*S. micrantha*	*S. persica*	*S. calocephala*	*S. olivieri*	*S. flavida*
Pollen type	Tricolpate	Triporate	Triporate	Triporate	Triporate	Triporate
Polar axis (P)	80.71 ± 3.24	28.44 ± 2.86	27.01 ± 2.02	22.04 ± 2.46	93.55 ± 4.12	90.32 ± 3.59
Equatorial axis (E)	58.07 ± 5.23	68.26 ± 3.61	98.42 ± 4.82	87.65 ± 4.46	121.39 ± 6.13	117.44 ± 5.15
(P/E)	1.38	0.41	0.27	0.25	0.77	0.76
Pollen shape	Prolate	Preoblate	Preoblate	Preoblate	Suboblate	Suboblate
Polar view	Circular	Circular-triangular	Triangular	Triangular	Circular-triangular	Circular-triangular
Aperture	Linear	Circular	Elliptic	Elliptic	Circular	Circular
Aperture diameter	40.67 ± 2.96	14.33 ± 1.27	15.69± 1.56	15.19 ± 2.05	15.02 ± 0.92	15.89 ± 1.08
Exine ornamentation	Spinulate-Spinuloid	Spinulate-Spinuloid	Gemmate-Spinuloid	Gemmate-Spinuloid	Spinulate-Spinuloid	Spinulate-Spinuloid

There are two major types of pollen grain apertures. It varies from tricolpate in *S. columbaria* to triporate in the rest studied species (Fig1) as *S.persica* and *S. calocephala* have elliptical pore and *S. micrantha*, *S. flavida* and *S. olivieri* have circular pores (Table 2).

The size of the pollen grain of studied taxa ranged from 28 × 68 µm (*P* × *E*) in *S. micrantha* to 93 × 121 µm (*P* × *E*) in *S. olivieri*. The ratio between the mean polar axis (*P*) and the mean equatorial diameter (*E*) can be used to assign the pollen grains to shape classes as of Erdtmann (1952). The shape of the pollen grain in equatorial view varied from preoblate to prolate among investigated taxa. The pollen shape in polar view varied from triangulate in *S. persica* and *S. calocephala*, circular in *S. colembaria* to circular-triangular in the rest of the species (Fig. 1).

Two types of exine ornamentation found in studied taxa. Gemmate type was observed in *S. persica* and *S. calocephala* as so that, between the gemmate are numerous low irregularly spaced spinuloid (Fig 2). Spinulate type was observed in *S. colembaria*, *S. flavida*, *S. micrantha* and *S. olivieri* as beset with numerous similar small conical spinuloid (Fig. 2).

The taxa investigated were separated from each other in a UPGMA tree (Fig. 3). Cluster analysis showed that species placed in two clusters. *S. olivieri* and *S. flavida* placed in one cluster and the remained taxa placed in other cluster. This cluster divided to two sub-clusters: subcluster I included *S. micrantha*, *S. persica* and *S. calocephala* and subcluster II contained *S. columbaria*.

Principal component analysis (PCA) shows that the polar axis (P), equatorial axis (E) and aperture diameter are main characters in grouping of species (Fig. 4).

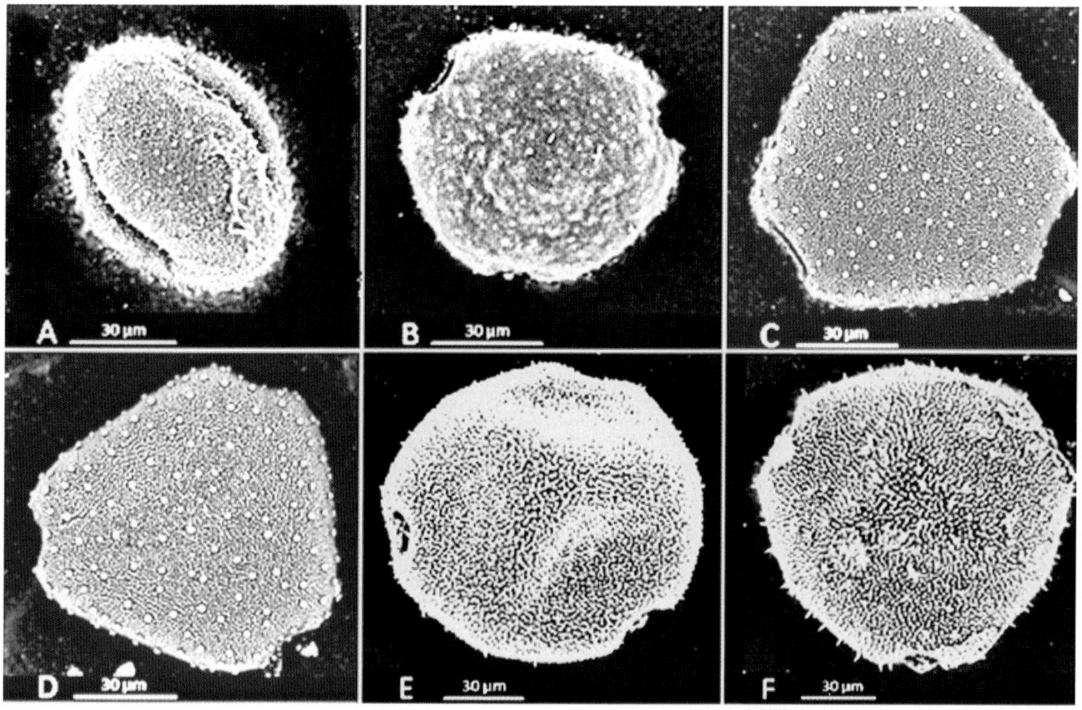

Fig. 1. Pollen electron micrograph of studied taxa. A: *S. columbaria*, B: *S. micrantha*, C: *S. persica*, D: *S. calocephala*, E: *S. olivieri* and F: *S. flavida*.

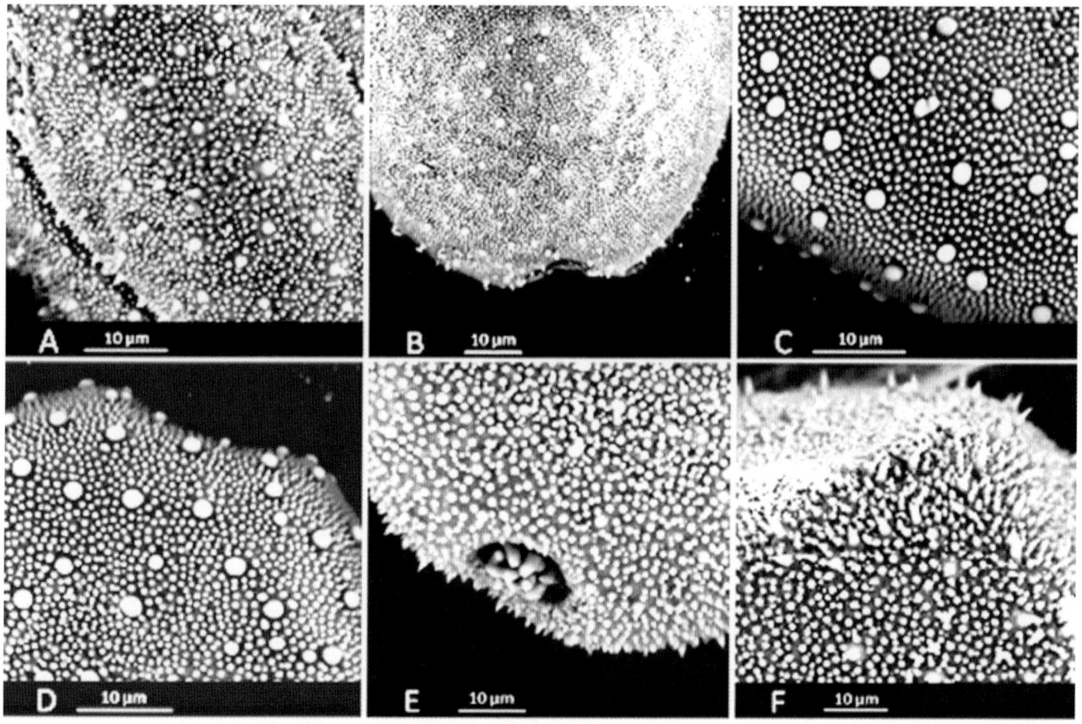

Fig. 2. Pollen surface ornamentations in studied taxa. A: *S. columbaria*, B: *S. micrantha*, C: *S. persica*, D: *S. calocephala*, E: *S. olivieri* and F: *S. flavida*.

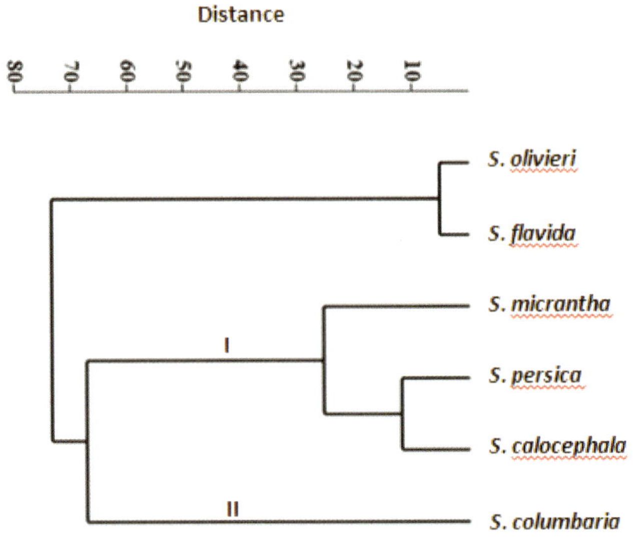

Fig. 3. UPGMA dendrogram showing the relationship among studied *Scabiosa* species based on palynological characters.

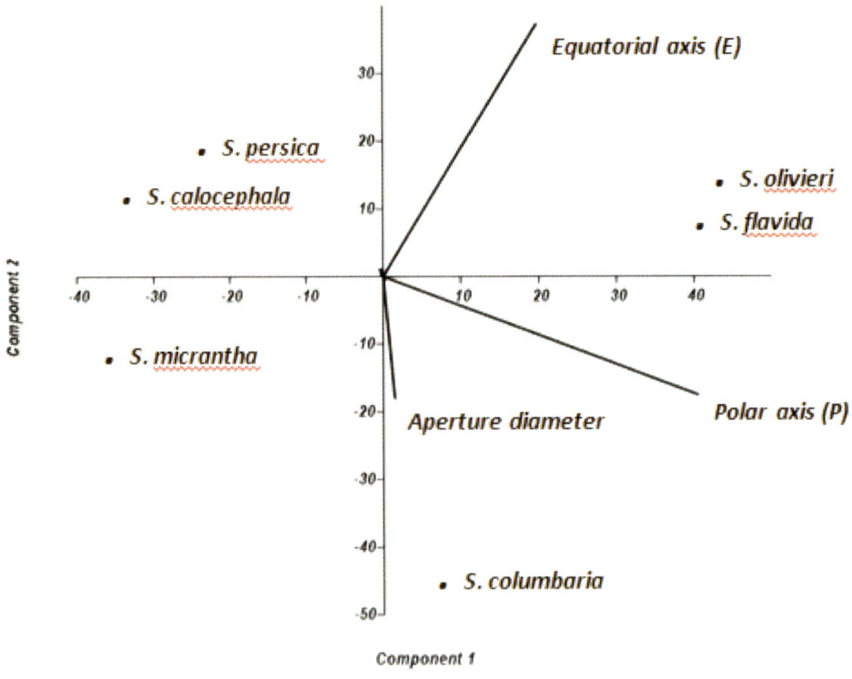

Fig. 4. Principal Component Analysis (PCA) among the studied *Scabiosa* species based on palynological characters.

Discussion

There are lots of debates on taxonomy of the genus *Scabiosa* and undoubtedly the genus has been the subject of the various taxonomic studies. Taxonomic classification of *Scabiosa* species is still unresolved. Because of morphological similarities between *Scabiosa* species are very high, so it seems that the application of pollen morphology can help us to resolve the taxonomical problems between the species. Previous studies show that pollen characters can be used for solving taxonomic problems (Khalik, 2010). In this study, palynological characters varied among the investigated taxa and were useful in identification of taxa. For example, *S. columbaria* can be distinguished from the other taxa based on aperture types. *P/E* ratio was useful in identification of *S. micrantha* from *S. olivieri* and *S. flavida*, where their exine ornamentation is very similar.

Various palynological studies on different species of the Dipsacaceae have confirmed that pollen characters were important to species identification. Perveen and Qaiser (2011) studied palynological characteristic in some species of Dipsacaceae from Pakistan. Their results indicated that pollen characters may be used to delimit the species.

Aperture types are the most significant pollen characters. On the basis of aperture types, *Scabiosa* is divided into two pollen types including tricolpate and triporate. *S. columbaria* was tricolpate while the rest studied taxa were triporate. *Scabiosa* is divided into two subgenera, including *Scabiosa* and *Asterocephalus*, based on morphological characters (such as presence of Groove or pit on epicalyx tube) (Rechinger, 1989) and according to present study, they have tricolpate and triporate apertures, respectively. Perveen and Qaiser (2011) divided the Pakistani species of the family Dipsacaceae into three pollen types on the basis of aperture viz., *Dipsacus inermis*–type, *Pterocephalus gedrosiacus*–type and *Scabiosa candollei*–type. Our results are in agreement with those of Perveen and Qaiser (2011) who have also observed triporate pollen in Pakistani *Scobiosa* species. *S. columbaria* species complex has long posed for a complex taxonomic issue. Hybridization is common and, as a result, the number of reported species (and subspecies) has varied widely (Bobrov, 1957; Grossman, 1975; Jasiewicz, 1976). A revision of the species limits in *S. columbaria* is much needed.

The UPGMA dendrogram based on palynological characters is in concordant with morphological classification. Mayer and Ehrendorfer (1999) investigated differentiation of the epicalyx, the corona, the epi-diaphragm, calyx, pollen, chromosome number and their taxonomic importance in genus *Scabiosa s.l* and concluded that these features could be helpful in circumscribing of this genus.

S. persica and *S. calocephala* have similar characters including: number of corona vein, epicalyx tube pits (oblong shape), presence of sulcus between pits, hidden bristle rather to corona. These two species are different in leaves, leaflets, number of involucre bracts (Jamzad, 1993).

S. olivieri and *S. flavida* have small dipsacaceous- head and thus they are different from other species. These two species have triangle pits, short corona, evident bristles rather to corona, lack of sulcus between pits. There is radiant flower in *S. flavida* but not in *S. olivieri*.

S. micrantha have oval shape dipsacaceous- head in contrast with other species. This specieshas oblong shape pits, sulcus between pits, and evident bristles rather to corona.

Despite other species (five above mentioned species), *S. columbaria* has 8-grooves along epicalyx tube (Jamzad, 1993). Morphological similarity between these taxa is approved by pollen characteristics (Fig 3). *S. micrantha* have some characters of *S. olivieri* group (*S. olivieri* and *S. flavida*) and some of *S. persica* group (*S. persica* and *S. calocephala*). Hence, on the basis of pollen features, *S. micrantha* located between them.

Taxonomic status of *S. calocephala* is unresolved but some data suggest that it is synonymous with *Lomelosia calocephala*. *S. calocephala* and *S. micrantha* based on the presence of pits on epicalyx tube, are placed in sect. *Asterocephalus*. The UPGMA dendrogram showed that this species were very similar to *S. persica*. Therefore, taxonomic status of *S. calocephala* does not change according to pollen morphology.

S. olivieri and *S. flavida* were previously classified within the genus *Scabiosa*, based on the calyx characters (Tackholm, 1974; Rechinger, 1989; Boulos, 2000). Mayer and Ehrendorfer

(1999) separated all species of *Scabiosa* set. *Olivierianae* into *Lomelosia*. In general, our results agree with those of Mayer and Ehrendorfer. Caputo *et al*. (2004) studied phylogenetic relationships among 17 species of Dipsacaceae and they divided Dipsacaceae into two major clades: one including *Lomelosia* and *Pycnocomon*, both in a sister group relationship with a clade including *Scabiosa*, *Sixalix* and *Pterocephalus*, and the other including the rest of species.

Unlike other studies, in this work exine ornamentation was not useful to distinguish among closely related genera such as *Scabiosa* and *Lomelosia*. The exine ornamentation of *S. micrantha* is very similar to *S. olivieri* and *S. flavida* while our results showed that *S. micrantha* was grouped in sect. *Asterocephalus* in accordance with Flora Iranica classification.

Feng *et al*. (2000) studied pollen of 17 species of the genus *Dipsacus* and divided the genus into three pollen types based on exine ornamentation viz., dispinulate-reticulate, spinulate-foveolate and dispinulate or rarely smooth. He further reported that pollen morphology was little helpful at specific level.

The results of this study showed that pollen morphology provided facile and reliable characters for taxonomic studies of *Scabiosa*.

References

Adanson, M. 1763. Familles des plantes. Vincent, Paris. **2**: 152.

Bobrov, E.G. 1957. Dipsacaceae. *In*: Flora of the USSR, Vol. **24** (Ed. by B.K. Shishkin). Akademii Nauk SSSR, Moscow.

Bonet, M.A. and Vallès, J. 2007. Ethnobotany of Montseny biosphere reserve (Catalonia, Iberian Peninsula): plants used in veterinary medicine. Journal of Ethnopharmacology **110**(1):130–147.

Boulos, L. 2000. Flora of Egypt. **2**. Cairo.

Caputo, P. and Cozzolino, S. 1994. A cladistics analysis of Dipsacaceae (Dipsacales). Plant System.Evol. **189**: 41–61.

Caputo, P., Cozzolino, S. and Moritti, A. 2004. Molecular phylogenetics of Dipsacaceae reveals parallel trends in seed dispersal syndromes. Plant System. Evol. **246**: 163–175.

Cronquist A. 1981. An Integrated System of Classification of Angiosperms. Columbia University Press, New York.

Devesa, J. A. 1984a. Revisidn del genero *Scabiosa* en la Peninsula Ibrrica e Islas Baleares. Lagascalia. **12**: 143-212.

Devesa, J. A. 1984b. *Pseudoscabiosa,* genero nuevo de *Dipsacaceae.* Lagascalia **12**: 213-221.

Erdtmann, G. 1952. Pollen morphology and plant taxonomy. Chronica Botanica Co., Massachusettes; Copenhagen.

Feng, X.F., Ai, T.M. and Xu, H.N. 2000.A study on pollen morphology of *Dipsacus*. Zhongguo Zhong Yao ZaZhi. **25**(7): 394-401.

Greuter, W. and Raus, T. 1985. Conservatoire et Jardinbotaniques de la Ville de Gene`ve, Gene`ve. Med-Checklist Notulae: 11. Willdenowia **15**: 61–64.

Grossman, F. 1975. Morphologisch-o¨kologischeUntersuchungenan *Scabiosa columbaris* L. s.l. immittleren und westlichen Alpengebiet. Vero¨ffentlichungen des Geobotanischen Institutes der ETH, Stiftung Ru¨bel. **25**: 1–125.

Haller, A. 1768. Historiastirpiumindigenarum Helvetiae, Societatis Typhographicae, Bern. **1**: 87.

Hammer Ø, Harper DAT, Ryan PD. 2001. PAST: Paleontological Statistics Software Package for Education and Data Analysis. Palaeontologia Electronica 4: 9. http://palaeo-electronica.org/ 2001_1/past/issue1 _01.htm.

Jamzad, Z. 1993. *Scabiosa* L. *In:* Assadi, M., Khatamsaz, M. and Maassoumi, A.A. (Eds.) Flora of Iran, Vol. **8**. Islamic Republic of Iran, Ministry of Jahad-e Sazandegi, Research Institute of Forests and Rangelands, Tehran, pp. 63–106.

Jasiewicz, A. 1976. Scabiosa. *In:* T.G. Tutin, V.H. Heywood, N.A. Burges, D.H. Valentine, S.M. Walters and D.A. Webb (Eds). Flora Europaea, Vol. **4**. Cambridge University Press, Cambridge, pp. 68–74.

Khalik, K.A. 2010. A palynological study of the family Dipsacaceae in Egypt and its taxonomic significance. J. Bot. Taxonomy & Geobotany **121**(3-4): 97-111.

Linnaeus, C. 1753. Species plantarum, exhibentesplantas rite cognitas, ad genera relatas, cum Differentiis Specificis, Nominibus Trivialibus, Synonymis Selectis,Locis Natalibus, Secundum systema sexualedigestas, Impensis Laurentii Salvii, Stockholm **1**: 98.

Lopez-Gonzalez, G. 1987. *Pterocephalidium,* unnuevogrneroibrrico de la familia *Dipsacaceae.* Anales Jard. Bot. Madrid. **43**: 245-252.

Mayer, V. and Ehrendorfer, F. 1999. Fruit differentiation, palynology, and systematics in the Scabiosa group of genera and Pseudoscabiosa (Dipsacales). PI. Syst. Evol. **216**: 135-166.

Nair, P.K.K. 1980. Pollen Morphology of Angiosperms.Vikas Publications, New Delhi.

Nowicke, J.W. and Skvarla, J.J. 1979. Pollen morphology: The potential influence in higher order systematics. Ann. Mo. Bot. Gard. **66**: 633-699.

Perveen, A. and Qaiser, M. 2011. Pollen flora of Pakistan-Dipsacaceae LXVIII., Pak. J. Botany **42**(6): 2825-2827.

Rechinger, K. H. 1989: *Scabiosa* sect. Olivierianae, sect. nova (Dipsacaceae). Willdenowia **19**: 137–151.

Rechinger, K. H and Lack, H. W. 1991. Dipsacaceae. *In*: Rechinger, K. H., (Ed.): Flora Iranica. 168, pp. 1-67.

Reveal, J. L. and Chase, M. W. 2011. APG III: Bibliographical information and synonymy of Magnoliidae. Phytotaxa **19**: 71–134.

Tackholm, V. 1974. Students' Flora of Egypt. Beirut, Cairo University.

The Plant List 2013. Version 1.1. Published on the Internet; <www.theplantlist.org>.accessed 1st January.

Ting, S.U. 1949. Illustration of pollen grains of some Chinese plants. Bot. Notiser. **4**: 277-282.

6

STOMATAL AND TRICHOME DIVERSITY IN *SENNA* MILL. FROM BANGLADESH

AYESA BEGUM, MD. OLIUR RAHMAN[1] AND MOMTAZ BEGUM

Department of Botany, University of Dhaka, Dhaka 1000, Bangladesh

Keywords: Senna; Stomata; Trichomes; Taxonomy; Bangladesh.

Abstract

Foliar epidermal characters of ten species of *Senna* Mill. found in Bangladesh are investigated. Anisocytic, anomocytic, paracytic, tetracytic and haxacytic stomata are found across the species. Anisocytic and paracytic stomata are common in all species. In addition to anisocytic and paracytic types, anomocytic stomata are found in *Senna sophera* and hexacytic stomata are observed in *S. auriculata*. Anticlinal wall is straight in *S. alata, S. hirsuta, S. occidentalis, S. siamea* and *S. tora*, curved in *S. auriculata, S. sophera* and *S. siamea*, and undulate in *S. hirsuta* and *S. occidentalis*. Glandular trichomes are observed in *S. hirsuta* and *S. occidentalis*, while non-gladular trichomes are predominant in the remaining species.

Introduction

Senna Mill. is a large, widespread genus and exhibits a high diversity of habits including herbs, shrubs, treelets, tall trees and lianas. The genus comprises 350 species and about 80% of its species occur in the American continent, while most of the remaining members are found in tropical Africa, Madagascar and Australia, and only a few species occur in South-eastern Asia and Pacific Island (Irwin and Barneby, 1982; Marazzi *et al*., 2006). Species of *Senna* were formerly included in *Cassia* L. *s.l.* (Irwin and Turner, 1960). Subsequent taxonomic treatments subdivided the *Cassia s.l.* into three distinct genera, *viz., Cassia s. str., Chamaecrista* Moench and *Senna* Mill. (Irwin and Barneby, 1981, 1982). *Senna* Mill. are characterized by presence of extrafloral nectaries, ebracteolate pedicels, straight or simply incurved filaments, flattened or cylindrical, irregularly dehiscent pods and areolate seeds, while in *Cassia* L., no extrafloral nectaries present, pedicels 2-bracteolate, filaments sigmoidally curved towards the base, pod indehiscent, and there is no areole on seed surface. *Chamaecrista* Moench is distinct by ciliolate anther-thecae along the suture, elastically dehiscent pod, smooth or pitted seed coat and absence of aerole on seed surface (Irwin and Barneby, 1982). The separation of *Senna* Mill. from *Cassia* L. was further established by taxonomic (Singh, 2001), phenetic (Boonkerd *et al*., 2005) and molecular studies (Acharya *et al*., 2011). In Bangladesh, *Senna* Mill. is represented by 11 species (Rahman *et al*., 2013).

Taxonomic relevance of vegetative anatomy in delimitation of taxa and establishment of itergeneric or interspecific relationships is well reported (Tomblinson, 1961; Stace, 1965; Kotresha and Seetharam, 2000; Tschan and Denk, 2012). Foliar anatomical features play an important role in distinguishing different groups of plants. Leaf is considered as the most varied organ anatomically in angiosperms which provides a variety of anatomical features that can be employed as useful taxonomic characters (Metcalfe and Chalk, 1950; Metcalfe, 1968; Stace, 1984). Foliar anatomical characters such as stomata and trichomes have been found instrumental in solving taxonomic problems. The taxonomic and phylogenetic significance of stomata and trichomes have long been recognized by various workers (Dilcher, 1974; Naik and Nirgude, 1981; Stace, 1984; Devi *et al*., 2013).

*Corresponding author. Email: prof.oliurrahman@gmail.com

Despite several studies based on foliar micromorphology have been made in defferent genera, *viz., Polygonum* (Lersten and Curtis, 1992), *Eugenia* (Fontenelle *et al.*, 1994), *Bauhinia* (Kotresha and Seetharam, 1995), *Hibiscus* (Shaheen *et al.*, 2009), *Fagopyrum* (Yasmin *et al.*, 2010) and *Cynoglossum* (Akçin *et al.*, 2012), the genus *Senna* Mill. received very little attention in this regard (Ogundipe *et al.*, 2009). Therefore the present study was undertaken in order to explore the features of stomata and trichome in the genus *Senna* Mill. occurring in Bangladesh and to evaluate these characters for species delimitation and interspecific relationship.

Materials and Methods

Ten species of *Senna* Mill. used in this study are *Senna alata* (L.) Roxb., *S. auriculata* (L.) Roxb., *S. hirsuta* (L.) Irwin & Barneby, *S. obtusifolia* (L.) Irwin & Barneby, *S. occidentalis* (L.) Link, *S. siamea* (Lam.) Irwin & Barneby, *S. sophera* (L.) Roxb., *S. surattensis* (Burm. f.) Irwin & Barneby, *S. timoriensis* (DC.) Irwin & Barneby and *S. tora* (L.) Roxb. The localities of these species along with the voucher numbers are appended in Table 1. Plant materials collected from different parts of the country as well as herbarium specimens deposited in Dhaka University Salar Khan Herbarium (DUSH) and Bangladesh National Herbarium (DACB) were studied.

Leaf samples were cut into small square pieces and soaked in a petridish containing KOH solution for 7-8 days. Then they were washed with tap water followed by treatment with hydrogen peroxide for 3-4 days till clearing of leaves. After that they were treated with 50% alcohol for 2 hours and then with safranin for 16-18 hours. The leaf tissues were treated again with 70% alcohol for 2 hours. Finally, the samples were treated with a mixture of alcohol and xylol in a 3:1 ratio and mounted in 70% glycerin. The samples were prepared in clean, dry and oil free slides and observed under compound microscope. Photomicrograph of each species was taken using Nikon Eclipse 200 camera in addition to the line drawing of stomata.

Table 1. List of species of *Senna* Mill. along with voucher specimens used in the present study.

No.	Species	Voucher specimens
1	*Senna alata* (L.) Roxb.	Dhaka: Dhaka University campus, 23.12.2011, Ayesa 65 (DUSH)
2	*S. auriculata* (L.) Roxb.	Dhaka: Sher-e-Bangla Agricultural University compound, 26.1.2011, Ayesa 07 (DUSH)
3	*S. hirsuta* (L.) Irwin & Barneby	Gazipur: Gazipur, 30.6.2011, Ayesa 40 (DACB)
4	*S. obtusifolia* (L.) Irwin & Barneby	Cox's Bazar: Teknaf, Mouchuni, 24.4.2011, Ayesa 32 (DUSH)
5	*S. occidentalis* (L.) Link	Dhaka: Dhaka University campus, 26.12.2010, Ayesa 02 (DUSH)
6	*S. siamea* (Lam.) Irwin & Barneby	Dhaka: Tejgaon, Old Airport, 27.12.2011, Ayesa 74 (DUSH)
7	*S. sophera* (L.) Roxb.	Dhaka: Dhaka University Campus, 30.4.11, Aeysa 33 (DUSH)
8	*S. surattensis* (Burm. f.) Irwin & Barneby	Dhaka: Dhaka University campus, 20.12.2011, Ayesa 47 (DUSH)
9	*S. timoriensis* (DC.) Irwin & Barneby	Chittagong Hill Tracts: Ruma P.S., Changnakra, 25.1.1965, M. S. Khan 1166 (DACB).
10	*S. tora* (L.) Roxb.	Dhaka: Dhaka University Botanical garden, 26.12.2011, Ayesa 69 (DUSH)

Results and Discussion

The genus *Senna* offers different types of stomata and trichomes. The shape of epidermal cells, nature of anticlinal walls, types of stomata and number of stomata per microscopic field are presented in Table 2. The epidermal cells are irregular or polygonal in outline. Polygonal cells are observed in *S. obtusifolia, S. siamea* and *S. tora*, while irregular types of cells are evident in *S. sophera*. The remaining species bear both irregular and polygonal cells (Table 2). The anticlinal walls are straight, undulate and curved across the genus. Straight anclinal walls are found in *S. obtusifolia, S. siamea* and *S. tora*; anticlinal walls are curved in *S. auriculata, S. sophera* and *S. surattensis,* and they are undulate, curved and straight in *S. hirsuta* and *S. occidentalis*. Stomata are mostly paracytic and anisocytic, however, tetracytic, anomocytic and hexacytic stomata have also been documented. Paracytic and anisocytic stomata are observed in all species, while anomocytic stomata are found only in *S. sophera*, and hexacytic stomata only in *S. auriculata* along with other types. Paired anisocytic and paired tetracytic stomata are present only in *S. auriculata.* Paired paracytic stomata are observed in *S. sophera* and *S. hirsuta* (Table 2; Figs 1&2). Amphistomatic stomata have been observed in all species employed, while hypostomatic stomata is found only in *S. siamea*.

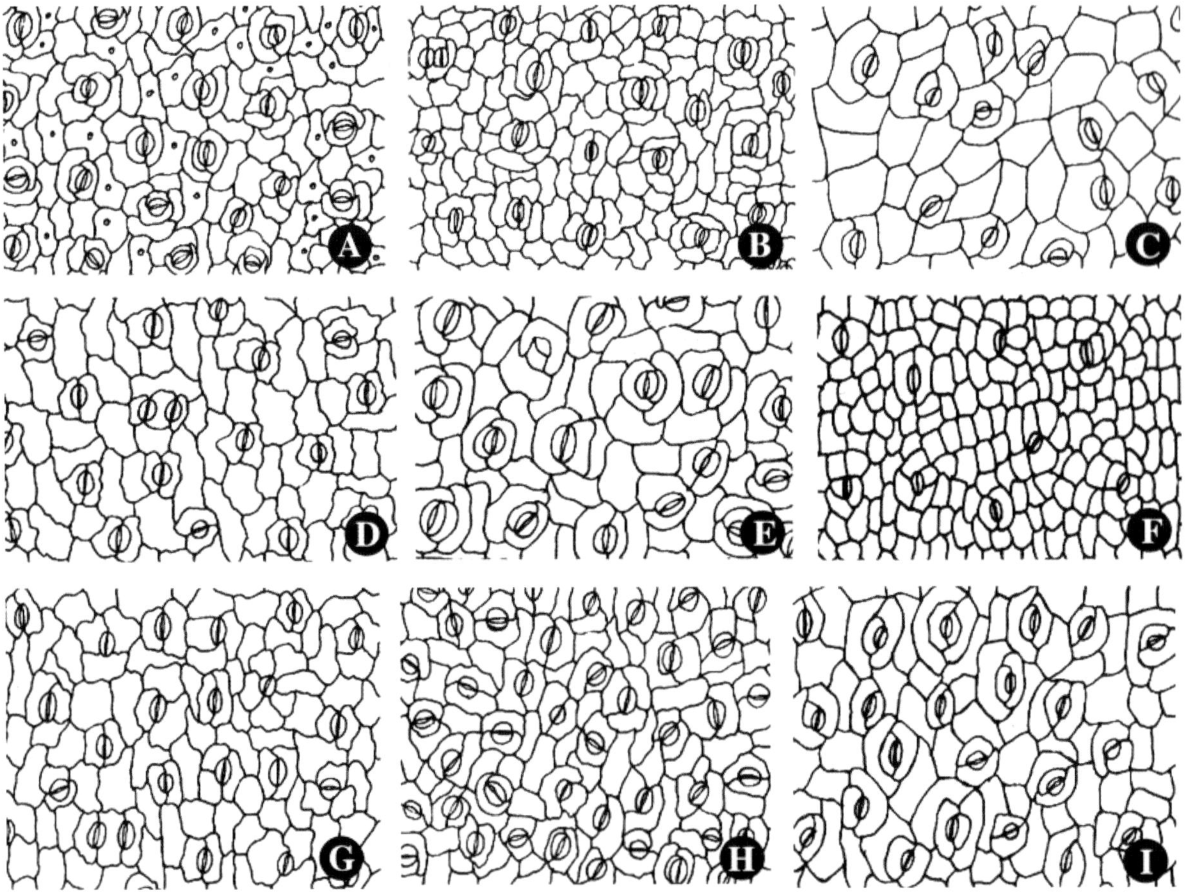

Fig. 1. Line drawings of different types of stomata found in *Senna* Mill., A) *Senna alata*, B) *S. auriculata*, C) *S. hirsuta*, D) *S. obtusifolia*, E) *S. occidentalis*, F) *S. siamea,* G) *S. sophera*, H) *S. surattensis*, I) *S.tora*.

Table 2. Foliar stomatal variation in *Senna* Mill. species used in the present study.

Species	Epidermal cell		Anticlinal wall		Types of stomata		Number of stomata/ microscopic field	
	Adaxial	Abaxial	Adaxial	Abaxial	Adaxial	Abaxial	Adaxial	Abaxial
Senna alata	Polygonal	Irregular	Straight	Straight, Curved	Paracytic, Anisocytic	Paracytic	29	50
S. auriculata	Polygonal, Irregular	Polygonal, Irregular	Curved	Curved	Paracytic, Anisocytic, Paired tetracytic	Paracytic, Anisocytic, Tetracytic, Hexacytic, Paired anisocytic	44	39
S. hirsuta	Polygonal, Irregular	Polygonal, Irregular	Straight, Undulate	Straight, Curved	Paracytic	Paracytic, Anisocytic, Paired paracytic	35	39
S. obtusifolia	Polygonal	Polygonal	Straight	Straight	Paracytic	Paracytic, Anisocytic	32	37
S. occidentalis	Polygonal, Irregular	Polygonal, Irregular	Straight, Undulate	Straight, Curved	Paracytic	Paracytic, Anisocytic, Tetracytic	21	46
S. siamea	Polygonal	Polygonal	Straight	Straight	-	Paracytic, Anisocytic, Tetracytic	-	0-7
S. sophera	Irregular	Irregular	Curved	Curved	Paracytic, Anisocytic, Anomocytic, Tetracytic, Paired paracytic	Paracytic, Anisocytic, Tetracytic	28	58
S. surattensis	Polygonal, Irregular	Polygonal, Irregular	Curved	Curved	Paracytic, Anisocytic, Tetracytic	Paracytic, Anisocytic	54	51
S. tora	Polygonal	Polygonal	Straight	Straight	Paracytic, Anisocytic	Paracytic, Anisocytic	32	25

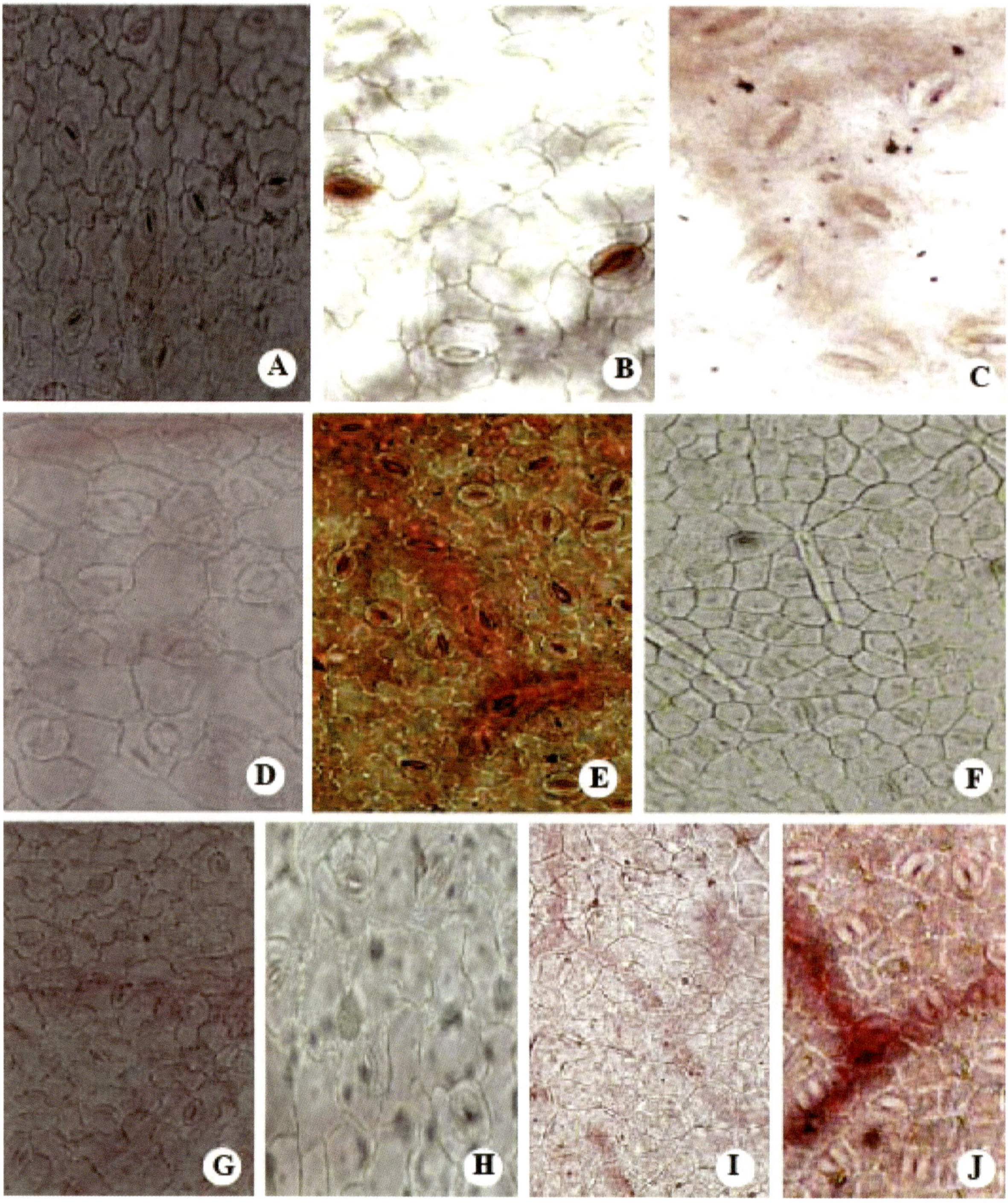

Fig. 2. Different types of stomata in *Senna* Mill. A) *Senna alata*, B) *S. auriculata*, C) *S. hirsuta*, D) *S. obtusifolia*, E) *S. occidentalis,* F) *S. siamea*, G) *S. sophera*, H) *S. surattensis*, I) *S. tora*, J) *S. timoriensis*.

The study reveals that both glandular and non-glandular trichomes are found in *Senna*. The trichomes observed in *Senna* vary in structure, form and distribution. Glandular multicellular trichomes are evident only in the abaxial surface of *S. hirsuta* and *S. occidentalis*. The number of cells varies from 5 to 7. However, no glandular unicellular trichomes are found in any species employed in the present study (Table 3). Non-glandular trichomes are evident in all species except *S. occidentalis*. Non-glandular multicellular trichomes are distinct in *S. hirsuta*, *S. obtusifolia* and *S. tora*, whereas, non-glandular unicellular trichomes are observed in eight species. In *S. alata, S. hirsuta* and *S. siamea* unicellular trichomes are sparsely noticed. Non-glandular unicellular trichomes may be conical or papilose. Very rarely uncinate trichomes are found only in *S. alata*. Non-glandular multicellular trichomes are consisted of 2-8 cells, which also vary greatly in size, shape and number of cells.Verrucose type of trichome wall are documented in *S. tora, S. obtusifolia, S. siamea, S. surattensis* and smooth walled trichomes are seen in *S. hirsuta, S. auriculata, S. timoriensis* (Fig. 3). The longest type of multicellular trichomes are present with up to 8 cells, comprising two or more basal cells ended by a much elongated apical cell in *S. hirsuta*. The pedastal cells might vary from 4 to 7. Multicellular verrucose trichome present in *S. tora* and *S. obtusifolia.* The cells of the stalk are almost equal in length. Smooth walled non-glandular multicellular trichomes are present in *S. hirsuta*, consisting of 2-6 cells and the cells may be equal or unequal in size. The upper cells always larger than the basal ones and become narrower towards the apex. *S. occidentalis* and *S. hirsuta* can easily be distinguished from the other species of *Senna* by presence of glandular multicellular trichomes.

Table 3. Trichome variation in *Senna* Mill. species employed in the present study.

Species	Glandular trichome				Non-glandular trichome			
	Unicellular		Multicellular		Unicellular		Multicellular	
	Adaxial surface	Abaxial surface	Adaxial surface	Abaxial surface	Adaxial surface	Abaxial surface	Adaxial surface	Abaxial surface
Senna alata	–	–	–	–	+ (very few)	+ (very few)	–	–
S. auriculata	–	–	–	–	++	++	–	–
S. hirsuta	–	–	–	+ (very few)	–	+ (very few)	+++ (2-6 celled)	+++ (2-4 celled)
S. obtusifolia	–	–	–	–	++	–	–	++ (1-7 celled)
S. occidentalis	–	–	–	++ (5-7 celled)	–	–	–	–
S. siamea	–	–	–	–	+ (very few)	+ (very few)	–	–
S. sophera	–	–	–	–	++	++	–	–
S. surattensis	–	–	–	–	++	++	–	–
S. timoriensis	–	–	–	–	++	+++	–	–
S. tora	–	–	–	–	–	–	–	++ (2-4 celled)

– = absent, + = sparsely present, ++ = moderately present, +++ = densely present.

The taxonomic value of leaf epidermal characters have been received much attention in the recent past, even the taxonomic monographs are now considered incomplete without micromorphology of the epidermis (Rejdali, 1991). In this context, however, little is known in the genus *Senna*. Recently Ogundipe *et al.* (2009) and Shaheed and Illoh (2010) studied foliar micromorphology of six species of *Senna* from Nigeria. The present study bridges the gaps in our knowledge of the some additional species of the genus *Senna* after Ogundipe *et al.* (2009) and Shaheed and Illoh (2010).

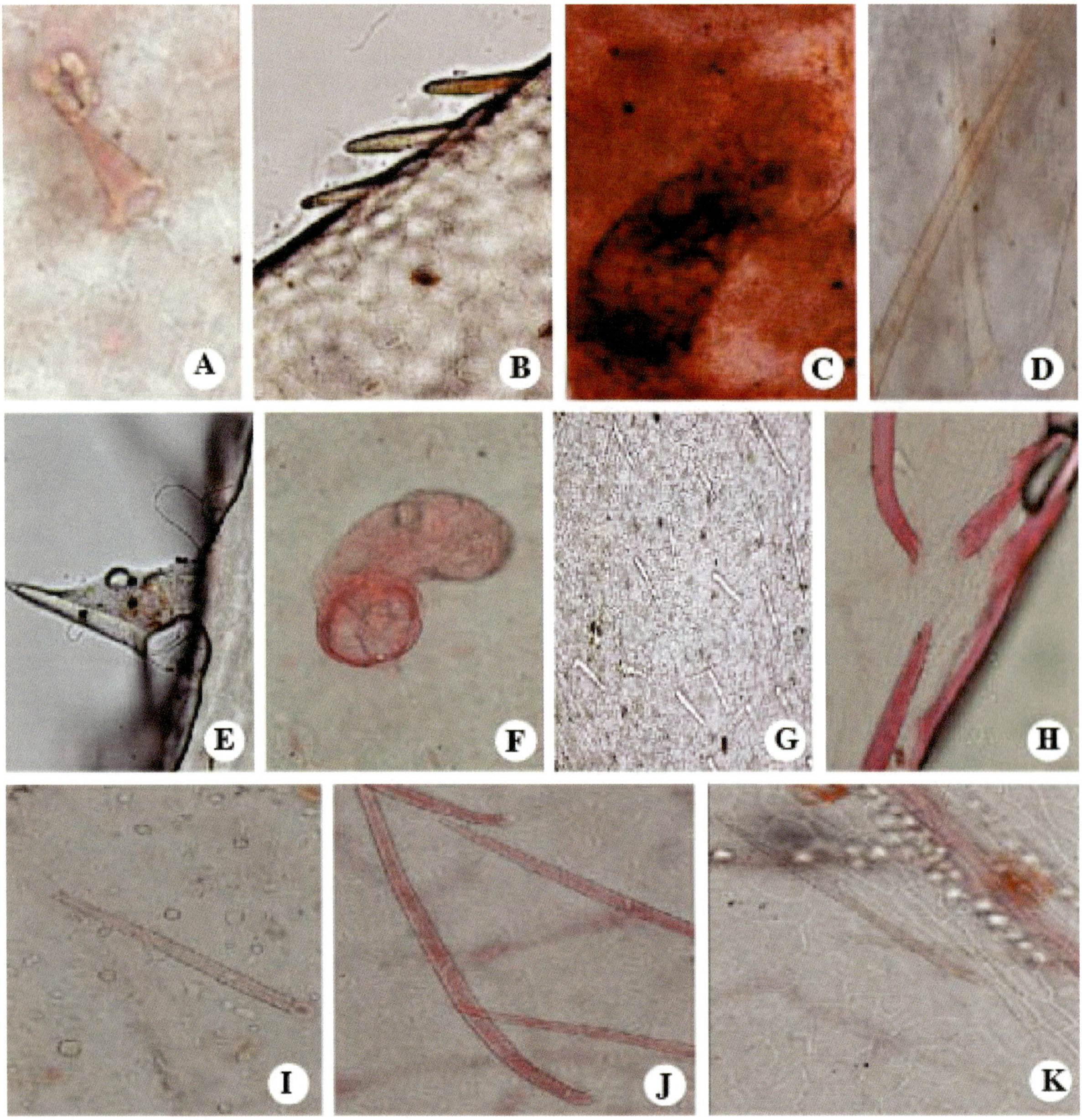

Fig. 3. Trichome diversity in *Senna* Mill., A) *Senna alata*, B) *S. auriculata*, C,D) *S. hirsuta*, E) *S. obtusifolia*, F) *S. occidentalis*, G) *S. siamea*, H) *S. sophera*, I) *S. surattensis*, J) *S. tora*., K) *S. timoriensis*.

The present study reveals that the foliar epidermal characters are important in taxonomic identification and species relationships in the members of the *Senna* examined. In our study *S. obtusifolia* and *S. tora* have been found closely related as evidenced by their epidermal cells which are polygonal in shape, and anticlinal walls are straight both in *S. obtusifolia* and *S. tora*. Moreover, paracytic and anisocytic stomata are common in these two species indicating a close relationships among them. In a morphometric study of the genus *Senna* Rahman *et al.* (2013) showed that *S. obtusifolia* and *S. tora* are very closely related as attested by the following shared characters: leaflets obovate, stipules linear, falcate, inflorescence short-racemose, axillary, ovary ribbed, style glabrous, stigma truncate and pod linear or subtetragonous, which has been found consitent with our study. Morphologically *S. alata* and *S. auriculata* are closely allied, however, the epidermal features do not support a close association between them. *S. auriculata* is characterized by presence of hexacytic stomata, while *S. sophera* is distinct by presence of anomocytic stomata. Saheed and Illoh (2010) observed that in *S. alata*, the adaxial epidermal cells are polygonal or irregular, while the abaxial cells are irregular in shape, and this species exhibited non-glandular trichomes. Our study presenting similar types of epidermal cells and non-glandular trichomes in *S. alata* support the work of Saheed and Illoh (2010). *S. obtusifolia* is characterized by presence of polygonal epidermal cells, straight anticlinal walls, and paracytic and anisocytic stomata, however, no anomocytic stomata were found in this species as documented by Saheed and Illoh (2010).

The study also unveils a close relationship between *S. hirsuta* and *S. occidentalis* as they share the following common characteristics: both polygonal and irregular epidermal cells, straight and undulate anticlinal walls in the adaxial surface, straight and curved walls in the abaxial surface, paracytic stomata in the adaxial surface, and both paracytic and anisocytic stomata in the abaxial surface (Table 2). *S. hirsuta* and *S. occidentalis* can also be differentiated from the remaining species by presence of multicellular glandular trichomes as observed in the abaxial surface (Table 3). While studying on the Nigerian *Senna* species Ogundipe *et al.* (2009) showed that *S. hirsuta* and *S. occidentalis* are closely allied as they bear paracytic and anisocytic types of stomata, and the epidermal cells are polygonal and irregular in both species. Results obtained from the present study have been found congruent with that of Ogundipe *et al.* (2009). The close affinity between *S. hirsuta* and *S. occidentalis* is also evidenced by cytological investigation where the somatic chromosome number 2n=28 was reported for these species (Bir and Kumari, 1980).

Acknowledgement

We would like to thank Prof. Dr. Md. Abul Hassan, Department of Botany, University of Dhaka for his cooperartion during the course of the study. Thanks are also due to the authority of Bangladesh National Herbarium for allowing us to examine the herbarium materials.

References

Acharya, L., Mukherjee, A.K. and Panda, P.C. 2011. Separation of the genera in the subtribe *Cassiinae* (Leguminosae: Caesalpinioidae) using molecular markers. Acta Botanica Brasilica **25**(1): 223-233.

Akçin, O.E., Çoşkunçelebi, K. and Şenel, G. 2012. Foliar anatomy of *Cynoglossum* L. (Boraginaceae) from north Anatolia, Turkey. Bangladesh J. Plant Taxon. **19**(2): 101-108.

Bir, S.S. and Kumari, S. 1980. Cytological evolution of the Leguminous flora of the Punjub plain. *In*: Bir, S.S. (Ed.), Recent Researches in Plant Science. Kalyani Publishers, Ludhiana, India, pp. 261-271.

Boonkerd, T., Pechsri, S. and Baum, B.R. 2005. A phenetic study of *Cassia s.l.* (Leguminosae-Caesalpinioideae: Cassieae: Cassiinae) in Thailand. Plant Syst. Evol. **252**: 153-165.

Devi, N.J., Padma, Y., Narasimhudu, C.L. and Raju, R.R.V. 2013. Diversity of stomata and trichomes in *Euphorbia* L. – I. Bangladesh J. Plant Taxon. **20**(1): 27-38.

Dilcher, D.L. 1974. Approaches to the identification of angiosperm leaf remains. Bot. Rev. **40**: 1-157.

Fontenelle, G.B., Costa, C.G. and Machado. R.D. 1994. Foliar anatomy and micromorphology of eleven species of *Eugenia* L. (Myrtaceae). Bot. J. Linn. Soc. **116**: 111-133.

Irwin, H.S. and Turner, B.L. 1960. Chromosomal relationships and taxonomic considerations in the genus *Cassia*. Am. J. Bot. **47**: 309-318.

Irwin, H.S. and Barneby, R.C. 1981. Tribe 2. *Cassiae* Bronn (1822). *In*: Pohlhill, R.M. and Raven, P.H. (Eds), Advances in Legume Systematics, Part 1. Royal Botanic Gardens, Kew, UK, pp. 97-106.

Irwin, H.S. and Barneby, R.C. 1982. The American Cassiinae. Memoir. New York Bot. Gard. **35**: 1-918.

Kotresha, K. and Seetharam, Y.N. 1995. Epidermal studies in some species of *Bauhinia* L. (Caesalpinioideae). Phytomorphology **45**(1&2): 127-137.

Kotresha, K. and Seetharam, Y.N. 2000. Epidermal micromorphology of some *Cassia* L. (Caesalpiniaceae). Phytomorphology **50**(3&4): 229-237.

Lersten, N.R. and Curtis, J.D. 1992. Foliar anatomy of *Polygonum* (Polygonaceae): Survey of epidermal and selected internal structures. Plant Syst. Evol. **182**(1-2): 71-106.

Marazzi, B., Endress, K.P., de Queiroz, L.P.and Conti, E. 2006. Phylogenetic relationships within *Senna* (Leguminosae, Cassiinae) based on three chloroplast DNA regions: Patterns in the evolution of floral symmetry and extrafloral nectaries. Am. J. Bot. **93**(2): 288-303.

Metcalfe, C.R. and Chalk, L. 1950. Anatomy of the Dicotyledons, Vol. **2**. Oxford, pp. 1014-1024.

Metcalfe, C.R. 1968. Current development in systematic plant anatomy. *In*: Heywood, V.H. (Ed.), Modern Methods in Plant Taxonomy. Academic Press, London, New York, pp. 45-57.

Metcalfe, C.R. and Chalk, L. 1979. Anatomy of the Dicotyledons, Second edition, Vol. **1.** Clarendon Press, pp. 63-75.

Naik, V.N. and Nirgude, S.M. 1981. Anatomy in relation to taxonomy of *Chlorophytum* (Liliaceae). Indian J. Bot. **4**(2): 48-60.

Ogundipe, O.T., Kadiri, A.B. and Adekanmbi, O.H. 2009. Foliar epidermal morphology of some Nigerian species of *Senna* (Caesalpiniaceae). Indian J. Sci. & Tech. **2**(10): 5-9.

Rahman, M.O., Rahman, M.Z. and Begum, A. 2013. Numerical taxonomy of the genus *Senna* Mill. from Bangladesh. Bangladesh J. Plant Taxon. **20**(1): 77-83.

Rejdali, M. 1991. Leaf micromorphology and taxonomy of North African species of *Sideritis* L. (Lamiaceae). Bot. J. Linn. Soc. **107**: 67-77.

Saheed, S.A. and Illoh, H.C. 2010. A taxonomic study of some species in *Cassiinae* (Leguminosae) using leaf epidermal characters. Not. Bot. Hort. Agrobot. Cluj **38**(1): 21-27.

Shaheen, N., Ajab, M., Hayat, M.Q. and Yasmin, G. 2009. Diversity of foliar trichomes and their systematic relevance in the genus *Hibiscus* (Malvaceae). Int. J. Agric. Biol. **11**: 279-284.

Singh, V. 2001. Monograph on Indian subtribe *Cassiinae* (Caesalpiniaceae). Scientific Editions, Jodhpur, India.

Stace, C.A. 1965. Cuticular studies as an aid to plant taxonomy. Bull. Br. Mus. Nat. Hist. **4**: 1-78.

Stace, C.A. 1984. The taxonomic importance of the leaf surface. *In:* Heywood, V.H. and Moore, D.M. (Eds), Current Concepts in Plant Taxonomy. Academic Press, London, pp. 67-94.

Tomblinson, P.B. 1961. Anatomical approach to the classification of the Musaceae. Bot. J. Linn. Soc. **55**: 779-809.

Tschan, G.F. and Denk, T. 2012. Trichome types, foliar indumentum and epicuticular wax in the Mediterranean gall oaks, *Quercus* subsection Galliferae (Fagaceae): implications for taxonomy, ecology and evolution. Bot. J. Linn. Soc. **169**: 611-644.

Yasmin, G., Khan, M.A., Shaheen, N. and Hayat, M.Q. 2010. Micromorphological investigation of foliar anatomy of *Fagopyrum* Mill. and *Rumex* L. of Polygonaceae. Pak. J. Bot. **42**(1): 47-57.

7

FOLIAR EPIDERMAL, STEM AND PETIOLE ANATOMY OF MEGHALAYAN *DIOSCOREA* L. (DIOSCOREACEAE) AND ITS SYSTEMATIC IMPLICATION

NILOFER SHEIKH[1] AND YOGENDRA KUMAR

Botany Department, North Eastern Hill University, Shillong-22, India

Keywords: Anatomy; *Dioscorea* ; Cluster Analysis; Principal Component Analysis; Systematics.

Abstract

A comparative anatomical study of eight species of *Dioscorea* L. from Meghalaya, North East India was carried out in order to evaluate the taxonomic significance of anatomical characters to differentiate the species. Characters were coded and analyzed by PCA and cluster analysis. The combination of selected qualitative and quantitative anatomical characters of foliar epidermis, stem and petiole were significant for identification of species. The characters that contributed most to the separation of the species were type of stomata, length of stomata, stomatal index, leaf epidermal hairs, stem epidermal hairs, layer of stem sclerenchyma, number of vascular bundle in outer ring and inner ring of stem, paired or unpaired metaxylem, presence of phloem at both ends or at one end, presence or absence of starch grain in stem, petiole epidermal hair and presence or absence starch grain and crystal in petiole. An indented dichotomous key based on anatomical characters was constructed to distinguish and identify the species.

Introduction

Dioscorea L., with about 602 species (Coursey, 1967) in the tropical and subtropical regions, regarded as the core genus of the family Dioscoreaceae is a pivotal taxon in the evolution of Liliopsida and occupies a basal position among all extant monocotyledonous plants (Chase *et al.*, 2006; Dahlgren, 1989). About 50 species of *Dioscorea* in India (Anonymous, 1952) and approximately 28 species are distributed in North East India (Sharma and Hore, 1995). Meghalaya, one of the eight hot spots of North East India has wide range of distribution of *Dioscorea* species in wild habitat. Only a few species of *Dioscorea* are consumed as food and also used as medicines by the indigenous people of this area in spite of its wide range of distribution. Identification of *Dioscorea* species has always presented a challenge to taxonomists due to its morphological diversity, dioecy and small flowers. Kunth (1924) divided *Dioscorea* into four subgenera based on seed morphology. Burkill (1960) divided the genus into 23 sections based on seed characters, tuber characters and male inflorescence morphology. Coursey (1967) divided the genus into 70 sections based on underground tuber. Several authors studied the floristics of the genus including Miege (1968), Milne-Redhead (1975), N'Kounkou (1993), Tellez and Schubert (1994), Miege and Sebsebe (1998) and Ding and Gilbert (2000).

Morphological approaches provide a workable system of taxonomy, yet it cannot be denied that data from other field like anatomy, cytology, palynology, molecular studies etc., provide evidences for the accurate delimitation of the species. Anatomical parameters play an important role in plant taxonomy (Metcalfe and Chalk, 1957). Anatomical characters have proved to be more useful for delimitation of higher taxonomic ranks, such as genera and families. Few researchers studied foliar epidermal anatomy of different species of *Dioscorea* (Abdulrahman *et al.*, 2009;

[1]Corresponding author. Email: nilofersheikh83@gmail.com

Aina and Atumeyi, 2011; Shah and Gopal, 1972), however, no concrete data of anatomical analysis used in taxonomic delimitation of this genus have been conducted earlier. Therefore, in the present study an attempt has been made to investigate the anatomical features of foliar epidermis, stem and petiole of different species of Meghalayan *Dioscorea* and to single out distinctive anatomical characters potentially useful for infrageneric classification.

Materials and Methods

Plant material

Fresh material of eight species of *Dioscorea viz., D. pentaphylla* L. (NEHU-11946), *D. alata* L. (NEHU-11944), *D. belophylla* (Prain) Haines (NEHU-11950), *D. glabra* Roxb. (NEHU-11937), *D. pubera* Bl. (NEHU-11949), *D. oppositifolia* L. (NEHU-11941), *D. lepcharum* Prain *et* Burk. (NEHU-11942) and *D. bulbifera* L. (NEHU-11935) were collected from wild habitats of Meghalaya. Voucher specimens were deposited in the Herbarium of Botany Department, North Eastern Hill University, Shillong.

Foliar epidermal anatomy

For foliar epidermal study, fresh leaf epidermal peelings from both the surfaces (adaxial and abaxial) and transverse section of the leaves for epidermal study were made by hand with the help of sharp razor and forceps. Ten individual per species were studied. The epidermal peelings and the transverse section of the leaves were cut into suitable size, taken on a clean slide, stained with 5% aqueous safranine, mounted in 50% glycerin and sealed the margins of cover slips with DPX. The prepared slides were observed under a light microscope at x40 magnification and photographs were taken. To calculate the stomatal index, following formula was used:

$$I = {}^{S}/_{E+S} \times 100$$

Where I = stomatal index, S = number of stomata per unit area, and E = number of epidermal cells per unit area. The terminology adopted by Metcalfe and Chalk (1950) and Metcalfe (1961) was followed to describe the stomatal types.

Stem and petiole anatomy

For the anatomical study of stem and petiole, transverse section of stem and petiole of 10 individual plants per species were made using sharp blade from the fresh material. Temporary slides were prepared following O'Brien *et al.* (1964) using Toluidine blue. The stained sections were observed under a light microscope at x10 magnification and Photographed. Presence and absence of starch grain in stem and petiole was separately analyzed by Lugol's solution (Jensen, 1962). The anatomical characters observed under microscopes were recorded.

Anatomical data recording and statistical analysis

A total of 27 characters including both qualitative and quantitative were observed (Table 1). The characters or traits that appeared in more than one state and differed among the examined species were coded and assessed through principal component analysis (PCA). Cluster analysis (CA) was also performed in order to group the species based on the similarity of anatomical characters or traits. Both PCA and CA were performed using XLSTAT ver. 2015.4.01.22283 statistical software.

Table 1. List of anatomical characters of stomata, stem and petiole of different *Dioscorea* species.

Serial No.	Traits acronym	Characters/ descriptors	Score code-descriptor code
		Stomata	
1	PAR	Absence/ presence of paracytic	Absent-0; present-1
2	ANS	Absence/ presence of anisocytic	Absent-0; present-1
3	TER	Absence/ presence of tetracytic	Absent-0; present-1
4	ANA	Absence/presence of anomocytic	Absent-0; present-1
5	Nosto	No of stomata per mm²	1-(95-124); 2-(125-154);3-(155-184)
6	Noepi	No of epidermal cell per mm²	1-(36-45);2-(46-55);3-(56-65);0->65
7	EpiH	Epidermal hairs	0-absent; 1-present
8	StoI	Stomatal index (%)	1-(10-15.5); 2-(16-20.5);3-(21-25.5)
9	Lsto	Length of stomata(μm)	1-(21-30μm);2-(31-40μm);3->40μm
10	Bsto	Breadth of stomata(μm)	1-(1-1.4μm);2-(1.5-1.9μm);3-(2-2.4μm)
		Stem	
11	STO	Stem outline	1-Wavy;2-angular; 3- round
12	SEpiH	Epidermal hair	1-present; 0-absent
13	OR	V.B in outer ring	1-(6-7 rings);2-(8-9rings);3-(10-11rings)
14	IR	V.B. in inner ring	1-(4-5 rings);2-(6-7rings);3-(8-9rings)
15	LC	Layer of cortex	1:-6 layers; 2:-7 layers
16	LScl	Layer of sclerenchyma	1:-4 layers; 2:-5 layers;3:-6 layers
17	StG	Stem starch grain	0-absent; 1-present
18	Mxl	Metaxylem	1-paired; 2-unpaired
19	Ph	Phloem	1-both ends; 2-one ends
20	Scr	Stem crystals	1-present; 0-absent
		Petiole	
21	PO	Petiole outline	1-pentagonal; 2-round; 3-others
22	PepiH	Petiole epidermal hair	0-absent; 1-present
23	Pcol	Petiole cortical layer	1:-3 layers; 2:-4 layers; 3:-5 layers
24	PSL	Petiole scherenchyma layer	1:-3 layers; 2:-4 layers; 3:-5 layers
25	V.B.	vascular bundle	1-6V.B.; 2-8V.B.; 3- 9 V.B.
26	Pcr	Petiole crystals	0-absent; 1-present
27	Pst	Petiole starch grain	0-absent; 1-present

* μm= micrometer.

Results

Foliar epidermal anatomy

The upper (adaxial) and lower (abaxial) epidermis has one layer of cells and each is covered by a cuticle. The outer epidermis is also characterized by the presence of hairs in some species. Stomata are restricted only to the lower surface (hypostomatic). Four types of stomata complex were identified namely anamocytic, tetracytic, paracytic and anisocytic (Fig. 1). *D. alata* possessed anisocytic, anamocytic and tetracytic stomata; *D. pubera* possessed paracytic and tetracytic stomata; *D. belophylla, D. pentaphylla* and *D. bulbifera* had paracytic, anisocytic and anamocytic types of stomata; *D. glabra* with anamocytic and tetracytic stomata; *D. lepcharum* and

D. oppositifolia possessed all the four types of stomata. In the present analysis, the stomatal index ranges from 15.3 in *D. alata* to 24.6 in *D. bulbifera*.

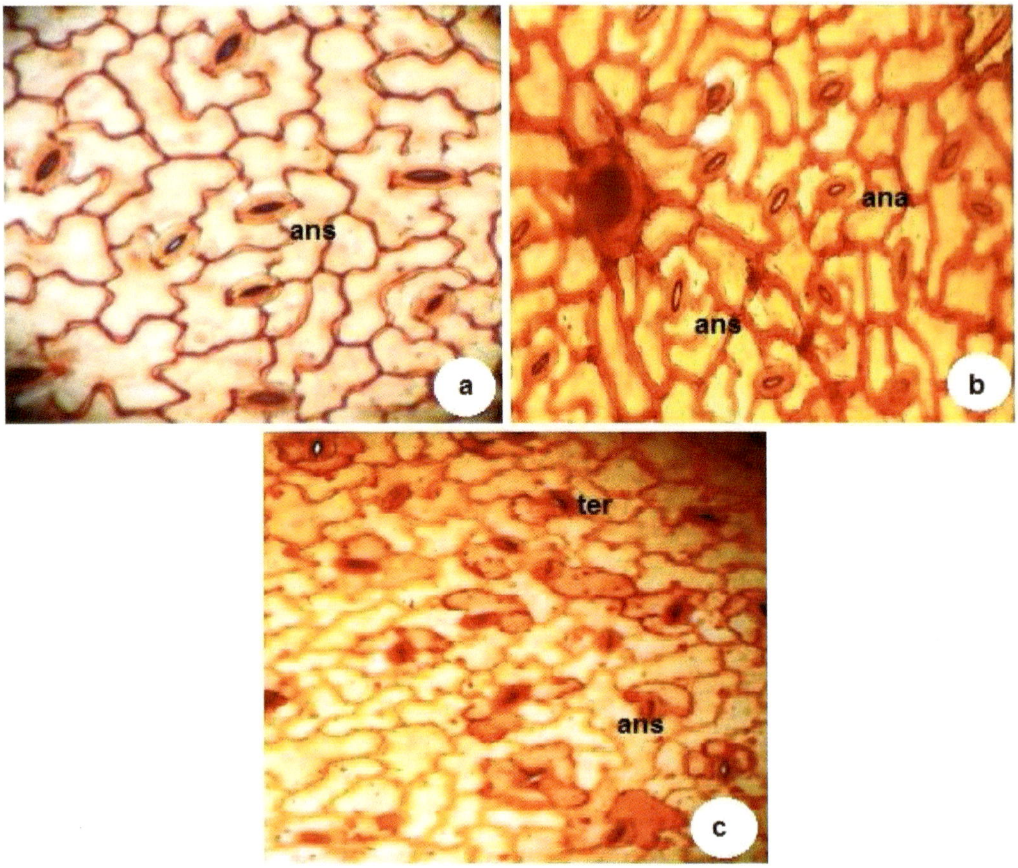

Fig.1. Foliar epidermal structures of *Dioscorea* species. **a)** *D.bulbifera*, **b)** *D. pubera*, **c)** *D. alata. (ana* - anomocytic stomata, *ans* - anisocytic stomata, *ter* - tetracytic stomata).

Stem anatomy

Transverse sections of stems are generally circular with longitudinal ridges or wings (Fig. 2). The epidermis consists of thin-walled rectangular, cuboidal or rounded cells. The epidermal cells are 1-cell layer thick. The cuticle is generally thin. In *D. pubera* some epidermal cells contained numerous hairs. The cortex lying just beneath the epidermis is composed of 3-7 layers of cells. The cortical cells are of various size and shapes. The inner boundary of cortex is located at a zone of sclerenchyma with 2-many cells. The pith occupies the central position and composed of thin-wall hexagonal parenchyma cells. The vascular bundles of the stem are arranged in two concentric circles. The vascular bundles of the outer circle are smaller than the inner, with 2 metaxylem vessels together with 1 phloem unit at the middle in the bundles. The bundles of inner circle have mostly paired metaxylem vessels with phloem unit present at both ends in all species except but in *D. bulbifera*, where the bundles of inner circle have unpaired metaxylem with phloem unit at one end. The number of vascular bundles in transverse section varies within the species.

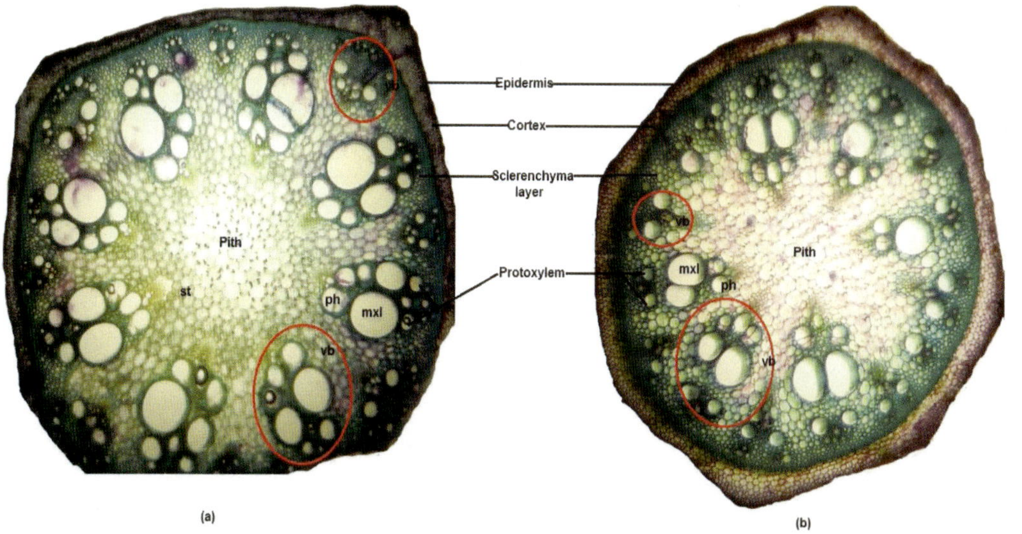

Fig. 2. Anatomy of the stem of *Dioscorea* species at 4X magnification. **a)** *D.alata*, **b)** *D.belophylla*. (*st* - starch grains, *ph* - phloem, *mxl* - metaxylem, *vb*- vascular bundle).

Petioles anatomy

The cortex is composed of collenchymatous tissues. The vascular bundles are arranged in a ring and are basically collateral with the presences of 2-3 phloem units in each bundle. The number of vascular bundles in each petiole is also variable but constant in each species (Fig. 3).

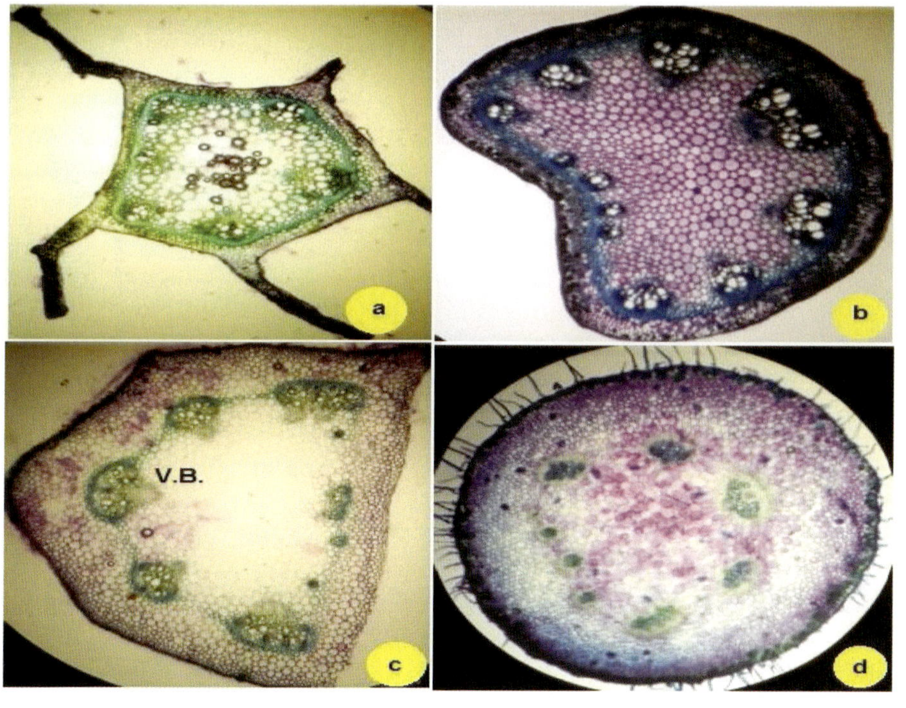

Fig.3. Anatomy of the petiole of *Dioscorea* species at 4X magnification. **a)** *D.alata*, **b)** *D. lepcharum*, **c** *D. belophylla*, **d)** *D. pubera.(V.B.* vascular bundle).

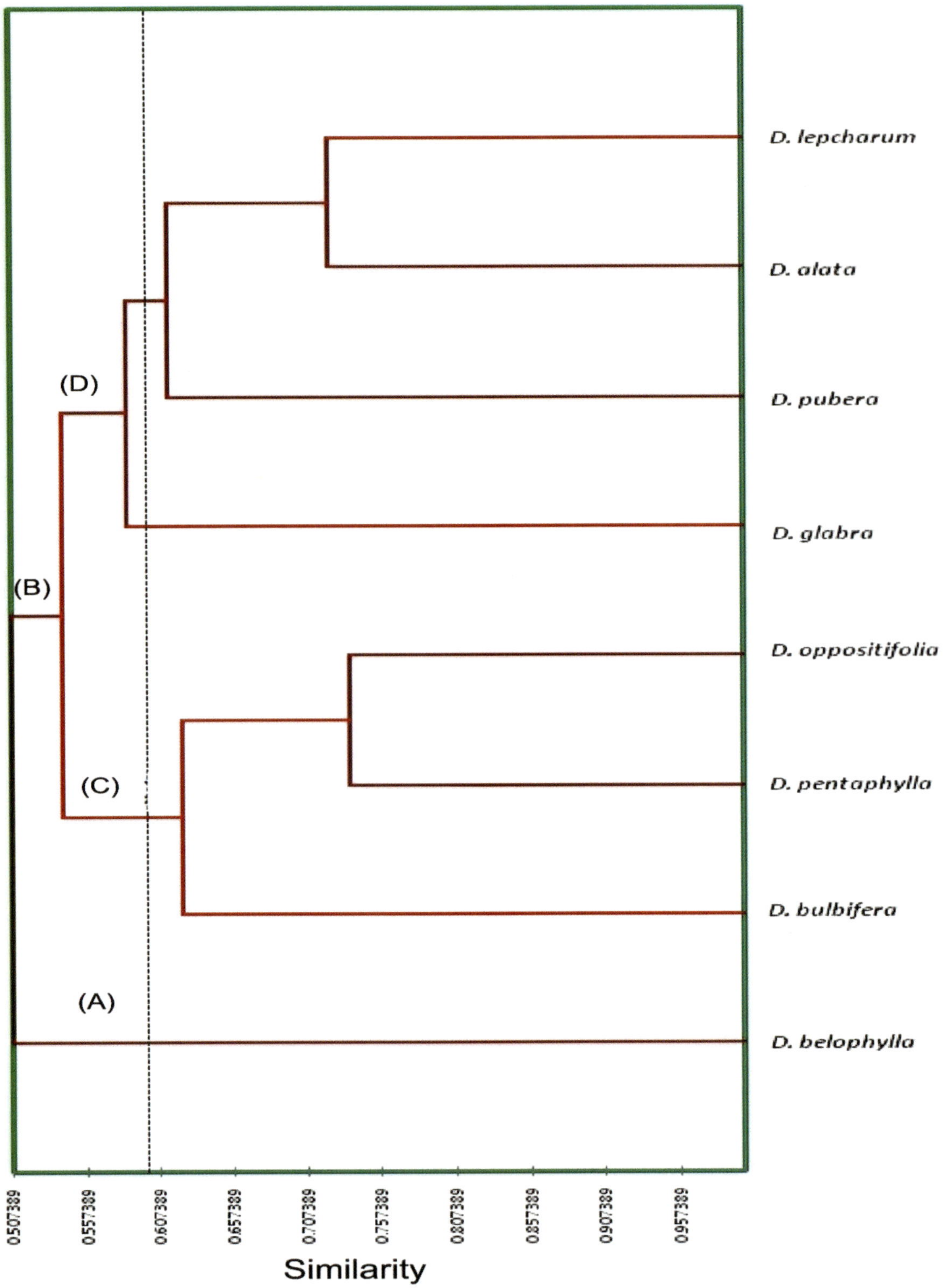

Fig.4. Dendrogram showing relationships among eight Dioscorea species based on anatomical characters.

Cluster analysis

The result of cluster analysis based on unweighted pair group average method (UPGMA) initially produced two main clusters, cluster A with a single species *D. belophylla* and the other cluster B at 0.507 level of similarity (Fig. 4). Within the cluster B, two subcluster C and D are separated at similarity level of 0.54. Subcluster C with *D. bulbifera* is separated from *D. pentaphylla* and *D. oppositifolia*. *D. pentaphylla* and *D. oppositifolia* forms a clade at 0.74 level of similarity. Similarly, subcluster D with *D. glabra* and *D. pubera* are separated from *D. alata* and *D. lepcharum* which forms a clade at 0.72 level of similarity.

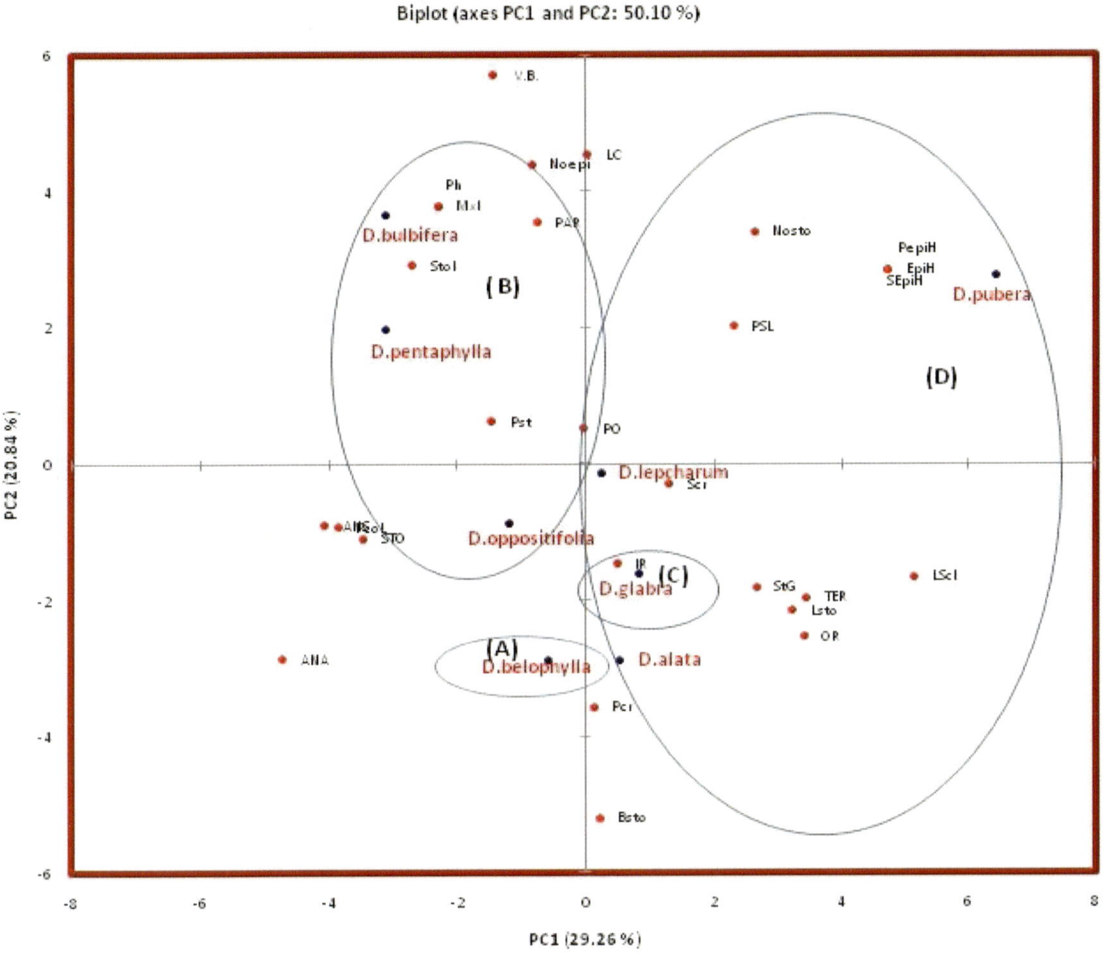

Fig. 5. A biplot based on first and second principal components of anatomical characters for eight *Dioscorea* species.

Principal component analysis

The first seven principal components explained about 100% of the variation, while the first component (PC1) alone counted for 29.26% of the variability (Table 2). Characters such as tetracytic type of stomata, presence or absence of leaf epidermal hair, length of stomata, presence of stem epidermal hair, number of vascular bundle present in outer ring of stem, number of schlerenchyma of in stem, presence of epidermal hair in petioles has the highest loading on PC1. The second component (PC2) explained 20.84% of the total variation was highly correlated with paracytic type of stomata, number of stomata, layers of cortex in stem, presence of paired or unpaired metaxylem in stem, presence of phloem at one end or at both ends and number of vascular bundles in petioles. The remaining component explained less variability. To assess the scores of the individual species with its characters or traits, PC1 and PC2 were plotted (Fig. 5). The group of species on the left side of the biplot is separated from the species on the right side of biplot due to certain combination of anatomical characters, *viz.*, anamocytic and paraytic types of stomata, presence of starch in petiole, number of vascular bundle in petiole, presence of metaxylem paired or unpaired in stem, presence of phloem at one end or at both ends in stem,

Table 2. Eigenvalues, variance, cumulative variance and component scores (eigenvectors) of the first 7 principal components (PC) for anatomical characters of *Dioscorea* species.

	Component scores						
	PC1	PC2	PC3	PC4	PC5	PC6	PC7
Eigenvalue	7.900	5.626	4.836	3.554	2.746	1.372	0.965
Variance (%)	29.260	20.836	17.912	13.162	10.171	5.083	3.576
Cumulative %	29.260	50.096	68.008	81.170	91.341	96.424	100.000
PAR	-0.138	**0.548**	-0.090	-0.734	0.078	0.270	-0.236
ANS	-0.742	-0.138	-0.471	-0.360	-0.275	0.043	-0.042
ANA	-0.863	-0.439	0.043	0.175	-0.109	0.056	-0.124
TER	**0.625**	-0.302	0.169	0.274	-0.522	0.349	0.143
Nosto	0.483	**0.525**	0.500	0.400	0.094	-0.138	-0.231
Noepi	-0.150	**0.678**	0.437	**0.518**	-0.131	-0.187	-0.081
EpiH	**0.863**	0.439	-0.043	-0.175	0.109	-0.056	0.124
StoI	-0.493	0.450	0.147	-0.357	**0.559**	-0.079	-0.294
Lsto	**0.586**	-0.331	-0.530	-0.259	-0.142	-0.197	-0.373
Bsto	0.041	-0.799	0.371	0.379	0.013	0.237	-0.147
STO	-0.634	-0.169	-0.091	-0.583	0.217	-0.062	0.413
SEpiH	**0.863**	0.439	-0.043	-0.175	0.109	-0.056	0.124
OR	**0.619**	-0.389	-0.566	-0.027	-0.017	0.161	-0.343
IR	0.089	-0.225	-0.351	**0.848**	0.187	0.249	0.047
LC	0.006	**0.701**	**0.571**	0.116	-0.224	0.341	-0.052
LScl	**0.937**	-0.255	-0.002	-0.132	-0.005	0.200	0.013
StG	0.484	-0.278	0.392	-0.207	**0.564**	0.374	0.183
Mxl	-0.418	**0.584**	-0.430	0.400	0.297	0.217	-0.057
Ph	-0.418	**0.584**	-0.430	0.400	0.297	0.217	-0.057
Scr	0.236	-0.044	-0.802	-0.238	-0.055	0.486	-0.068
PO	-0.003	0.081	**0.856**	-0.114	-0.459	0.184	-0.052
PepiH	**0.863**	0.439	-0.043	-0.175	0.109	-0.056	0.124
PcoL	-0.704	-0.143	0.303	-0.272	-0.529	0.171	-0.095
PSL	0.419	0.312	-0.470	0.081	-0.613	-0.316	-0.155
V.B.	-0.259	**0.880**	-0.171	-0.047	-0.124	0.332	-0.028
Pcr	0.022	-0.548	0.346	0.138	**0.686**	-0.071	-0.290
Pst	-0.265	0.096	-0.742	**0.545**	-0.044	-0.104	0.246

Coefficient in bold indicate descriptors that are highly correlated with the corresponding principal component.

whereas species on the right side of the plot is represented with characters such as presence of foliar epidermal, stem epidermal and petiole epidermal hairs, number of petiole schlerenchyma layers, number of stomata per unit area etc. The finding was consistent with the separation of species into four major groups by UPGMA clustering (Fig. 4).

Discussion

The use of anatomical characters or traits for taxonomic studies has proved useful for identification of fragmented plant and herbarium specimens (Metcalfe and Chalk, 1957). Anatomy can provide useful information for establishing interrelations between taxa at the species and supra species levels. Sometimes it can also help in individual identifications. The internal structure of leaf is more affected by environmental factors and thus is of little value for delimiting taxonomic groups. Other characters of leaf, such as the epidermis and stomata have proved to be much more reliable for taxonomic consideration in many genera (Uphof, 1962; Dickison, 2000; Yang and Lin, 2005; Strgulc-Krajsek *et al.,* 2006). The petiole structure is of considerable taxonomic importance in many genera, since it is less affected by environmental changes (Metcalfe and Chalk, 1957). The result of the present study allows the selection of some diagnostic anatomical characters for the identification of Meghalayan *Dioscorea* species. Stebbin and Khush (1961) and Ayensu (1972) reported that all stomata in *Dioscorea* species were anomocytic which was found in consistent with Shah and Gopal (1972) and Abdulrahaman *et al.* (2009). Our results showing presence of different types of stomata including paracytic, tetracytic and anomocytic etc were found congruent with those of Shah and Gopal (1972) and Abdulrahaman *et al.* (2009). The present study reveals an anamocytic character trait contributes much towards variability. Presence of starch grain in stem and petiole is also an important character trait for differentiating among the species (Onwuene, 1978). From the PCA analysis, the character trait i.e. presence of starch grain in petiole was another important trait which shows high rate of variability. Hence the foregoing analysis and presented illustration clearly show the importance of anatomical data as an additional tool in the taxonomy of the genus and also contributes to the identification of species. The results provided some useful features for future phylogenetic and taxonomic studies. Based on these features, an artificial indented dichotomous key is presented hereunder to delimit the species.

Dichotomous key to the Meghalayan species of *Dioscorea* based on anatomical characters.

1.	Leaf, stem and petiole pubescent	***D. pubera***
-	Leaf, stem and petiole glabrous	2
2.	Absence of tetracytic stomata	3
-	Presence of tetracytic stomata	5
3.	Layer of schlerenchyma in stem is more than 5; absence of starch grain in stem; vascular bundle in outer ring in stem is more than 9	***D. belophylla***
-	Layer of schlerenchyma in stem is less than 5; presence of starch grain in stem; vascular bundle in outer ring in stem is less than 9	4
4.	Inner ring of vascular bundle in stem is less than 6; metaxylem in stem is paired; phloem on both ends of vascular bundle	***D. pentaphylla***
-	Inner ring of vascular bundle in stem is more than 6; metaxylem unpaired; phloem at one end of vascular bundle	***D. bulbifera***

5.	Presence of starch grain in stem; outer ring of vascular bundle more than 9	6
-	Absence of starch grain in stem; outer ring of vascular bundle less than 9	7
6.	Inner ring of vascular bundle more than 9; petiole outline mainly pentagonal; layer of cortex in petiole less than 4	***D. alata***
-	Inner ring of vascular bundle less than 9, petiole outline crescent shape; layer of cortex in petiole more than 4	***D. lepcharum***
7.	Inner ring of vascular bundle in stem less than 8; presence of paracytic stomata; absence of petiole crystal, layer of cortex in petiole is more than 4	***D. oppositifolia***
-	Inner ring of vascular bundle more than 8; absence of paracytic stomata; presence of petiole crystal, layer of cortex in petiole is less than 4	***D.glabra***

Acknowledgements

The authors are thankful to the Head of Botany Department, North Eastern Hill University, Shillong for providing the facilities in the department and also to UGC for providing financial support under the scheme of Maulana Azad minority national fellowship. Our sincere thanks to the Joint Director, Botanical Survey of India eastern circle, Shillong for giving us the facilities during the tenure of this research programme.

References

Abdulrahaman, A.A., Egbedo, F.O. and Oladele, F.A. 2009. Stomatal complex types, stomatal density and the stomatal index in some species of *Dioscorea*. Arch. Biol. Sci., Belgrade **61**(4): 847–851.

Aina, O.D. and Atumeyi, S. 2011. Foliar epidermal anatomy of four species of *Dioscorea*. Advances Appl. Sci. Res. **2**(4): 21–24.

Anonymous, 1952. The Wealth of India. Vol. **3**. D-E, CSIR, New Delhi.

Ayensu, E.S. 1972. Dioscoreales. *In:* Metcalfe, C. (Ed.), Anatomy of the Monocotyledons. Vol. **6**. Clarendon Press, Oxford, pp. 1–182.

Burkill, I.H. 1960. The organography and the evolution of the Dioscoreaceae, the family of the yams. Bot. J. Linn. Soc. **56**: 319–412.

Chase, M.W., Fay, M.F., Devey, D.S., Maurin, O., Ronsted, N., Davies, T.J., Pillon,Y., Petersen, G., Seberg, O., Tamura, M.N., Asmussen, C.B., Hilu, K., Borsch, T., Davis, J.I., Stvenson, D.W., Pires, J.C., Givnish, T.J., Sytsma, T.J., McPherson, M.A., Graham, S.W. and Rai, H.S. 2006. Multigene analysis of monocot relationships, a summary. Aliso **22**: 63–75.

Coursey, D.G. 1967. Yams: An account of the nature, origins, cultivation and utilization of the useful members of the Dioscoreaceae: Tropical Agricultural series. Longmans, Green and Co. Ltd., London, 230 pp.

Dahlgren, G. 1989. An undated angiosperm classification. Bot. J. Linn. Soc. **100**: 197–203.

Ding, Z. and Gilbert, M.G. 2000. Dioscoreaceae. *In*: Wu, Z. and Raven, P.H. (Eds), Flora of China. Vol. **24**. Flagellariaceae through Marantaceae. Science Press: Beijing, and Missouri Botanical Garden Press, St. Louis, pp. 276–296.

Dickison, W.C. 2000. Integrative Plant Anatomy. Academic Press, San Diego.

Jensen, W.A. 1962. Botanical Histochemistry: Principles and Practice. San Francisco: W.H. Freeman & Company.

Kunth, R. 1924. Dioscoreaceae. *In*: Engler, H.G.A. (Ed) Das Pflanzenreich, 87(IV. 43). Leipzig: H. R. Engelmann (J.Cramer). pp. 1–387.

Metcalfe, C.R. 1961. The anatomical approach to systematics: general introduction with special reference to recent work on monocotyledons. Rec. Adv. Bot. **1**: 146–150.

Metcalfe, C.R. and Chalk, L. 1950. Anatomy of Dicotyledons, Vols.□ and □. Clarendon Press, Oxford.

Metcalfe, C.R. and Chalk, L. 1957. Anatomy of the Dicotyledons, Vol **I**. Clarendon Press, Oxford, pp. 502–535.

Miege, J. 1968. Dioscoreaceae. *In*: Flora of West Tropical Africa, 2nd ed. F. N. Hepper. London: Crown Agents, pp.144–154.

Miege, J. and Sebsebe, D. 1998. Dioscoreaceae. *In:* Edwards, S., Sebsebe, D. and Hedberg, I. (Eds), Flora of Ethiopia and Eritrea. Vol. **6**. Addis Ababa: The National Herbarium, Biology Department, Science Faculty, Addis Ababa University; Uppsala: Department of Systematic Botany, Uppsala University, pp 55–62.

Milne-Redhead, E. 1975. Dioscoreaceae. *In*: Polhill, R.M. (Ed), Flora of Tropical East Africa. London. Crown Agents.

N'Kounkou, J.S. 1993. Les Dioscoreaceae du Congo. Fragmenta Floristica Geobotanica, Supplementa **2**: 139–182.

O'Brien, T.P., Feder, N. and Mccully, M.E. 1964. Polychromatic staining of plant cell walls by toluidine blue O. Protoplasma **5**(2): 368–373.

Onwueme, I.C. 1978. The Tropical Tuber Crops; Yams, Cassava, Sweet Potato and Cocoyams. John Wiley & Sons, New York.

Shah, G.L. and Gopal, B.V. 1972. Some six species observations on diversity of stomata and trichomes in *Dioscorea* (Yams) Ann Bot. (London) **36**(148): 997–1004.

Sharma, B.D. and Hore, D.K. 1995. Genetic resource of yams in N.E. India with special reference to Garo Hills, Meghalaya. Indian J. Hill Farming **8**: 145–151.

Stebbins, G.L. and Khush, G.S. 1961. Variation in the organization of the stomatal complex in the leaf epidermis of monocotyledons and its bearing on their phylogeny. Ann. J. Bot. **48**: 51–90.

Strgulc-Krajsek, S., Dermastia, M. and Jogan, N. 2006. Determination key for Central European *Epilobium* species based on trichome morphology. Bot. Helv. **116**: 169–178.

Tellez, V.O. and Schubert, B.G. 1994. Dioscoreaceae. *In*: Davidse, G. Sousa, M. and Chater A.O. (Eds). Flora Mesoamericana. Volumen **6**. Universidad Nacional Autonoma de Mexico, pp. 54–65.

Uphof, J.C. 1962. Plant Hairs, Encyclopedia of Plant Anatomy IV. Borntraeger, Berlin.

Yang, Z.R. and Lin, Q. 2005. Comparative morphology of the leaf epidermis in *Schisandra* (Schisandraceae). Bot. J. Linn. Soc. **148**: 39–56.

8

MORPHOLOGICAL AND ANATOMICAL STUDIES OF THE NEWLY RECORDED GENUS *PLAGIOPTERON* GRIFF. (CELASTRACEAE) FROM VIETNAM

VU TIEN CHINH, RITESH KUMAR CHOUDHARY[1], TRAN THE BACH, BUI HONG QUANG, DO VAN HAI, SANGMI EUM[2] AND JOONGKU LEE[3*]

Institute of Ecology and Biological Resources,
Vietnam Academy of Sciences and Technology, Hanoi, Vietnam

Keywords: Anatomy; Celastraceae; New generic record; *Plagiopteron*; Vietnam.

Abstract

Plagiopteron Griff. (Celastraceae), collected from Bac Giang Province of Vietnam is reported here as a new generic record for the flora of Vietnam. It is supposedly an early derived member of the subfamily Hippocrateoideae (composed of tribes Campylostemoneae and Hippocrateeae), within Hippocrateaceae. The genus differs from other members of Celastraceae by its many bicyclically arranged stamens. Leaf anatomical studies have been performed to understand various features such as venation, vascular bundles, collenchyma and stomata. Taxonomic notes, description, and photographs are provided, together with short notes on the distribution, ecology and phenology of the species.

Introduction

The family Celastraceae is represented by 98 genera and 1264 species and distributed across tropics and subtropics, some in temperate regions of the world (Simmons, 2001). In Vietnam, they are represented by 13 genera and 80 species (Ban, 2003). During a recent field work in An Lac Village of Bac Giang Province in Vietnam (Fig. 1), an interesting plant belonging to Celastraceae was collected. After examining all the specimens housed in Vietnamese herbaria (HN, HNU, VNM) the collected specimens did not match with any known Vietnamese genus. Critical morphological and anatomical observation, consultation of type specimens and relevant literature studies were carried out and conclusively it was identified as *Plagiopteron suaveolens* Griff. The genus differs from other members of the family Celastraceae by its many bicyclically arranged stamens. This monotypic genus was hitherto known to occur in South China, Myanmar, Bangladesh and Thailand. The present report is a new distributional record of this genus in Vietnam.

Leaf anatomical studies have been proved useful in the solution of certain taxonomic and phylogenetic problems and delimitation of various genera (Robinson, 1969; Cutler *et al.,* 2007). The importance of anatomical characters for the Celastraceae taxonomy was confirmed by Smith & Robinson (1971), Den Hartog *et al.* (1978) and Mennega (1997). Smith & Robinson (1971) used leaf epidermal features in the definition of species Hippocrateaceae occurring in Santa Catarina. Den Hartog *et al.* (1978) supported the expansion of the Celastraceae family concept based on the types of stomata and crystal-cells in the leaf epidermis of Celastraceae *s.l.* Considering the importance of anatomical studies in Celastraceae, we carried out the same for *Plagiopteron.*

[1]Biodiversity & Palaeobiology Group, Agharkar Research Institute, G.G. Agarkar Road, Pune, India.
[2]International Biological Material Research Center, Korea Research Institute of Bioscience and Biotechnology, Daejeon, Republic of Korea
[3]Department of Environment and Forest Resources, Chungnam National University, Daejeon, Republic of Korea
*Corresponding author.

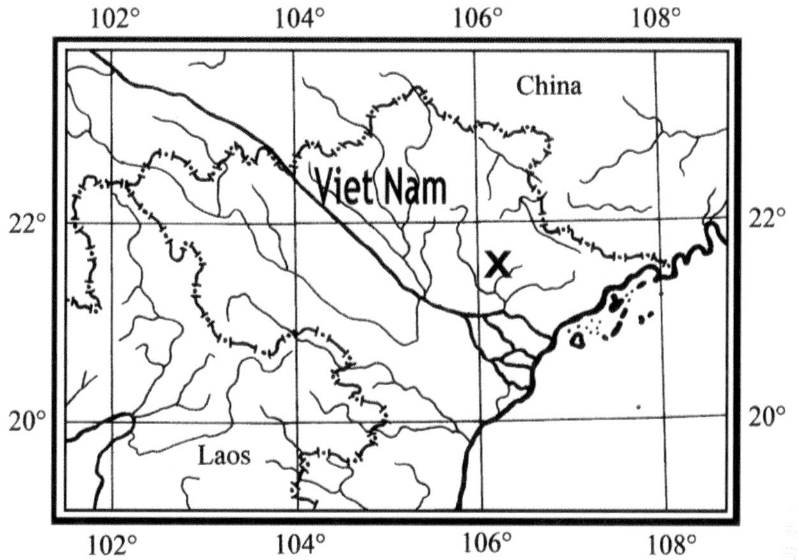

Fig. 1. Distribution map showing location of *Plagiopteron suaveolens* in Vietnam.

Materials and Methods

The specimens were collected from the natural habitats, thoroughly processed using standard herbarium techniques (Jain and Rao, 1977) and deposited in Herbarium of Institute of Ecology & Biological Resources (HN) Vietnam and Korea Research Institute of Bioscience & Biotechnology (KRIB). Inflorescences were fixed in FAA (5 ml of Formalin: 5 ml of Glacial Acetic acid: 90 ml of Ethyl Alcohol) solution and detailed morphological studies were carried out under Nikon Eclipse Ci Stereo Microscope. Photographs were taken using Nikon DS-Fi2 Camera.

Leaves were fixed in FAA and then stored in 70% ethanol to minimize hardening. Samples were taken from the main vein and the region between the midrib and margin. Cross-sections were performed in a rotary microtome, colored in methylene blue and carmine red, and then assembled according to standard techniques of plant anatomy (Johansen, 1940). The anatomical analysis and photographic documentation pertaining to light microscopy were performed under light microscope Nikon Eclipse Ci with DS-Fi2 Camera.

Results

Taxonomic Treatment

Plagiopteron Griff. Calcutta J. Nat. Hist. 4: 244, 1844. t. 13.

Type: China, *s.n.* (BR photo!).

Scandent shrubs, woody, laticiferous. Branches and branchlets densely stellate-pubescent. Leaves opposite, simple, pinnately nerved, hairy. Inflorescence axillary, paniculate, multi-flowered, densely stellate hairy. Flowers bisexual, small. Calyx (2–) 4 (or 5)-partite, segments unequal in length. Petals 3, rarely 4, sepal-like, reflexed. Stamens numerous, on a disk; filaments filiform, free; anthers dehiscing transversely. Gynoecium 3-carpellary, syncarpous. Ovary superior, 3-loculed, sessile, densely tomentose; ovules 2 per locule; style 1, subulate; stigma minutely 3-lobed. Fruit a turbinate septicidal capsule with 3 spreading wings ultimately separating into 3 cocci. Seeds ovate.

Plagiopteron suaveolens Griff., Calcutta J. Nat. Hist. 4: 244 (1843); Ya Tang & Sue Zmarzty, Fl. China 11: 439 (2007). *P. chinensis* X.X. Chen, Acta Bot. Yunnan 2(3): 331 (1980); *P. fragrans* Griff. Calcutta J. Nat. Hist. 4: 244 (1843) t.13. **(Fig. 2)**.

Type: China, *s.n.* (BR photo!).

Fig. 2. *Plagiopteron suaveolens* A. Habit; B. & C. Adaxial and abaxial surface of leaf; D. A flowering twig; E. Flowers; F. Infructescence; G. Fruits.

A large scandent shrub; young parts rusty stellate-pubescent. Petiole c. 1.0 - 1.5 cm long, densely tomentose; leaf blade elliptic, ovate-elliptic, ovate-oblong or nearly round, 3.0 - 13.5 × 2.5 - 7.5 cm, papery, abaxially densely brown stellate tomentose, adaxially finally pubescent on veins only, midvein strongly prominent abaxially, raised adaxially, lateral veins 5 or 7 pairs, base rounded or slightly cordate, less often acute to obtusely cuneate, or apparently finely crenate in dried material; margin entire; apex acute to acuminate. Inflorescence paniculate, axillary, toward twig apex, usually shorter than leaf blade; peduncle densely brown stellate tomentose, lower unbranched part of peduncle 4.5 - 5.0 cm long. Pedicel flower c. 5.0 - 6.5 mm long, tomentose; bracteoles lanceolate, 2.0 - 3.0 mm long. Sepals 4, spreading, hairy, unequal in size, lanceolate, 1.0 - 1.5 mm long, tomentose. Petals 3 or 4, nearly round or, narrowly ovate, 2.5 - 3.5 mm long, outside tomentose, yellow, inside with sparser weak hairs. Stamens many, on a disk, slightly united at base, much longer than petals; filaments filiform, slightly dilated above, 2.0 - 3.0 mm long, glabrous; anthers globose or pyriform, minute, glabrous, dehiscent by apical pores. Ovary superior, densely brown pubescent, 3-loculed with 2 basal erect ovules in each locule; style subulate, simple, c. 2.0 mm long, slender, tapering to apex, nearly as long as filaments, tomentose at base; stigma minutely 3-lobed. Capsules turbinate, woody, expanded at apex into 3 spreading wings and ultimately separating into 3 cocci; wings spathulate, 2.0 - 4.0 cm long, unequal with subparallel veins, sparsely stellate hairy. Seeds ovate c. 0.3 - 0.5 mm long.

Flowering: May to June; *fruiting*: July.

Specimens examined: Vietnam, Bac Giang Province, Son Dong district, An Lac Village, 18 July 2015, c. N21°22'11.1" E106°56'02.8", 166 masl, *T.T. Bach*, *T.D. Binh*, *V.T. Chinh*, *D.V. Hai*, *B.H. Quang*, *D.H. Son*, VK-6488 (HN, KRIBB).

Ecology: Plagiopteron suaveolens was found growing in the shade along the streams and at the forest borders of limestone hills at about 166 m asl. However, it has also been reported to occur in the open evergreen forest, secondary forest (evergreen forest type), open deciduous forest, or sometimes at the edge of cultivation from Thailand at the altitudinal range of 80-430 m asl (Phonsena, 2004).

Distribution: Vietnam (Bac Giang Province, Son Dong District, An Lac village). Bangladesh (Silhet), China (Guangxi), Myanmar (Tenasserim) and Thailand (Lampang, Kalasin, Nakhon Ratchasima, Sa Kaeo).

Leaf anatomical features

Leaf anatomical characteristics of *Plagiopteron suaveolens* are described under two headings. **(Fig. 3)**.

Central vein

The epidermis cells cover the entire leaf surface and are continuous with the surface of stem to which the leaf is attached. The upper epidermal cells (B) are single layered with some unicellular-hairs (A); the lower epidermis consists of ordinary cells and guard cells (I). The guard cells occur in pair, separated by an opening or pore (stoma). The epidermal cells are continued by plenty of collenchyma-cell layers (C).

The transverse section shows the vascular systems in central veins. The vascular bundles extend throughout the leaf. The conducting elements are xylem (E) and phloem (D).

Some sclerenchyma cells (F) were observed in the Central part.

Leaf blade

The upper and lower epidermis cells are continued with those which occur in Central veins.

Mesophyll of leaf consists of 3 parts: the sub-epidermis (G); the palisade parenchyma (H) and the spongy parenchyma. The upper epidermis cells are followed by two layers of sub-epidermis and the palisade parenchyma layer with narrow cells (H). The spongy parenchyma (J) extends from the palisade parenchyma (K) to the lower epidermis. Cells of the spongy parenchyma are irregular in shape and loosely arranged.

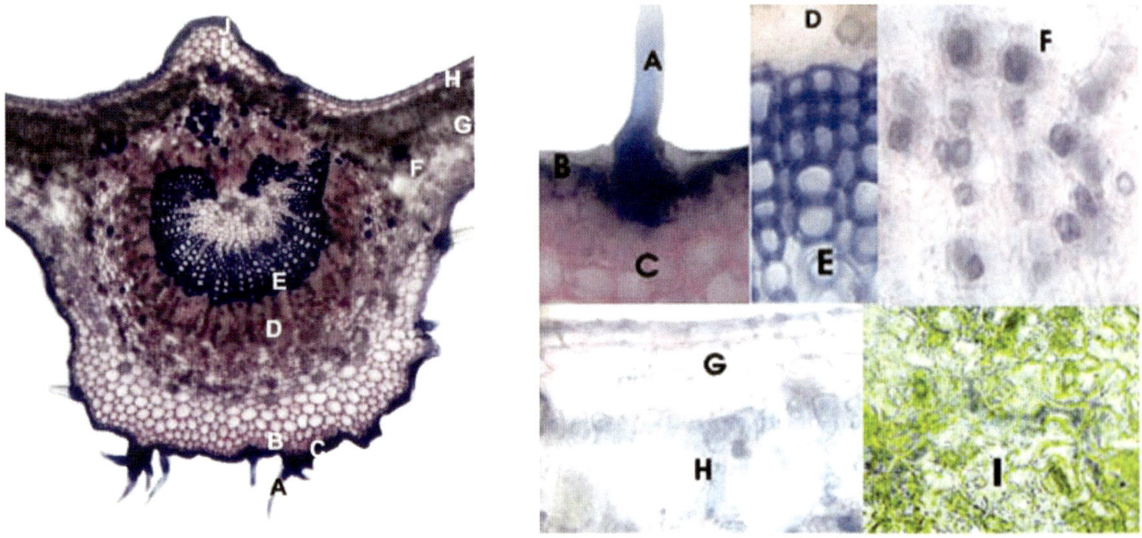

Transverse Section (T.S.) Magnified view of T.S.

Fig. 3. Detailed leaf anatomical characteristics of *Plagiopteron suaveolens* as seen in transverse section. A. Unicellular hair; B. Upper Epidermis; C. Collenchyma cells; D. Phloem; E. Xylem; F. Sclerenchyma cells; G. Sub-epidermis; H. Palisade parenchyma; I. Guard cells; J. Spongy parenchyma; K. Palisade parenchyma; L. Sub-epidermis; M. Upper collenchyma; N. Upper epidermis.

Taxonomic Notes

The genus *Plagiopteron* was earlier placed under the monotypic family *Plagiopteraceae* (Bass *et al.*, 1979) however, its woody climbing habit, the presence of large opposite leaves, small flowers in many-flowered terminal cymes, and fruit with Acer-like wings pointed out its affinity with Celastraceae. Further similarities were found in the elastic material of the leaf midrib vascular cylinder. Nonetheless, the pollen characters did not support its placement in Celastraceae (Tang, 1994). Recent systematic studies using molecular data placed *Plagiopteron* in the Celastraceae, close to the Hippocrateoid genera. The genus is supposedly an early derived member of the subfamily Hippocrateoideae (composed of tribes Campylostemoneae and Hippocrateeae), within Hippocrateaceae. This relationship has been supported by the molecular studies using 26S *nrDNA*, *rbcL*, *atpB* and *phyB* loci (Simmons *et al.*, 2001b; Savolainen *et al.*, 2000, 2000b; Soltis *et al.*, 2000). This relationship is also consistent with recognition of the embryological similarities between *Plagiopteron* and Celastraceae (Tang, 1994) and simultaneous analysis of *rbcL* and morphological characters (Nandi *et al.*, 1998). Synapomorphies of *Plagiopteron* with Hippocrateaceae are extrose, transversely dehiscent stamens that are not versatile. *Plagiopteron*, however differs from other members of Celastraceae by its many bicyclically arranged stamens. A synapomorphy of *Plagiopteron* with Hippocrateoideae is the distinctive capsular fruits that are deeply lobed between the locules (Simmons *et al.*, 2001a). The present report will help to prioritize conservation of this species in Vietnam, Bangladesh, China, Thailand and Myanmar.

Acknowledgments

We are thankful to the Institute of Ecology and Biological Resources (IEBR), Vietnam Academy of Science and Technology Hanoi (VAST), Chungnam National University (CNU), International Biological Material Research Center (IBMRC), Korea Research Institute of Bioscience and Biotechnology (KRIBB), Korea and The National Foundation for Science and Technology Development (NAFOSTED-106-NN.99-2015.26) for permitting us to carry out this study. RKC (2nd author) acknowledges to the MoU between Agharkar Research Institute, India and IBMRC, KRIBB which facilitated this research. Thanks are also due to the anonymous reviewers for their constructive comments which helped the manuscript to improve further.

References

Ban, N.T. 2003. Celastraceae, *In*: Checklist of Plant Species of Vietnam. Vol. **2**. Agricultural Publishing House, Vietnam, pp. 1120–1134.

Baas, P., Geesink, R., van Heel, W.A. and Muller, J. 1979. The affinities of *Plagiopteron suaveolens* Griff. (Plagiopteraceae). Grana. **18**: 69–89.

Cutler, D.F., Botha T. and Stevenson D.W. 2007. Plant Anatomy: an applied approach. Blackwell Publishing, Malden, pp. 1–302.

Den Hartog, R.M., Tholen, V. and Bass, P. 1978. Epidermal characters of the Celastraceae *sensu lato*. Acta Bot. Neerl. **27**: 355–388.

Jain, S.K. and Rao, R.R. 1977. A Handbook of Field and Herbarium Methods. Today and Tomorrow's Printers and Publishers, New Delhi, pp. 1–157.

Johansen, D.A. 1940. Plant microtechnique. New York, Mc. Graw Hill Book.

Mennega, A.M.W. 1997. Wood Anatomy of the Hippocrateoideae (Celastraceae). IAWA-Journal. **18**(4): 331–368.

Nandi, O.I., Chase, M.W. and Endress, P.K. 1998. A combined cladistic analysis of angiosperms using rbcL and non-molecular data sets. Ann. Mo. Bot. Gard. **85**: 137–212.

Phonsena, P. 2004. *Plagiopteron suaveolens* (Plagiopteraceae): an emended description. Thai Forest Bulletin (Botany), **32**: 123–131.

Robinson, H. 1969. A monograph on foliar anatomy of the genera *Connelia*, *Cottendorfia*, and *Navia* (Bromeliaceae). Smithsonian Contrib. Bot. **2**: 1–41.

Savolainen, V., Fay, M.F., Albach, D.C., Backlund, A., van der Bank, M., Cameron, K.M., Johnson, S.A., Lledo, M.D., Pintaud, J.-C., Powell, M., Sheahan, M.C., Soltis, D.E., Soltis, P.S., Weston, P., Whitten, M., Wurdack, K.J., and Chase, M.W. 2000a. Phylogeny of the eudicots: A nearly complete familial analysis based on rbcL gene sequences. Kew Bull. **55**: 257–309.

Savolainen, V., Chase, M.W., Hoot, S.B., Morton, C.M., Soltis, D.E., Bayer, C., Fay, M.F., De Brujin, A., Sullivan, S., and Qiu, Y.-L. 2000b. Phylogenetics of flowering plants based upon a combined analysis of plastid atpB and rbcL gene sequences. Syst. Biol. **49**: 306–362.

Simmons, M.P. 2001. Celastraceae. *In*: Kubitzki K. (Ed.), The Families and Genera of Vascular Plants. Vol. **6**. Springer-Verlag, Berlin, pp. 29–64.

Simmons, M.P., Clevinger, C.C., Savolainen, V., Archer, R.H., Mathews, S. and Doyle, J.J. 2001b. Phylogeny of the Celastraceae inferred from phytochrome B and morphology. Am. J. Bot. **88**: 313–325.

Smith, L.B. and Robinson, H.E. 1971. *Hippocrateaceas. In*: Reitz, R. (ed.), Flora Ilustrada Catarinense. Itajai, Herbario Barbosa Rodrigues, pp. 1–33.

Soltis, D.E., Soltis, P.S., Chase, M.W., Mort, M.E., Albach, D.C., Zanis, M., Savolainen, V., Hahn, W.H., Hoot, S.B., Fay, M.F., Axtell, M., Swensen, S.M., Nixon, K.C. and Farris, J.S. 2000. Angiosperm phylogeny inferred from a combined data set of 18S rDNA, rbcL, and atpB sequences. Bot. J. Linn. Soc. **133**: 381–461.

Tang, Y. (1994). Embryology of *Plagiopteron suaveolens* Griffith (Plagiopteraceae) and its systematic implications. Bot. J. Linn. Soc. **116**: 145–157.

9

POLLEN MORPHOLOGY AND NUMERICAL ANALYSIS OF *TAMARIX* L. (TAMARICACEAE) IN EGYPT AND ITS SYSTEMATIC IMPLICATION

AHMED ELKORDY[1] AND AHMED FARIED[2]

Botany & Microbiology Department, Fac. of Science, Sohag University, 82524, Egypt

Keywords: Tamarix; Pollen Morphology; Numerical Taxonomy; UPGMA Cluster; PCO, Egypt.

Abstract

A palynological and multivariate study of six species of *Tamarix* L. distributed in Egypt was carried out. Pollen morphology was examined by Light Microscope (LM) and Scanning Electron Microscope (SEM). The systematic study of these species was conducted by means of numerical analysis using UPGMA clustering and PCO analysis based on 33 morphological characters, including life form, vegetative parts, fruits, floral characters, seeds and pollen grains. Pollen grains were found to be monads, radially symmetrical, isopolar, small-sized and homocolpate. *Tamarix nilotica* can be easily separated from other taxa by its subprolate pollen shape. Two major clades were identified by multivariate analysis of morphological characters; one of them included three species, namely *T. amplexicaulis*, *T. passerinoides* and *T. macrocarpa*. Our results indicate that there is a wide range of morphological similarity among the species of section Polyadenia. The other clade included *T. aphylla*, *T. tetragyna* and *T. nilotica* from section Tamarix and section *Oligadenia*.

Introduction

Tamarix L. is one of the four genera of Tamaricaceae, and consists of halophytic shrubs and dwarf trees native to Europe, Asia, southern and northern Africa (Baum,1978).The taxonomy of *Tamarix* is notoriously complex (Bunge, 1852; Zohary, 1972; Baum, 1978; Villar *et al.*, 2014). Most of its members show few distinctive external features (Baum, 1978) and most taxa are difficult to distinguish in the vegetative state (Crins, 1989). Hybridization is considered as one of the most important factors which may play a great role in this taxonomic interference (Rusanov, 1949; Wilken, 1993). The most recent global revision of the genus *Tamarix* by Baum (1978) includes three featured sections, viz. *Tamarix, Oligadenia* and *Polyadenia*, separated primarily by petal length, number of stamens, shape of androecial disk and attachment of filament insertion on the androecial disk. Furthermore, these sections are split into nine series based on several floral and vegetative characters. Intermediate forms have been observed for many characters which are used in identification of taxa within the genus, and these characters usually vary seasonally on the same individual (Rusanov,1949).

Species of *Tamarix* cannot be identified without the presence of floral and fruit characters (Crins, 1989). According to Baum (1978) and Crins (1989), morphological characters such as life form and growth morphology, leaf characters (vaginate vs. sessile), number of floral parts and morphology of androecial disk can be used for distinguishing certain species. Other characters such as shape of petal and sepal, presence or absence of hairs on the raceme rachis and the nature of filament attachment to the androecial disc are discussed by Gaskin and Schaal (2003). According to Venturella *et al.* (2007) most taxa of *Tamarix* L. have more or less the same morphological and ecological similarity, infraspecific variability and teratology of floral elements.

[1]Correspondent author. Email: elkordy3000@yahoo.com
[2]Botany & Microbiology Department, Faculty of Science, Assiut University, 71515, Egypt.

The difficulties in the identification of *Tamarix* species sometimes give rise to inaccurate descriptions and problems in analytical keys.

Täckholm (1974) reported five species of *Tamarix* from Egypt, namely *T. aphylla*, *T. tetragyna*, *T. nilotica*, *T. amplexicaulis* and *T. passerinoides* Hosni (2000) added new three taxa to the flora of Egypt, viz. *T. macrocarpa, T. mannifera* and *T. arborea.* Boulos (2000) in his last issue of Flora of Egypt reduced the number of species to six and treated *T. mannifera* and *T. arborea* as synonyms of *T. nilotica*. Moreover, he reported *T. macrocarpa* from Egypt, but still some authors consider *T. macrocarpa* to be conspecific with *Tamarix passerinoides*; others treat it as a variety of the latter species. Baum (1978) treated *T. mannifera, T. arabica, T. gallica,* and *T. arborea* as separate good species; while Boulos (2000) treated all of them as synonyms of *T. nilotica*. Baum (1978) treated *T. meyeri* as good separate species, while Boulos (2000) considered it as a synonym of *T. tetragyna*. Concerning *T. effusa* and *T. deserti,* the two species are considered by Baum (1978) and Boulos (2000) as synonyms of *T. tetragyna* (Table 1).

The objectives of this work are to investigate suitability of pollen morphology for distinguishing Egyptian *Tamarix* and by using 33 characters, how do multivariate analyses agree with the sectional placement of the genus *Tamarix*.

Materials and Methods

Palynological study

Pollen grains examined for *Tamarix* species in this study were collected from herbarium specimens given in Appendix 1. Pollen grains were acetolized and mounted on a metallic stub in a few drops of ethanol according to methods outlined in Moore *et al.* (1991). Light microscopy of pollen grains was carried out using an Olympus type BH-2 research microscope. The measurements were based on at least 20 pollen grains per specimen. Photomicrographs were taken under an Olympus photomicroscope microscope. For scanning electron microscopy (SEM), Pollen grains were prepared according to the procedure given by Baum *et al.* (1970). The specimens were coated with gold in an Apolaron E1100 ion sputtering device, then viewed at 25–30 kv in a JOEL JSM 5300 scanning electron microscope. The means of Polar axis (*P*) and equatorial diameter (*E*) were measured and the *P*/*E* ratio was calculated over all specimens. The main morphological characters and terminology and concepts of pollen grains are based on Huysmans *et al.* (2003); Punt *et al.* (2007); Erdtman (2013).

Multivariate analysis of 33 morphological characters

Plant materials: The multivariate study is based on herbarium specimens borrowed from the following herbaria: CAI, CAIM, ASTU, W, and SHG. In addition, fresh materials of the most species were collected, field observations were made from several localities in Egypt. Species constituted the OTU (Operational Taxonomic Unit) are presented in Table 3.

In order to broadly sample the morphological variation, the OTUs consist of a number of collections/accessions (either herbarium specimens or fresh material or both) from different localities in Egypt. For some taxa, materials from Egypt were not available or limited, so specimens from other countries were used (OTUs 2 and 4.)

Morphological character observations: Morphological characters and character states scored for plant, seed, and pollen are presented in Table 4. A total of 33 characters were measured on each specimen, comprising 15 quantitative and 18 qualitative characters; 3 of the qualitative characters were scored as binary and the rest were scored as multi-state characters.

Plant morphology, flower and fruit characters: The measurements for all specimens of a taxon were averaged into one OTU scored for each of the characters. OTU scores for quantitative characters were averages of measurements of at least 10 specimens (where possible). Because herbarium specimens cannot be considered to be a random sample of species, we followed Wieringa (1999) by calculating the mean of the minimum and maximum measurement of all specimens for species. The complete data matrixes for the Numerical Taxonomy study, including specimen citations are available as appendixes from the Department of Botany and Microbiology, Sohag University, Faculty of Science, Egypt.

Table 1. Synopsis of the Egyptian *Tamarix* (according to Baum 1978 and Boulos 2000).

No.	Taxa	Baum (1978)			Boulos (2000)
		Sect.	Series	Rank	
1.	*Tamarix nilotica*				Good species
2.	*T. mannifera*		Leptostachyae	Good sp.	
3.	*T. arborea*			Good sp.	Syns. of *T. nilotica*
4.	*T. arabica*			Good sp.	
5.	*T. gallica*	Tamarix	Gallica	Good sp.	
6.	*T. aphylla*				Good species
7.	*Thuja aphylla*		Vagianates		
8.	*T. orientalis*				Syns. of *T. aphylla*
9.	*T. articulata*				
10.	*T. tetragyna*				Good species
11.	*T. meyeri*	Oligadenia	Anisandrae	Good sp.	Syn. of *T. tetragyna*
12.	*T. effuse*				Syns. of *T. tetragyna*
13.	*T. deserti*				
14.	*T. passerinoides*				
15.	*T. macrocarpa*				Good species
16.	*T. amplexicaulis*	Polyadenia	Pleiandrae		

Data analysis

Two types of analyses were performed with PAST (Paleontological Statistics Version 3.15) (Hammer *et al.,* 2001). First, we performed a cluster analysis using average taxonomic distance and UPGMA clustering (procedures RHO and HAMMING similarity index) to reduce the effects of different scales of measurement for different characters. Secondly a principal coordinates analysis (PCO) was performed, using the product-moment correlation as a coefficient. The procedure (RHO similarity index) was used to calculate the distance matrix and PCO based on STAND data.

Results

Pollen grains in Tamarix species are free (monads), tricolpate, prolate to subprolate *Pollen grains:* in equatorial view to spheroidal in polar view; ranging from 11.92 to 18.56 µm in polar axis length, 7.82 to 12.77 µm in equatorial diameter. The ratio of polar axis to the equatorial diameter *(P/E)* is between 1.3 and 1.63; they all have small, fine to coarse reticulate sculpturing grains, luminae polygonal in shape. Apertures are 3, and simple (Table 2).

All pollen grains of studied species found to be isopolar and radially symmetrical. They are suboblate, oblate sphenoidal, spheridal, prolate, prolate spheroidal and subprotate in shape (Figs 1-

6). The ratio between the mean polar axis (*P*) and the mean equatorial diameter (*E*) can be used to assign the pollen grains to shape classes as follows (Punt *et al.*, 2007).

Pollen grains suboblate when P/E = 0.75–0.875; Pollen grains oblate spheroidal when P/E= 0.875–1.0; Pollen grains spheroidal when P/E = 1.0; Pollen grains prolate spheroidal when P/E = 1.1–1.14; Pollen grains subprolate when P/E = 1.14–1.33; Pollen grains prolate when P/E= 1.33–2.0

For each studied species, the mean of polar axis is plotted against the mean equatorial diameter (Fig. 7). Oblate and peroblate grains would lie above this 45° line, and oblate spheroidal and suboblate above but near this line. Perfectly spheroidal grains must lie along the 45° line in Fig. 1, with perprolate and prolate grains below this line and subprolate and prolate spheroidal below but near the line. Concerning the studied species, there was a tendency for pollen grains to have a prolate shape in all species examined except *T. nilotica* issubprolate (Fig. 3A). The pollen grains seem to be mostly circular in polar view, often slightly lobed due to sunken colpi.

Pollen grains of studied taxa are comparatively small; the average polar axis value ranges from 11.96 µm in *T. nilotica* to 16.84 in *T. passerinoides*; the average *E* value varies from 9.14 µm in *T. nilotica* to 11.41 µm in *T. aphylla* (Table 2). All pollen grains are characterized by simple apertures and are zonocolpate; the colpi are narrow to slit-like after acetolysis, they are usually widest at the equatorial view and gradually narrow towards the poles. The number of colpi is always three.

Pollen grain have the same ornamentation pattern with reticulate tectum, circular to polygonal laminae in shape, 0.46 - 0.91 µm in length (Figs 1B-6B).

Table 2. Morphological data of pollen grain characters of genus *Tamarix*.

Sl. No.	Species	Polar axis (P µm)		Equatorial axis (E µm)		P/E	Pollen shape	Aperture no.	Surface ornamentation
		Range	Mean	Range	Mean				
1.	*T. aphylla*	13.30 –18.56	15.91	10.11–12.77	11.41	1.39	Prolate	3	Reticulate
2.	*T. tetragyna*	14.38 –17.62	16.23	7.82–11.99	10.32	1.57	Prolate	3	Reticulate
3.	*T. nilotica*	11.92 –11.99	11.96	8.64–9.94	9.14	1.30	Subprolate	3	Reticulate
4.	*T. amplexicaulis*	14.49 –16.1	15.54	10.71–10.97	10.81	1.40	Prolate	3	Reticulate
5.	*T. passerinoides*	14.75 –17.91	16.84	8.92–10.90	10.26	1.64	Prolate	3	Reticulate
6.	*T. macrocarpa*	14.79 – 17.10	15.64	10.92 \–9.18	10.23	1.52	Prolate	3	Reticulate

Equatorial diameter (*E*); Polar axis (*P*); the ratio between the Polar and Equatorial (*P/E*)

Multivariate analysis of 33 morphological characters

Cluster analysis

Figure 8 shows the dendrogram of all OTUs studied, clustered by the UPGMA method. The cophenetic correlation of distance matrix and tree matrix was 0.9512, indicating a good fit of the dendrogram to the distance matrix, see Rohlf (1990) .

Two clades were identified, namely clade A and Clade. Clade A is divided into two subgroups: subgroup (I) comprising *T. amplexicaulis,* and subgroup (II) comprising *T. passerinoides* and *T. macrocarpa.* Clade B is also divided into two subgroups: subgroup (III) comprising *T. nilotica and T. tetragyna*, while sub group (IV) comprising only *T. aphylla.*

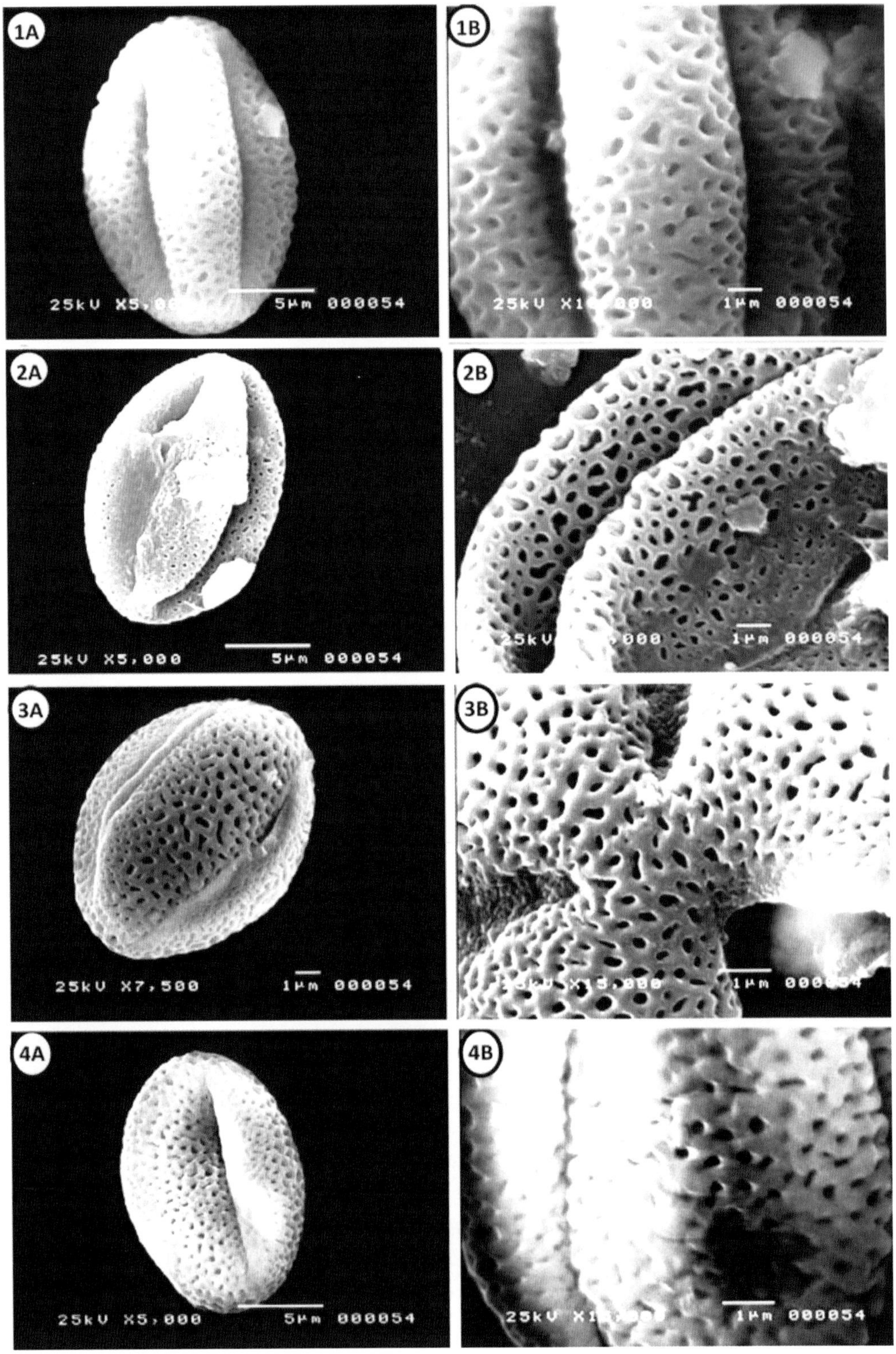

Figs 1-4. Pollen grains photographs of studied species under Scanning electron microscope: A- entire pollen grains; B - enlargement part of pollen grain exine: 1. *T. aphylla,* 2. *T. tetragyna,* 3. *T. nilotica*, 4. *T. amplexicaulis.*

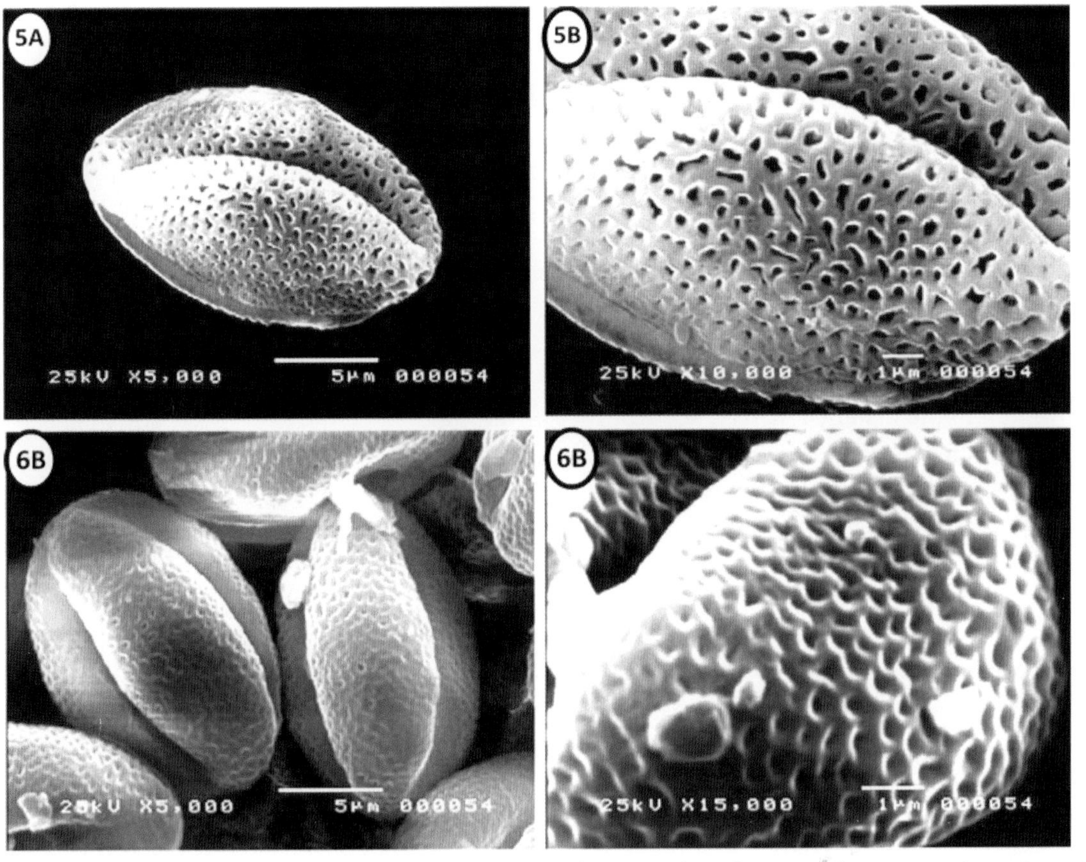

Figs 5-6. Pollen grains photographs of studied species under Scanning electron microscope: A- entire pollen grains; B - enlargement part of pollen grain exine: 5. *T. passerinoides,* 6. *T. macrocarpa.*

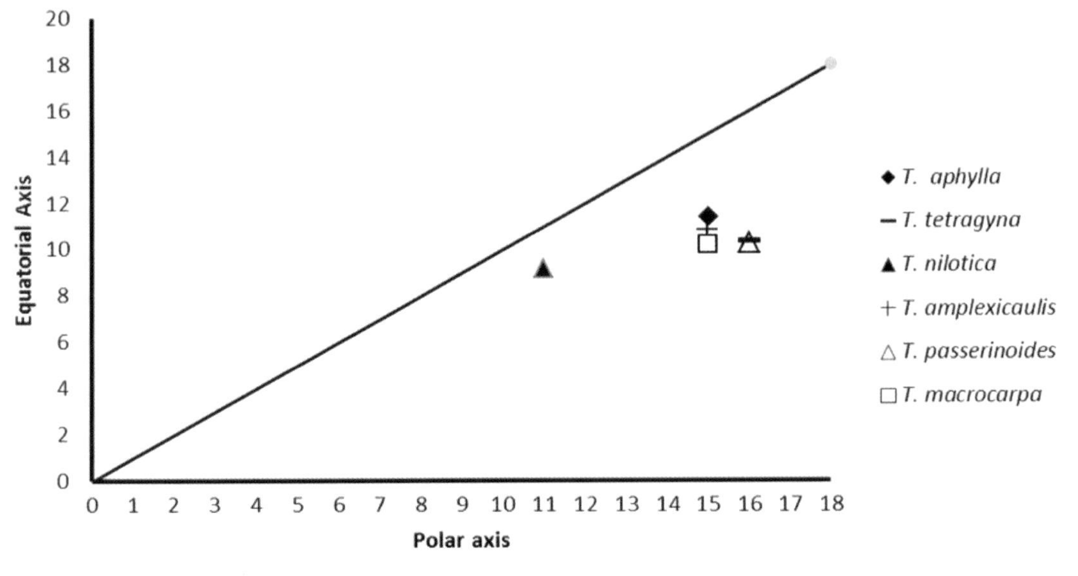

Fig. 7. Pollen mean equatorial vs. polar measurements.

Principal components analysis (PCO)

The plot of 6 OTU's on the first three principal components is shown in Figs 9 and 10. These components interpret 87.788 % of the total observed variation.

On the first component (44.523 % of the total variation in Figs. 9 and 10) a segregation is demonstrated between two groups. 1) *T. amplexicaulis, T. passerinoides* and *T. macrocarpa;* 2) *T. nilotica, T. tetragyna* and *T. aphylla.*

Multivariate analysis of 33 morphological characters

Cluster analysis

Figure 8 shows the dendrogram of all OTUs studied, clustered by the UPGMA method. The cophenetic correlation of distance matrix and tree matrix was 0.9512, indicating a good fit of the dendrogram to the distance matrix, see Rohlf (1990) .

Two clades were identified, namely clade A and Clade. Clade A is divided into two subgroups: subgroup (I) comprising *T. amplexicaulis,* and subgroup (II) comprising *T. passerinoides* and *T. macrocarpa.* Clade B is also divided into two subgroups: subgroup (III) comprising *T. nilotica and T. tetragyna*, while sub group (IV) comprising only *T. aphylla.*

Principal components analysis (PCO)

The plot of 6 OTU's on the first three principal components is shown in Figs 9 and 10. These components interpret 87.788 % of the total observed variation.

On the first component (44.523 % of the total variation in Figs. 9 and 10) a segregation is demonstrated between two groups. 1) *T. amplexicaulis, T. passerinoides* and *T. macrocarpa;* 2) *T. nilotica, T. tetragyna* and *T. aphylla.*

The main characters explaining this separation (characters with high factor loading 0.6) are Life cycle, Color of plant bark, Leaf shape, Leaf base, Leaf apex, Inflorescence type, Inflorescence length, Bract apex, Sepal length, Sepal shape, Petal length, Petal colour, Stamen number, Disc presence, filament attachment position to the androecial disk, Style number, Capsule length, Capsule shape, Seed length and Seed shape.

The second (28.688% of the total variation in Figs 9&10) and third components (14.577 % of the total variation) do not reveal a divide between studied groups of the OTU's.

Table 3. List of OTU's for the *Tamarix* species used for the studies arranged by section and series according to Baum (1978).

No.	Taxon	Origin	No. of individuals	Baum (1978)
1.	*T. aphylla*	Egypt	10	Section: Tamarix Series: Vagianates
2.	*T. tetragyna*	Egypt, Israel	4	Section: Oligadenia Series: Anisandrae
3.	*T. nilotica*	Egypt	15	Section: Tamarix Series: Leptostachyae
4.	*T. amplexicaulis*	Tunisia	5	Section: Polyadenia Series: Pleiandrae
5.	*T. passerinoides*	Egypt	3	Section: Polyadenia Series: Pleiandrae
6.	*T. macrocarpa*	Egypt	7	Section: Polyadenia Series: Pleiandrae

Table 4. Characters and character states used in morphometric analysis of the genus *Tamarix* in Egypt.

Characters	Character states
1. Life cycle	1. Shrubs
	2. Trees
2. Plant height	1. 1-5 m
	2. 5-12 m
	1. Glabrous
3. Stem indumentum	2. Papillose
	3. Papillose to glabrous
4. Color of plant bark	1. Reddish brown
	2. Purple to dark brown
	1. Sheath-like
5. Leaf shape	2. Oblong-lanceolate
	3. Ovate-deltoid
	4. Ovate-deltoid to lanceolate
	5. Amplexicaul
6. Leaf length	1. 1-3 mm
	2. 3-6 mm
	1. Sessile
7. Leaf base	2. Cordate and clasping
	3. Amplexicaul
	1. Acute
8. Leaf apex	2. Acute to acuminate and not spreading
	3. Acute to acuminate and spreading
	4. Short pointed apex
9. Inflorescence type	1. Raceme
	2. Simple compact spike
	3. Paniculate
	4. Simple raceme or paniculate
10. Inflorescence length	1. 3-5 cm
	2. 5-10 cm
11. Inflorescence width	1. 0.3-0.8 cm
	2. 0.8-1.2 cm
12. Bract length	1. 1-2 mm
	2. 2-3.5 mm
13. Bract shape	1. Ovate-deltoid
	2. Triangular
	3. Narrowly triangular
	4. Oblong-linear
	1. Acute
14. Bract apex	2. Acuminate
	3. Upper acuminate, lower obtuse
	1. 0 mm
15. Pedicel length	2. 0.5-1 mm
	3. 1-1.5 mm
16. Flower diamter	1. 2-4 mm
	2. 5-6 mm
17. Sepal length	1. 1-1.5
	2. 1.5-2

Table 4 contd.

Characters	Character states
18. Sepal shape	1. Ovate 2. Broadly ovate 3. Deltoid-ovate 4. Ovate-elliptic 5. Outer 2 smaller and broadly ovate to broadly elliptical, the inner larger, broadly elliptical to suborbicular
19. Sepal apex	1. Acute 2. Obtuse 3. Outer 2 acute, the inner obtuse
20. Petal shape	1. Oblong elliptic 2. Ovate elliptic 3. Obovate elliptic 4. Broadly elliptic to ovate 5. Oblong, elliptical-oblong to obovate
21. Petal length	1. 1.5-3.0 mm 2. 3.0-4.5 mm
22. Petal colour	1. White 2. Pinkish 3. Pink
23. Stamen number	1. 4-5 2. 6-10 3. 10-13
24. Disc presence	1. Present 2. Absent
25. Position of filament insertion on the androecial disk	1. Filament not arising from disc 2. Jointed with the deep sinuses of the disc 3. Inserted at the entire or retuse disc lobes 4. Exserted, inserted in the sinuses of the disc
26. Style no.	1. 3 2. 4
27. Capsule length	1. 3-6 mm 2. 6-13 mm
28. Capsule shape	1. Pyriform 2. Pyramidal 3- Pyramidal with 4-valved 4- Ovoid pyramidal
29. Seed length	1. 0.5 mm 2. 1-1.5 mm
30. Seed shape	1. Oblong 2. Ovoid oblong 3. Terete 4. Cylindrical
31. Pollen shape	1. Prolate 2. Subprolate
32. Mean of polar axis	1. 11.96 2. 15.54-16.84
33. Mean of equatorial axis	1. 9.14 2. 10.23-11.41

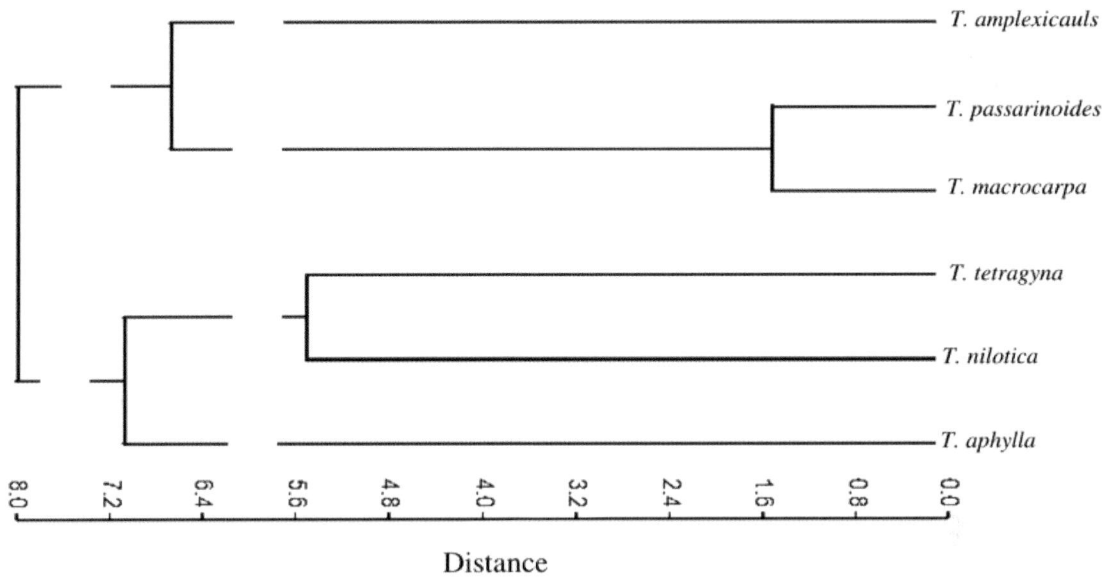

Fig. 8. UPGMA dendrogram of Tamarix speies showing interspecific relationship.method.

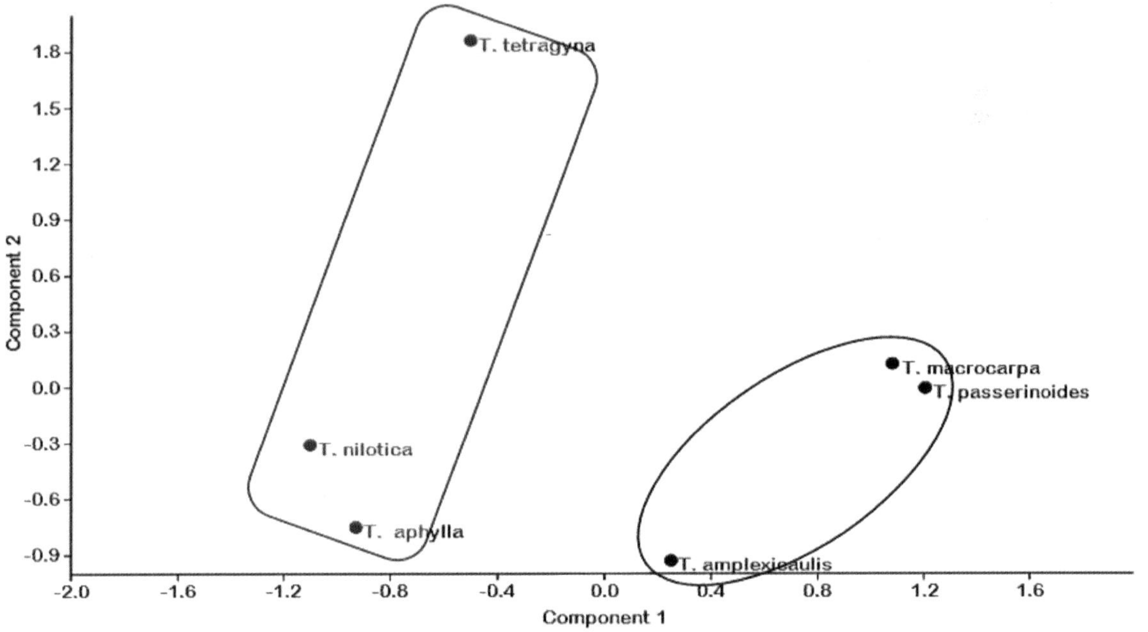

Fig.9. Scatter plot of the 6 OTUs plotted against the first principal component by the second principal component in *Tamarix* species.

Table 5. Vegetative and floral characters on the first three Principal coordinates axes showing highest factor loading. Factor loading values ≥ ±0.6 are shaded.

No.	Characters	Principal components		
		PC 1	PC 2	PC 3
		Factor loading		
1.	Life cycle	-0.7856	-0.4095	-0.062791
2.	Plant height	-0.45467	-0.36715	0.72195
3.	Stem indumentum	-0.20668	0.72258	-0.62667
4.	Color of plant bark	0.64978	0.72503	0.043484
5.	Leaf shape	0.71333	-0.051789	-0.66075
6.	Leaf length	-0.2465	0.91203	0.17145
7.	Leaf base	0.9793	-0.13841	-0.085181
8.	Leaf apex	-0.66548	-0.43643	0.5511
9.	Inflorescence type	0.7604	-0.29822	0.33339
10.	Inflorescence length	-0.9244	0.2937	0.068594
11.	Inflorescence width	-0.2465	0.91203	0.17145
12.	Bract length	-0.2465	0.91203	0.17145
13.	Bract shape	-0.45615	0.84115	-0.18956
14.	Bract apex	-0.74335	0.65303	0.13174
15.	Pedicel length	0.37491	0.71374	0.018628
16.	Flower diamter	-0.2465	0.91203	0.17145
17.	Sepal length	0.64978	0.72503	0.043484
18.	Sepal shape	-0.69836	0.28471	0.4677
19.	Sepal apex	-0.51942	0.74201	0.11933
20.	Petal shape	0.13981	0.10727	0.57966
21.	Petal length	0.64978	0.72503	0.043484
22.	Petal colour	0.73988	-0.10203	0.20584
23.	Stamen number	0.91476	-0.19773	-0.08085
24.	Disc presence	0.9244	-0.2937	-0.068594
25.	Position of filament insertion on the androecial disk	-0.82778	0.32416	-0.43225
26.	Style number	0.92759	0.051465	-0.0052335
27.	Capsule length	0.88406	0.047996	-0.089424
28.	Capsule shape	0.82411	0.47214	-0.26183
29.	Seed length	0.7856	0.4095	0.062791
30.	Seed shape	-0.74882	-0.59962	0.18855
31.	Pollen shape	-0.53904	-0.15084	-0.80138
32.	Mean of polar Axis	0.53904	0.15084	0.80138
33.	Mean of Equatorial axis	0.53904	0.15084	0.80138
	Percentage per PCO	44.523	28.688	14.577

Percentage total variation for the first three principal components is 87.788 %

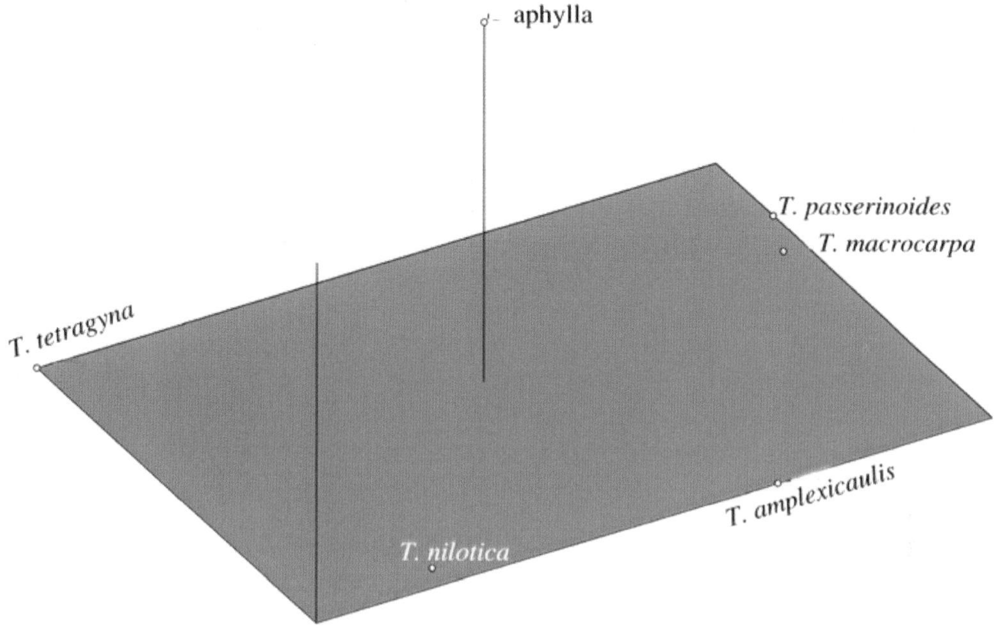

Fig. 10. Principal component analysis of *Tamarix* species explaining 87.788 % of the variation.

Discussion

Tamarix is known as a stenopalynous genus. The morphology of pollen grains is remarkably similar, especially together with the aspects of small tricolpate and reticulate tectum. These results show congruence with those of Gaskin *et al.* (2004) who showed that the genus *Tamarix* is a monophyletic group based on data from the nuclear ribosomal region 18S and chloroplast regions *rbc*L and *trn*A Ser (GCU)/ *trn*A Gly (UCC). The pollen grains shape is either prolate or subprolate; they are subprolate in only *T. nilotica* (Fig. 3A), and prolate in the remaining species (Figa 1A- 6A). These results agree with those of Baum *et al.* (1970).

The size of pollen grains overlaps for most of studied species, although *T. nilotica* is easily distinguished by their relatively small grains, against the rest of the species. The number of apertures is commonly 3, and this agree with the results of Baum *et al.* (1970). The ornamentation pattern of exine was found to be reticulate (Figs 1B-6B).

Morphological characters play an important role in taxonomy and to circumscribe taxa. Taxonomical problems appear when taxa show a huge amount of variability, due to ecological niches and phenotypic plasticity (Van den Berg and Groendijk Wilders, 1999).

Baum (1978) divided the genus *Tamarix* into Sections and Series (Table 1). Baum's study was based on a small number of morphological characters, either vegetative or floral such as the number of stamen, the length of petal, shape of androecial disk, and position of filament insertion on the androecial disk. In our present work, nearly all characters were scored and numerical methods (UPGMA and PCO) were applied to demonstrate the relationships between studied taxa and approximate the grade of variation among taxa. UPGMA analysis gives insight into degree of similarity among the OTU's and whether they form groups/clusters. PCO analysis reflects which characters are important on the axes, and indicates the significant characters based on the highest factor loading (Table 5).

Therefore it becomes clear which characters cause the separation between groups and can be useful to distinguish taxa. In general, the results display congruence between the UPGMA clustering and PCO analyses; two major clades were identified which have been given the names of A and B. Clade A included *T. amplexicaulis*, *T. passerinoides* and *T. macrocarpa*. Baum (1978) placed these species in Section *Polyadenia* series *Pleiandrae*. The results obtained from both UPGMA and PCO analysis confirmed that, the group of *T. amplexicaulis*, *T. passerinoides* and *T. macrocarpa* is a well-distinguished group characterized by amplexicaul or sessile leaves with auriculate bases, racemes 6-10 (-15) mm broad, pentamerous, stamens 6-15 (mostly 10), of these 5 antesepalous and with slightly longer filaments and androecial disc with no nectariferous lobes. Our results are congruent with that of Baum (1978).

Clade B included three species, *T. aphylla* and *T. nilotica* of section *Tamarix* series Vagianates and Leptostachyae respectively and *T. tetragyna* of section *Oligadenia* series Anisandrae. The results of both UPGMA and PCO confirmed that the group of *T. aphylla, T. nilotica and T. tetragyna* is a well-distinguished group with racemes 3-5 mm broad or in dioecious trees 5-7 mm broad, petals 1.0- 2.25 mm long, stamens usually 5 (antesepalous) and disc various. Baum (1978) placed *T. aphylla* and *T. nilotica* in section Tamarix, and *T. tetragyna* in section *Oligadenia*. Moreover, our results placed *T. nilotica* closer to *T. tetragyna* than to *T. aphylla*, which is contrary to Baum's sections. Arianmanesh *et al.* (2015) presented a phylogenetic analysis based on internal transcribed spacer (ITS) of 15 *Tamarix* samples recognized by recent taxonomic treatments from Iran. In addition, 19 previously ITS sequences from GenBank were used. The results of data analysis indicated that, the classification of the genus *Tamarix* into three sections based on some morphological characters by Baum (1978) is not supported by that molecular analyses as well as the classification of species according to the morphology of androecial disc, but the morphology of leaf and number of flower parts are useful for the classification and identification of *Tamarix* species. Both of morphological characteristics and molecular data will be most effective to determine the evolution of the genus *Tamarix*.

According to Arianmanesh *et al.* (2015), two main clades were found; Clade (B) included *T. nilotica, T. passerinoides T. tetragyna* var. *meyeri* and *T. aphylla* and clade (C) included *T. amplexicaulis* and *T. tetragyna* var. *deserti*. Our results are in matching with those of Arianmanesh *et al.* (2015), in which they support the presence of *T. nilotica, T. tetragyna* and *T. aphylla* in one section, and *T. amplexicaulis* in other section, and disagree with those of Baum (1978).

Our results using the UPGMA and PCO analysis to show the similarities between species indicate that there is some degree of similarity between the species of section Polyadenia.

Acknowledgements

We are grateful to the Directors and Curators of Naturhistorisches Museum Wien (Vienna) Herbarium (W), Kew Herbarium (K), Leiden Herbarium (L) Wageningen University Herbarium (WAG), Cairo University Herbarium (CAI), Agriculture Museum in Dokki (CAIM), and Assiut University Herbarium (ASTU) for the loan of specimens. Our great thanks are due to Dr. John Gaskin; Botanist/Research Leader PMRU, USDA ARS NPARL. Sidney, USA for going through the manuscript and making valuable suggestions.

References

Arianmanesh, R., Mehregan, I., Nejadsatari, T., Assadi and M. 2015. Molecular phylogeny of *Tamarix* (Tamaricaceae) species from Iran based on ITS sequence data. Eur. J. Exp. Biol. **5**: 44–50.

Baum, B. 1978. The genus *Tamarix*. Israel Academy of Sciences and Humanities.

Baum, B., Bassett, I. and Crompton, C. 1970. Pollen morphology and its relationships to taxonomy and distribution of *Tamarix*, series Vaginantes. Österreichische Botanische Zeitschrift **118**: 182–188.

Boulos, L. 2000. Flora of Egypt (Geraniaceae- Boraginaceae). Al Hadara Publishing, Egypt, pp. 124–130.

Bunge, A.G.V. 1852. Tentamen generis Tamaricum species accuratius definiendi. Dorpati: ex Officina Academica viduae JC Schuenmanni et C. Mattieseni.

Crins, W. 1989. The Tamaricaceae in the southeastern United States. J. Arnold Arboretum **70**: 403–405.

Erdtman, G. 2013. An Introduction to Pollen Analysis. Read Books Ltd.

Gaskin, J.F. and Schaal, B.A. 2003. Molecular phylogenetic investigation of US invasive *Tamarix*. Systematic Botany **28**: 86–95.

Gaskin, J.F., Ghahremani-nejad, F., Zhang, D.Y. and Londo, J.P. 2004. A systematic overview of Frankeniaceae and Tamaricaceae from nuclear rDNA and plastid sequence data. Ann. Miss. Bot. Gard. **91**: 401–409.

Hammer, Ø., Harper, D.A.T., and Ryan, P. D. 2001. PAST: Paleontological Statistics Software Package for Education and Data Analysis. Palaeontologia Electronica **4** (1): pp. 9. Available at: http://folk.uio.no/ohammer/past/.

Hosni, H. A. 2000. Tamaricaceae in the flora of Egypt. Taeckholmia **20**: 17–31.

Huysmans, S., Dessein, S., Smets, E. and Robbrecht, E. 2003. Pollen morphology of NW European representatives confirms monophyly of Rubieae (Rubiaceae). Rev. Palaeobot. and Palynol. **127**: 219–240.

Moore, P., Webb, J. and Collinson, M. 1991. Pollen Analysis, **2**nd. Blackwell, Oxford.

Punt, W., Hoen, P., Blackmore, S., Nilsson, S., Le Thomas, A. 2007. Glossary of pollen and spore terminology. Rev. Palaeobot. and Palynol. **143**: 1–81.

Rusanov, F. 1949. Sredniyeaziatskie Tamariksi. Tash-kent.[Tamarisks of Central Asia.].

Täckholm, V. 1974. Students' Flora of Egypt. Cairo University, Egypt.pp. 366–369.

Van den Berg, R.G. and Groendijk Wilders, N. 1999. Numerical analysis of the taxa of series Circaeifolia (*Solanum* sect. Petota). Solanaceae IV: advances in biology and utilization. Kew: The Royal Botanic Gardens, Kew,pp. 213–226.

Venturella, G., Baum, B. and Mandracchia, G. 2007. The genus *Tamarix* (Tamaricaceae) in Sicily: first contribution. Flora mediterranea **17**: 25–46.

Villar, J.L., Alonso, M.Á., Vicente, A., Juan, A. and Crespo, M.B. 2014. The genus *Tamarix* (Tamaricaceae) in Crete (Greece). Willdenowia **44**: 321–326.

Wieringa, J.J. 1999. Monopetalanthus exit. A systematic study of *Aphanocalyx, Bikinia, Icuria, Michelsonia* and *Tetraberlinia* (Leguminosae,Caesalpinioideae). Wageningen Agricultural University, Wageningen, Netherlands, pp. 320.

Wilken, D.H. 1993. Tamaricaceae. *In*: Hickman, J.C. (Ed.), The Jepson Manual: Higher Plants of California University of California Press, Berkeley, Los Angeles, London, pp. 1080.

Zohary, M. 1972. *Tamarix*. *In*: Zohary, M. and Feinbrun-Dothan, N.(Eds.), Flora Palaestina, Vol. **2.** Israel Academy of Sciences and Humanities, Jerusalem, pp. 351–364.

10

POLLEN MORPHOLOGY OF IRANIAN *DRACOCEPHALUM* L. (LAMIACEAE) AND ITS TAXONOMIC SIGNIFICANCE

MARYAM NADERIFAR[1], ALI SONBOLI[2] AND ABBAS GHOLIPOUR[1]

Department of Biology, Medicinal Plants and Drugs Research Institute, Shahid Beheshti University, Tehran, Iran

Keywords: Lamiaceae; *Dracocephalum*; Taxonomy; Pollen; Iran.

Abstract

Pollen morphology of 11 Iranian *Dracocephalum* L. species was investigated using light and scanning electron microscopy to evaluate their taxonomic significance for the infrageneric classification of the genus. Pollen grains of all examined taxa were isopolar, hexacolpate, circular in polar view and spheroidal to prolate in equatorial view (P/E = 1.0–2.0). The smallest pollen grains were observed in *D. aucheri* (P = 29.7 μm, E = 22.6 μm), while the largest pollen was found in *D. lindbergii* (P = 45.1 μm, E = 33.7 μm). The highest and lowest apocolpium index (AI) were measured in *D. aucheri* (AI = 0.27) and *D. surmandinum* (AI = 0.08), respectively. Colpus membrane was egranulate in all examined species except for *D. multicaule* and *D. ghahremanii*. The main exine ornamentation type was characterized as bireticulate including five different subtypes. The results revealed that the exine ornamentation is a diagnostic character useful for the classification of *Dracocephalum*.

Introduction

The genus *Dracocephalum* L. is the second largest genus in the subtribe Nepetinae, tribe Mentheae of Lamiaceae family. The genus is primarily of the Old World and consists of 71 species worldwide, of which 69 species are native to Eurasia (Budantsev, 1987, 1993; Kadereit, 2004), and North Africa and North America have one species each. Morphologically, *Dracocephalum* is a heterogeneous genus comprising species with calyx glabrous within and not gibbous at base; sinuses between lobes of calyx with swollen folds at base, bracteoles aristately toothed, nutlets elliptic to oblong and areole not curved. The genus *Dracocephalum* was divided into three subgenera by Budantsev (1987), *viz*., subg. *Dracocephalum* including seven sections (stamens included, anthers glabrous), *Fedtschenkiella* (Kudr.) Schischk. (stamens exserted, anthers glabrous), and *Ruyschiana* (Mill.) Briq. (stamens included, anthers pubescent). The members of this genus are well-known as medicinal plants with several uses, such as anti-hyperlipidemic, analgesic, antimicrobial, antioxidant, anticancer and oxidative stress protective activity (Sajjadi *et al*., 1998; Jahaniani *et al*., 2005; Sonboli *et al*., 2008).

Rechinger (1982) in his treatment for Flora Iranica area, recognized 18 *Dracocephalum* species of which eight species are growing in Iran. Since the publication of the Flora Iranica, one new species (*D. ghahremanii* Jamzad), a new record (*D. lindbergii* Rech.f.), and a resurrected species (*D. oligadenium* Bornm. & Gauba) have been added to the genus (Esfandiari, 1985; Jamzad, 2012). Considering new findings, the number of species now stands 11, of which five species (*D. ghahremanii* Jamzad, *D. kotschyi* Boiss., *D. oligadenium* Bornm. & Gauba, *D. polychaetum* Bornm. and *D. surmandinum* Rech. f.) are endemic.

---

[1]Department of Biology, Payame Noor University, 19395-4697, Tehran, Iran.

[2]Corresponding author. Email: a-sonboli@sbu.ac.ir

According to the aperture number and number of nuclei, Erdtman (1945) divided the Lamiaceae into two subfamilies. While Lamioideae subfamily comprised tricolpate and bi-nucleate pollen grains, Nepetoideae subfamily is characterized by hexacolpate and tri-nucleate pollen grains. Investigations of pollen morphology in the Lamiaceae have been essential as an aid to classification within this family (Erdtman, 1945; Harley *et al.*, 1992; Abu-Asab and Cantino, 1994).

Pollen morphology of different members of Lamiaceae and other plant families have been frequently investigated by several authors (Moon *et al.*, 2008 ; Ozler *et al.*, 2011; Badamtsetsg *et al.*, 2012; Sarwar and Takahashi, 2012, 2013; Jamzad and Hasani-Nejad, 2014) that show its taxonomic and phylogenetic importance. As far as our literature survey could ascertain, there are only a few publications on pollen morphology in *Dracocephalum.* Moon *et al.* (2008) investigated pollen morphology of 12 genera belonging to the tribe Mentheae, subtribe Nepetineae, including 13 species of *Dracocephalum*, one of which, *D. subcapitatum* (Kuntze) Lipsky, grows in northeast of Iran. Pollen morphology of three *Dracocephalum* species, *viz. D. foetidum* Bge., *D. grandiflorum* L. and *D. ruyschiana* L. from Mongolia have already been investigated (Badamtsetsg *et al.*, 2012).

The pollen grain morphology of *Dracocephalum* species growing in Iran have not been studied yet. Therefore, the main objectives of the present study are to provide a detailed account of the pollen morphology of 11 *Dracocephalum* species in Iran by light microscopy (LM) and scanning electron microscopy (SEM), and to evaluate its taxonomic significance for the classification of the genus.

Materials and Methods

Plant specimens of 11 *Dracocephalum* species were collected from their natural habitats between 2007 and 2013 and the voucher specimens were deposited in Medicinal Plants and Drugs Research Institute Herbarium of Shahid Beheshti University (MPH), Tehran (Table 1). For the LM studies, pollen grains were prepared following the method of Erdtman (1960). Pollen characteristics, including pollen shape, polar axis (P), equatorial axis (E), colpus length (Cl), exine thickness (Et), mesocolpium diameter (Me), and apocolpium index (AI), were measured from at least 20 mature pollen grains per sample under an Olympus BX–51 microscope. Results are provided as minimum, maximum and mean ± standard deviations.

For the SEM studies, pollen grains were transferred directly to stubs with double-sided adhesive tape and micrographs were obtained using KYKY–EM 3200 SEM at an accelerating voltage of 25 kV. Some characters, such as number of primary lumina per 100 μm^2, shape of primary lumina, diameter of primary lumina, number of perforation in primary lumina, presence or absence of granule on colpus membrane, presence or absence of 1–2 central large hole per primary lumina, and sculpturing type were measured and characterized. The pollen terminology follows Faegri and Iversen (1989), Harley *et al.* (1992) and Punt *et al.* (2007).

Results

The diagnostic pollen grain characters of 11 species of *Dracocephalum* investigated are presented in Tables 2 and 3. Photographs of some *Dracocephalum* species in their native habitats are presented in Figure 1. The LM and SEM micrographs of the studied pollen grains are illustrated in Figures 2–4. The measured parameters of the pollen grains are described below.

Size: The pollen grains were dispersed as monads. The mean size of the polar axis (P) ranges from 29.7 μm in *Dracocephalum aucheri* to 45.1 μm in *D. lindbergii*; the mean size of the equatorial axis (E) ranges from 22.6 μm in *D. aucheri* to 33.7 μm in *D. lindbergii* (Table 2).

Shape: The shape of the pollen grains in equatorial view varies from prolate to spheroidal, whereas their shape in polar view is more or less circular (Table 2; Figs 2–4). Often grain forms vary and co-exist between prolate and subprolate.

Table 1. Collection data and voucher information of *Dracocephalum* species studied.

No.	Species	Collection data
1.	*Dracocephalum aucheri* Boiss.	Tehran: Tuchal mountain, 24.06.2008. Gholipour 1281 (MPH)
2.	*D. moldavica* L.	Mazandaran: Sari, Cultivated, 10.06.2009. Gholipour 1693 (MPH)
3.	*D. thymiflorum* L.	Mazandaran: Siah Bishe, Karaj toward Chalus, Allamol, 30.06.2009. Sonboli & Gholipour 1643 (MPH)
4.	*D. multicaule* Montbr. & Auch. *ex* Benth.	West Azarbayejan: Khoy, Qotor, Habash, Arvin mountain, 01.07.2012. Gholipour 824 (SPNH)
5.	*D. kotschyi* Boiss. *	Tehran: Fasham, Shemshak toward Dizin, 30.06.2010. Sonboli, Hadian & Moridi 1631 (MPH)
6.	*D. lindbergii* Rech. f.	North Khorasan: Bojnourd, Rein, Aladagh mountain, 21.05.2007. Sonboli, Kanani & Gholipour 1737 (MPH)
7.	*D. oligadenium* Bornm. & Gauba *	Mazandaran: Siah Bishe, chalus road, Allamol village, 09.07.2009. Sonboli, Hadian & Moridi 1627 (MPH)
8.	*D. surmandinum* Rech. f. *	Esfahane: Semirom toward Shahreza, Surmand mountain, 18.06.2007. Sonboli, Kanani & Gholipour 1179 (MPH)
9.	*D. polychaetum* Bornm. *	Kerman: Babini village, Hazar mountain, 07.05.2008. Kanani, Gholipour & Mirtajadini 1276 (MPH)
10.	*D. subcapitatum* (Kuntze) Lipsky	Khorasan-e Razavi: Mashhad, Kalat, after Sandugh shekan pass, 10.06.2009. Sonboli & Gholipour 1626 (MPH)
11.	*D. ghahremanii* Jamzad *	Semnan: Shahmirzad, Chashm, Nizva mountain, 12.07.2007. Sonboli & Gholipour 1219 (MPH)

* Endemic to Iran

Apertures: The examined pollen grains were isopolar and hexacolpate. Colpi with their acute ends were distributed symmetrically. The colpus length ranged from 25.5 μm in *D. aucheri* to 40.2 μm in *D. lindbergii*, but mesocolpium varied from 7.6 μm in *D. ghahremanii* to 11.5 μm in *D. lindbergii* (Table 2). The highest apocolpium index (AI) was measured in *D. aucheri* (AI = 0.27), whereas the lowest AI was found in *D. surmandinum* (AI = 0.08) (Table 2). Colpus membrane was egranulate in all examined taxa except for *D. multicaule* and *D. ghahremanii* that were granulate (Table 3; Figs 3 & 4).

Exine sculpturing: The exine thickness varied from 0.8 μm in *D. kotschyi* and *D. thymiflorum* to 1.1 μm in *D. multicaule, D. oligadenium, D. surmandinum* and *D. ghahremanii* (Table 2). The main and common exine ornamentation type, examined with SEM in Iranian *Dracocephalum* species studied, was found to be bireticulate. A bireticulum consists of a non-congruent, two-layered reticulum. The main reticulum is referred to as the primary reticulum and the substratum as the secondary reticulum. It can be divided into five subtypes based on shape of primary lumen, the presence or absence of at least one large hole in each primary lumen and the number of secondary lumina per primary lumen (Table 3).

Table 2. Pollen morphological data on LM micrographs of *Dracocephalum* species studied.

Species	P (μm)	E (μm)	Shape				Cl (μm)	Et (μm)	Me (μm)	AI	LM figures
			P	Sp	Ps	S					
Dracocephalum aucheri	24.0 (29.7 ± 2.8) 33.4	19.0 (22.6 ± 2.0) 25.0	++	+	±	-	20.0 (25.5 ± 2.8) 29.0	0.4 (0.9±0.2) 1.2	6.0 (7.9 ± 1.3) 9.6	0.16 (0.20 ± 0.05) 0.27	2A, B
D. moldavica	29.0 (38.8 ± 4.0) 44.0	20.0 (27.5 ± 4.4) 36.0	++	+	±	-	20.2 (32.5 ± 5.3) 39.0	0.4 (0.9±0.3) 1.4	6.0 (8.1 ± 1.6) 11.0	0.15 (0.19 ± 0.04) 0.23	2C, D
D. thymiflorum	28.6 (36.9 ± 5.0) 44.0	19.2 (27.5 ± 5.8) 36.0	+	++	±	-	20.0 (29.1 ± 4.8) 35.0	0.4 (0.8±0.2) 1.2	7.2 (10.0 ± 1.6) 11.6	0.14 (0.20 ± 0.04) 0.25	2E, F
D. multicaule	35.2 (38.9 ± 2.5) 44.2	25.0 (31.4 ± 3.7) 38.4	++	+	±	+	29.0 (32.8 ± 2.7) 39.0	0.8 (1.1±0.3) 1.6	6.0 (7.8 ± 1.8) 13.6	0.12 (0.16 ± 0.03) 0.19	2G–I
D. kotschyi	36.4 (43.7 ± 3.7) 50.8	25.0 (33.5 ± 3.3) 38.6	+	++	±	-	30.0 (38.1 ± 3.6) 44.8	0.6 (0.8±0.1) 1.0	6.0 (11.1 ± 2.0) 13.6	0.10 (0.14 ± 0.02) 0.17	2J –L
D. lindbergii	37.0 (45.1 ± 4.1) 53.8	25.2 (33.7 ± 5.3) 39.2	+	++	-	-	33.0 (40.2 ± 3.8) 48.0	0.6 (0.9±0.3) 1.6	7.0 (11.5 ± 2.8) 16.0	0.14 (0.15 ± 0.01) 0.18	2M, N
D. oligadenium	31.0 (38.8 ± 3.7) 45.6	26.0 (31.3 ± 3.0) 36.0	+	++	+	±	22.0 (30.5 ± 4.3) 38.0	0.8 (1.1±0.2) 1.7	5.6 (8.5 ± 2.2) 13.0	0.12 (0.15 ± 0.02) 0.18	2O, P
D. surmandinum	27.0 (35.8 ± 4.8) 44.0	23.0 (29.5 ± 4.1) 38.0	+	++	+	±	24.0 (30.3 ± 3.8) 38.4	0.8 (1.1±0.2) 1.6	5.0 (7.7 ± 2.0) 12.0	0.08 (0.14 ± 0.05) 0.21	2Q, R
D. polychaetum	33.2 (43.1 ± 4.3) 49.2	21.0 (28.7 ± 5.2) 43.6	++	+	±	-	28.0 (36.9 ± 4.8) 43.0	0.6 (0.9±0.2) 1.4	6.0 (7.7 ± 1.1) 10.0	0.15 (0.18 ± 0.02) 0.20	2S, T
D. subcapitatum	35.0 (38.7 ± 2.2) 42.0	26.0 (30.3 ± 2.7) 36.0	+	++	±	-	25.0 (30.8 ± 3.7) 38.8	0.6 (0.9±0.1) 1.0	5.0 (7.9 ± 2.3) 13.0	0.11 (0.13 ± 0.02) 0.16	2U, V
D. ghahremanii	28.4 (34.0 ± 4.3) 43.0	20.0 (25.4 ± 4.0) 33.0	+	++	+	-	21.0 (27.4 ± 4.5) 36.0	0.8 (1.1±0.3) 1.7	5.0 (7.6 ± 1.7) 11.2	0.16 (0.19 ± 0.03) 0.23	2W, X

All size are in μm, except for AI. Numbers refer to minimum, mean±standard deviation and maximum; P = Polar axis, E = Equatorial diameter, Shape = based on P/E ratio. Cl = Colpus length, Et = Exine thickness, Me = Mesocolpium diameter, AI = Apocolpium index, P = Prolate, Sp = Subprolate, Ps = Prolate sphroidal, S = Sphroidal; (+) Presence, (-) Absence, (++) dominant, (±) rare.

Table 3. Pollen morphological data on SEM micrographs of *Dracocephalum* species studied.

Species	Number of primary lumina per 100 μm^2	Diameter of primary lumina (μm)	Thickness of primary muri (μm)	Colpus membrane	Sculpturing subtype*	SEM figures
D. aucheri	20	2.81	< 0.5	Egranulate	I - 1	3A, B
D. moldavica	5	3.29	< 0.5	Egranulate	I - 2	3C, D
D. thymiflorum	24	3.11	> 0.5	Egranulate	I - 3	3E–H
D. polychaetum	26	4.00	> 0.5	Egranulate	I - 4	3I, J
D. multicaule	30	2.37	< 0.5	Granulate	I - 5	3K, L
D. kotschyi	17	3.06	< 0.5	Egranulate	I - 5	4A, B
D. lindbergii	12	2.82	< 0.5	Egranulate	I - 5	4C, D
D. oligadenium	20	3.45	< 0.5	Egranulate	I - 5	4E, F
D. surmandinum	15	3.10	< 0.5	Egranulate	I - 5	4G, H
D. subcapitatum	11	4.00	> 0.5	Egranulate	I - 5	4I, J
D. ghahremanii	24	2.97	< 0.5	Granulate	I - 5	4K, L

*Sculpturing type = Bireticulate.

Fig. 1. *Dracocephalum* species in native habitats. A: *D. aucheri*; B: *D. surmandinum*; C: *D. kotschyi*; D: *D. ghahremanii*; E: *D. multicaule*; F: *D. polychaetum*.

Type I – 1: The exine ornamentation of *D. aucheri* is characterized by polygonal primary lumina shape, absence of large hole and the number of secondary lumina is less than ten per primary lumen (Fig. 3A, B).

Type I – 2: The shape of primary lumina is polygonal-elongate; without large hole and the number of secondary lumina are over ten per primary lumen which was observed in *D. moldavica* (Fig. 3C, D).

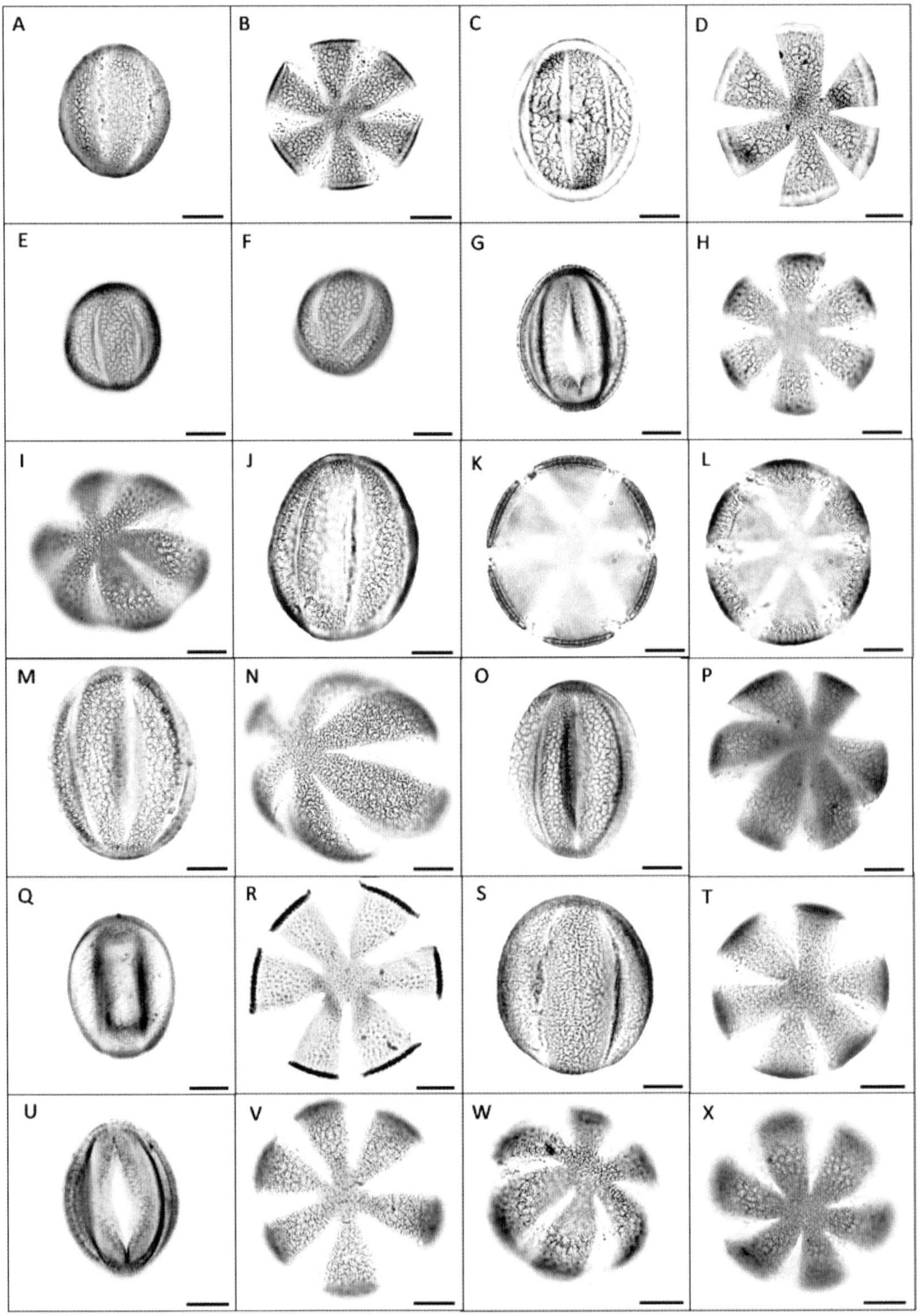

Fig. 2. LM micrographs of pollen grains of *Dracocephalum* L. A,B: *D. aucheri*; C,D: *D. moldavica*; E,F: *D. thymiflorum*; G–I: *D. multicaule*; J–L: *D. kotschyi*; M,N: *D. lindbergii*; O,P: *D. oligadenium*; Q,R: *D. surmandinum*; S,T: *D. polychaetum*; U,V: *D. subcapitatum*; W,X: *D. ghahremanii*. Scale bars = 10 μm.

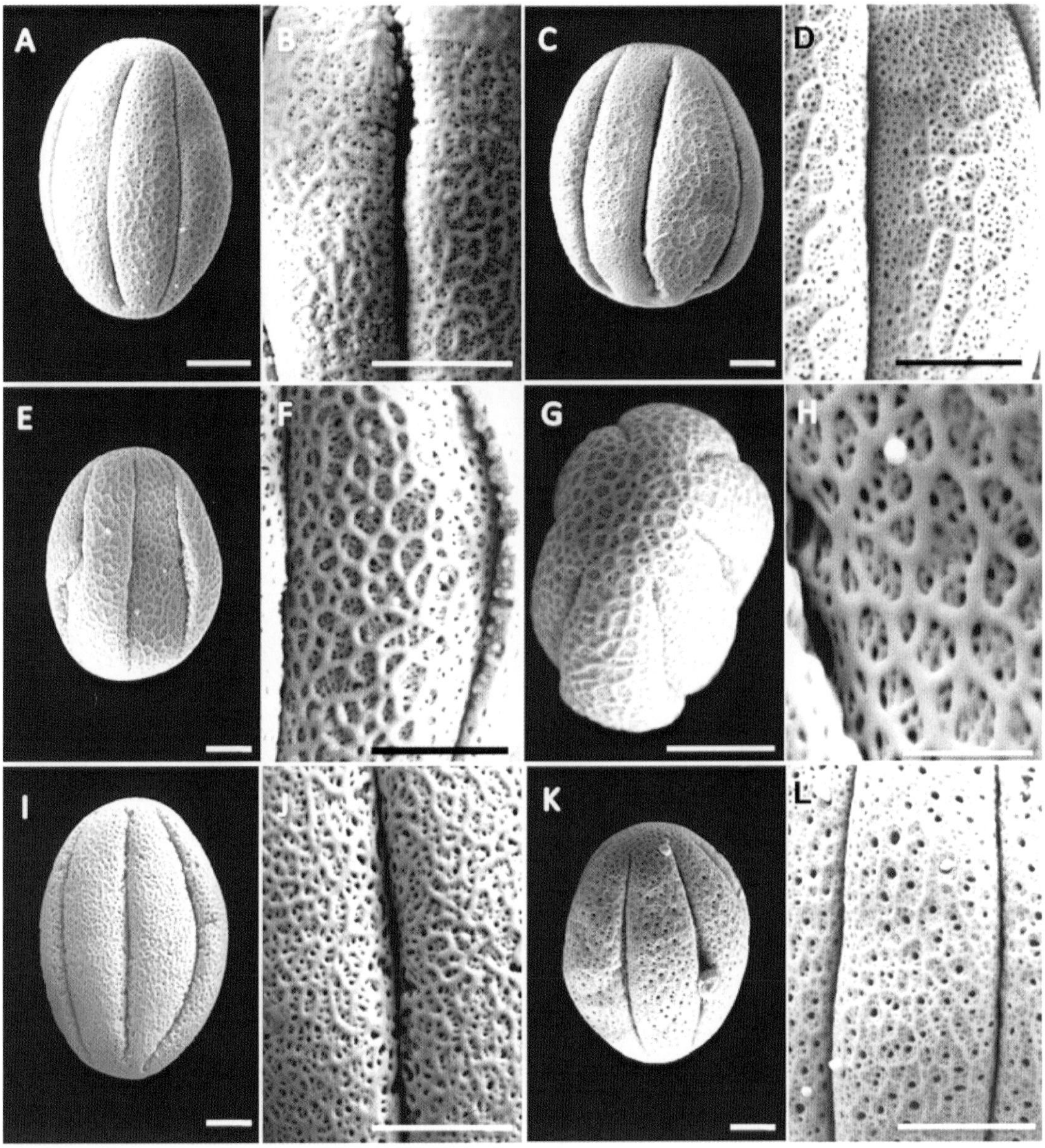

Fig. 3. SEM micrographs of pollen grains of *Dracocephalum* L.. A,B: *D. aucheri*; C,D: *D. moldavica*; E–H: *D. thymiflorum*; I,J: *D. polychaetum*; K,L: *D. multicaule*. Scale bars: A–G, I–L = 10 μm; H = 5 μm.

Type I – 3: In *D. thymiflorum* the shape of primary lumina was polygonal-rounded with strongly thickened primary muri. The number of secondary lumina is more than ten per primary lumen (Fig. 3E–H).

Type I – 4: The shape of primary lumen was polygonal-elongate, without large hole and the number of secondary lumina are less than ten per primary lumen which was observed in *D. polychaetum* (Fig. 3I, J).

Type I – 5: This subtype is frequent and common in a group of species, which are morphologically similar too. The primary lumen is polygonal and each primary lumen contains more than 10 secondary lumina and at least one large hole, which was observed in *D. ghahremanii*, *D. kotschyi*, *D. lindbergii*, *D. multicaule*, *D. oligadenium*, *D. subcapitatum* and *D. surmandinum* (Fig. 3K, L, 4A–L).

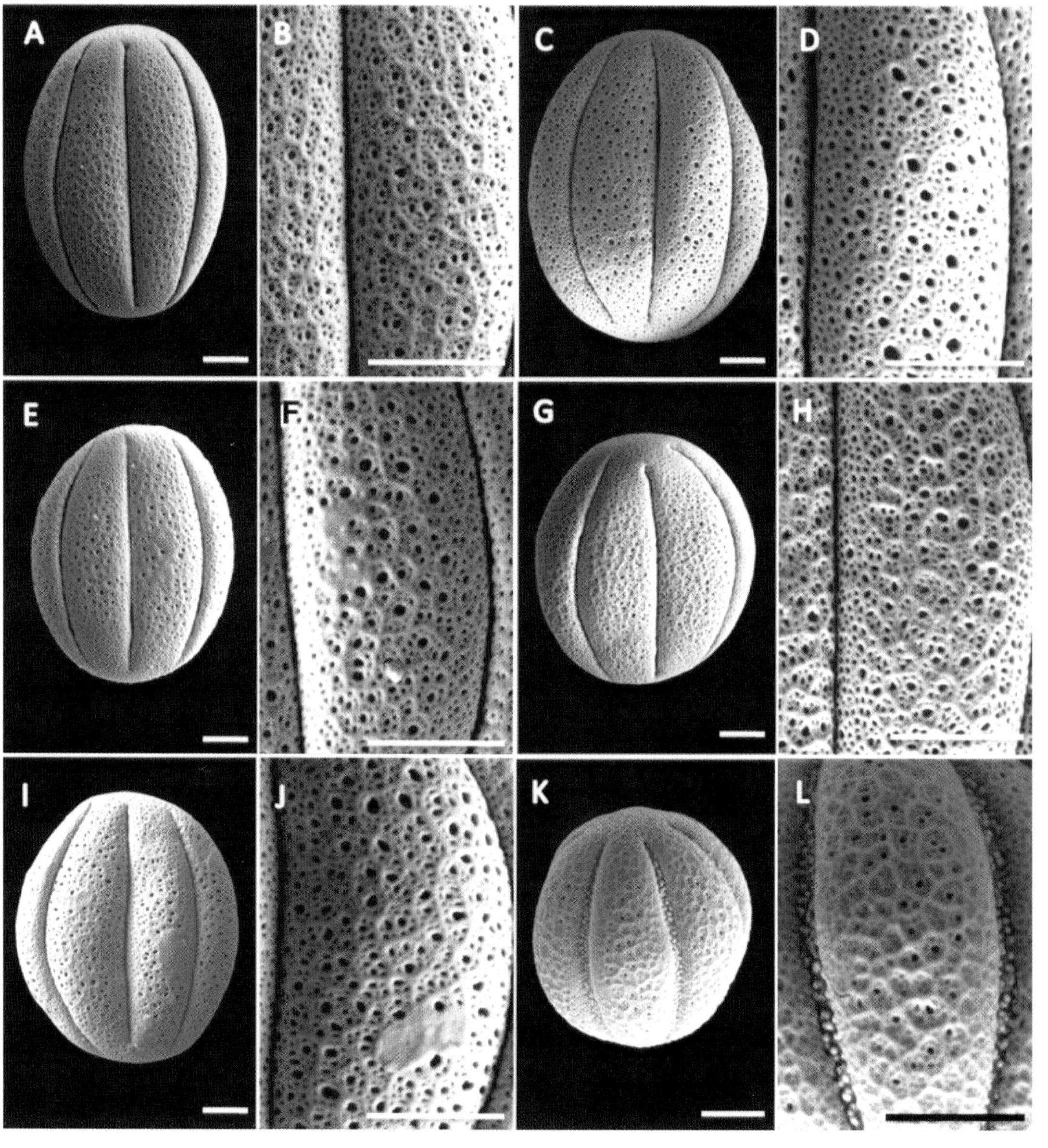

Fig. 4. SEM micrographs of pollen grains of *Dracocephalum* L. A,B: *D. kotschyi*; C,D: *D. lindbergii*; E,F: *D. oligadenium*; G,H: *D. surmandinum*; I,J: *D. subcapitatum*; K,L: *D. ghahremanii*. Scale bar = 10 µm.

Discussion

The present study revealed that all of the examined taxa had isopolar and hexacolpate pollen grains. Moon *et al.* (2008) examined the pollen morphology of 13 *Dracocephalum* species distributed mainly in Central Asia. Among them *D. imberbe* (P = 31.7 µm, E = 24.7 µm) and *D. grandiflorum* (P = 54.6 µm, E = 44.0 µm) were found as the smallest and largest taxa, respectively. In another study, size of the pollen grains of *D. grandiflorum* was found to be large (P = 52.63 µm, E = 29.88 µm), whilst *D. foetidum* and *D. ruyschiana* were characterized to be medium (Badamtsetsg *et al.*, 2012).

Polar axis and equatorial diameter of *D. subcapitatum* examined from Turkmenistan were reported to be P = 37.8 µm and E = 37.1 µm (Moon *et al.*, 2008), while in *D. subcapitatum* collected from northeast of Iran and presented in this study, mean polar axis and equatorial diameter are 38.7 µm and 30.3 µm, respectively. Subprolate (Sp) and prolate-spheroidal (Ps) were characterized as the most common pollen shape in the studied species of *Dracocephalum* by Moon *et al.* (2008), but in the Iranian *Dracocephalum* species studied here, subprolate (Sp) was found to be the frequent pollen shape followed by prolate (P) (Table 2). The colpus length (CL) varied from 25.5 µm in *D. aucheri* to 40.2 µm in *D. lindbergii* (Table 2). The mean colpus length in *D. subcapitatum* from Iran was 30.8 µm, whereas mean colpus length in *D. subcapitatum* from Turkmenistan (Moon *et al.*, 2008) was 32.7 µm. Apocolpium index (AI) in the present investigation ranged from 0.08 to 0.27 for Iranian *Dracocephalum* species, while it was found 0.08 to 0.31 for the species studied by Moon *et al.* (2008).

Colpus membranes were granulate in *D. multicaule* and *D. ghahremanii*, whilst the remaining taxa were egranulate (Table 3; Figs 3K,L, 4K,L). Microreticulate exine sculpturing was recorded as the dominant type in *Dracocephalum* species studied by Moon *et al.* (2008) except for *D. nutans*, *D. palmatum* and *D. stamineum*, in which bireticulate type was found as the exine ornamentation type. Bireticulate was also found as the exine ornamentation type in *D. foetidum*, while microreticulate was determined for *D. grandiflorum* and *D. ruyschiana* (Badamtsetsg *et al.*, 2012). It is noticeable that the microreticulate exine sculpturing in the taxa studied from Iran was not observed.

Taxonomic implication of pollen data

Our investigation on Iranian species of *Dracocephalum* revealed that several pollen characters can be of taxonomical value. In some cases, these characters supported the specific concept of formerly described species. For example, the separate taxonomic position of *D. oligadenium* is usually controversial. Esfandiari (1985), based on morphological characters, resurrected *D. oligadenium* as a distinct species from *D. kotschyi*. Recently, Jamzad (2012) considered *D. oligadenium* as a synonym of *D. kotschyi* in her treatment of the family Lamiaceae for *Flora of Iran*. Based on pollen data, *D. oligadenium* shows diagnostic features compared to *D. kotschyi* such as: smaller polar axis (38.8 µm vs 43.7 µm), colpus length (30.5 µm vs 38.1 µm), exine thickness (1.1 µm vs 0.8 µm), mesocolpium diameter (8.5 µm vs 11.1 µm) and apocolpium index (0.15 vs 0.14), which are useful for separating it from *D. kotschyi*. Considering the results obtained here from palynological data along with diagnostic morphological characters we may propose that *D. oligadenium* is a distinct species from *D. kotschyi* and could be resurrected as a separate taxon.

According to Flora Iranica (Rechinger, 1982), *D. subcapitatum* is distributed in Khorasan and Semnan provinces of Iran. Based on plants collected from Semnan province, which were previously identified as *D. subcapitatum*, Jamzad (Jamzad, 2012) introduced the new species *D. ghahremanii*. Previous study based on molecular RAPD marker showed that the Semnan population of *D. subcapitatum* is clearly different from other three populations of Khorasan (Sonboli *et al.*, 2011). The pollen morphological characteristics, i.e. number of primary lumina per 100 µm^2 (24 vs 11), diameter of primary lumina (2.97 µm vs 4.0 µm), granulate colpus membrane versus egranulate, provide strong evidence supporting the separate taxonomic position of *D. ghahremanii* from its close relative *D. subcapitatum*. Moreover, some species, *viz. D. moldavica, D. thymiflorum* and *D. aucheri* are morphologically well-characterized from the other species of *Dracocephalum* in Iran. These species can also be separated from other species of the genus based on pollen morphology (see Table 3). For example, polygonal-rounded lumen shape in *D. thymiflorum,* the number of primary lumina per 100 µm^2 in *D. moldavica* and exine sculpturing

(Type-I) in *D. aucheri* well-characterized these species from the rest, respectively. Finally, we can conclude that investigation into pollen morphology helps to characterize the taxa of *Dracocephalum*. Studies employing additional taxa of the genus are necessary which might further contribute to utilization of pollen micromorphological characters as significant attributes in classification of the species.

Acknowledgements

The authors thank the research council of Medicinal Plants and Drugs Research Institute Herbarium of Shahid Beheshti University (MPH) for financial support. They also express their gratitude to Dr. N. Olanj for her kind cooperation. The useful comments of an anonymous referee in improving the manuscript are highly appreciated.

References

Abu-Asab, M.S. and Cantino, P.D. 1994. Systematic implications of pollen morphology in subfamilies Lamioideae and Pogostemonoideae (Labiatae). Ann. Missouri Bot. Gard. **81**: 653–686.

Badamtsetsg, B., Myoung, L.S. and Yuon, L.H. 2012. Pollen morphology of the family Lamiaceae in Mongolia. J. Korean Nature **5**: 169–179.

Budantsev, A.L. 1987. The system of the genus *Dracocephalum* (Lamiaceae). Bot. Zh. **72**: 260–267 (in Russian).

Budantsev, A.L. 1993. A synopsis of the tribe Nepeteae (Lamiaceae) – the genera *Lophanthus*, *Dracocephalum*, *Cedronella*, *Schizonepeta* and *Agastache*. Bot. Zh. **78**: 106–115 (in Russian).

Erdtman, G. 1945. Pollen morphology and plant taxonomy. IV. Labiatae, Verbenaceae and Avicenniaceae. Svensk Bot. Tidskr. **39**: 279–285.

Erdtman, G. 1960. The acetolysis method, a revised description. Svensk Bot. Tidskr. **54**: 561–564.

Esfandiari, E. 1985. *Dracocephalum oligadenium* (Labiatae), a distinct species. Iran J. Bot. **3**: 75–76.

Faegri, K. and Iversen, J. 1989. Textbook of Pollen Analysis, 4th ed. John Wiley & Sons, New York.

Harley, M.M., Paton, A., Harley, R.M. and Cade, P.G. 1992. Pollen morphological studies in tribe Ocimeae (Nepetoidaea: Labiatae): I. *Ocimum* L. Grana **31**: 161–176.

Jahaniani, F., Ebrahimi, S.A., Rahbar-Roshandel, N. and Mahmoudian, M. 2005. Xanthomicrol is the main cytotoxic component of *Dracocephalum kotschyi* and a potential anticancer agent. Phytochemistry **66**: 1581–1592.

Jamzad, Z. 2012. *Dracocephalum* L. *In*: Assadi, M., Maassoumi, A.A. and Mozaffarian, M. (Eds.), Flora of Iran, No. **76**. Research Institute of Forests and Rangelands Press, Tehran, Iran, pp. 424–444.

Jamzad, Z. and Hasani-Nejad, M. 2014. Taxonomic implications of pollen exine morphology in infrageneric classification of *Scutellaria* (Lamiaceae). Nord. J. Bot. **32**: 233–244.

Kaderiet, J.W. 2004. Flowering Plants: Dicotyledons, Lamiales (except Acanthaceae including Avicenniaceae). *In*: Kubitzki K. (Ed.), The Families and Genera of Vascular Plants,Vol. **7**. Springer-Verlag, Berlin.

Moon, H.K., Vinckier, S., Smets, E. and Huysmans, S. 2008. Comparative pollen morphology and ultrastructure of Mentheae subtribe Nepetinae (Lamiaceae). Rev. Palaeobot. Palynol. **149**: 174–186.

Ozler, H., Pehlivan, S., Kahraman, A., Dogan, M., Celep, F., Baser, B., Yavru, A. and Bagherpour, S. 2011. Pollen morphology of the genus *Salvia* L. (Lamiaceae) in Turkey. Flora **206**: 316–327.

Punt, W., Hoen, P.P., Blackmore, S., Nilsson, S. and Le Thomas, A. 2007. Glossary of pollen and spore terminology. Rev. Palaeobot. Palynol. **143**: 1–81.

Rechinger, K.H. 1982. Labiatae *In*: Rechinger, K.H. (Ed.), Flora Iranica, No. **150**. Akademische Druck–u. Verlagsanstalt, Graz, Austria.

Sajjadi, S.E., Movahedian Attar, A. and Yektaeian, A. 1998. Antihyperlipidemic effect of hydroalcoholic extract, and polyphenolic fraction from *Dracocephalum kotschyi* Boiss. Pharm. Acta Helv. **73**: 167–170.

Sarwar, A.K.M.G. and Takahashi, H. 2012. Pollen morphology of *Kalmia* L. (Phyllodoceae, Ericaceae) and its taxonomic significance. Bangladesh J. Plant Taxon. **19**(2): 123−133.

Sarwar, A.K.M.G. and Takahashi, H. 2013. Pollen morphology of *Rhododendron* L. and related genera and its taxonomic significance. Bangladesh J. Plant Taxon. **20**(2): 185−199.

Sonboli, A., Gholipour, A., Mirjalili, M.H. and Amini Rad, M. 2011. Molecular characterization of Iranian *Dracocephalum* (Lamiaceae) species based on RAPD data. Acta Biol. Szeged. **55**: 227–230.

Sonboli, A., Mojarrad, M., Gholipour, A., Nejad Ebrahimi, S. and Arman, M. 2008. Biological activity and composition of the Essential oil of *Dracocephalum moldavica* L. grown in Iran. Nat. Prod. Commun. **3**: 1547–1550.

11

ACHENE MICROMORPHOLOGY OF SEVEN TAXA OF *ACHILLEA* L. (ASTERACEAE) FROM TURKEY

TULAY AYTAS AKCIN[1] AND ADNAN AKCIN[2]

Department of Biology, Faculty of Arts and Science, Ondokuz Mayıs University, Samsun, Turkey

Keywords: Achene micromorphology; *Achillea*; Slime cells; Turkey.

Abstract

Micromorphological characters of achenes in seven taxa of Turkish *Achillea* L. (Asteraceae) were investigated using stereomicroscope and scanning electron microscope (SEM). Some morphological descriptions of achenes were given for each species. *A.biserrata* Bieb. has the biggest (0.69±0.092 x 2.01±0.252 mm) and *A. grandiflora* Friv. has the smallest (0.30±0.018 x 1.12±0.058 mm) achenes. The achenes are oblong-lanceolate in *A.biserrata* and *A. teretifolia* Willd. and they are oblong in the remaining taxa. In surface sculpturing, the ornamentation and slime cell distribution varied among the taxa. However, *A. biebersteinii* Afan. has distinct slime cells forming groups scattered over the achene surface. Mature achenes are ribbed and glabrous in all studied taxa. *A. biserrata* has distinct carpopodium structure.

Introduction

The genus *Achillea* L. (Asteraceae) includes about 140 species distributed in South-west Asia and South-eastern Europe (Akyalcın *et al.*, 2011). According to recent studies, the genus *Achillea* is represented in Turkey by 48 species (54 taxa), 24 of which are endemic for Anatolia (Akyalcın *et al.*, 2011). *Achillea* L. is classified into five sections, namely sect. *Othantus* (Hoffmanns. & Link) Ehrend. & Y. P. Guo (one species), sect. *Babounya* (DC.) O. Hoffm. (30 species), sect. *Ptarmica* (Mill.) W.D.J. Koch (2 species), sect. *Anthemoideae* (DC.) Heimerl (2 species) and sect. *Achillea* (13 species) (Huber-Morath, 1975; Duman, 2000; Arabacı and Yıldız, 2006; Arabacı and Budak, 2009).

Achene micromorphological characters have been found useful in systematics of the family Asteraceae (Abid and Qaiser, 2007a, b; Shekhar *et al.*, 2011). Cypselar external morphology and anatomy in members of different tribes of Asteraceae are found important for delimitation of genera (Garg and Sharma, 2007; Pandey and Kumari, 2007). Abid and Qaiser (2002) studied cypselar morphology of *Dittrichia* Greuter, *Duhaldea* DC., *Inula* L., *Iphiona* Cass. and *Pentanema* Cass. (Asteraceae) from Pakistan and Kashmir and concluded that two distinct groups of taxa can be recognized. It was confirmed that most of the species of *Anaphalis* DC. (Asteraceae) were delimited due to their distinct micromorphological characters of cypsela (Abid and Qaiser, 2007a). Zhu *et al.* (2006) concluded that achene wall anatomy and surface sculpturing of *Lactuca* L. (Asteraceae) and related genera displayed variation within genera.

Slime cells are widespread in higher plants, especially in fruits and/or seeds in different families, *viz.,* Brassicaceae, Euphorbiaceae, Plantaginaceae, Linaceae, Malvaceae and Lamiaceae (Huang *et al.*, 2000; Western *et al.,* 2000). Slime cells are usually rectangular and they form ladder-like columns that are elongated in a parallel form to the long axis of the achene. Within the Asteraceae, slime cells have been reported in *Achillea* (Kreitschitz and Valles, 2007; Akcin and

---

[1]Corresponding author. Email: taytas@omu.edu.tr

[1]Department of Biology, Faculty of Arts and Science, Amasya University, Amasya 05100, Turkey

Akcin, 2010), *Anthemis* L., *Chrysanthemum* L. (Grubert, 1974) and *Artemisia* L. (Huang *et al.*, 2000; Yakovleva *et al.*, 2002). It is also suggested that slime plays an important role in controlling of germination, mostly in plants that grow in the condition of water deficiency in arid and semiarid environments (Kreitschitz and Valles, 2007). It can also help fruit or seed dispersal and defence against pathogens (Huang and Gutterman,1999; Huang *et al.*, 2000). Despite several studies on achene micromorphology of *Achillea* were carried out recently (Abid and Qaiser, 2009; Akcin and Akcin, 2010), however, there is very little information on the Turkish taxa. The main aim of the present work is to study the detailed achene morphology of some Turkish taxa belonging to the genus *Achillea* and to find out how useful these characters are in the systematics of the genus *Achillea*.

Materials and Methods

Achenes of *Achillea* L. taxa were studied by stereomicroscope (Leica T1A) and Scanning Electron Microscope (SEM) (JEOL-Neoscope JCM-5000). The studied plant materials were collected from different populations in Turkey. A list of specimens examined is given in Table 1. The specimens have been deposited at the Herbarium of the Department of Biology, University of Ondokuz Mayıs, Turkey (OMUB).

Table 1. List of *Achillea* L. taxa examined along with their vouchers.

Taxon	Vouchers
Sect. ***Achillea*** *s. lat.*	
1. *Achillea biebersteinii* Afan.	A5 Amasya: Vicinity of Amasya Education Faculty, road side, 550 m, 11.6.2010, T. Akcin, OMUB 6446
2. *A. coarctata* Poir.	B5 Kayseri: Kayseri to Avanos, road side, 1100 m, 10.6. 2008, T. Akcin, OMUB 6447
3. *A. grandifolia* Friv.	A6 Samsun: Kocadağ, 1200 m, 20.7.2008, A. Akcin, OMUB 6448
4. *A. millefolium* L. subsp. *millefolium*	A5 Kırşehir: Mucur, 900 m, 21.6.2009, A. Akcin, OMUB 6444
5. *A. millefolium* L. subsp. *pannonica* (Scheele) Hayek	A5 Amasya: Yemişen village, 800 m, 28.7.2010, A. Akcin, OMUB 6445
Sect. ***Babounya*** (DC.) O. Hoffm.	
6. *A. teretifolia* Willd. (endemic)	A5 Amasya: Yemişen village, road side, 750 m, 28.7.2010, T. Akçin, OMUB 6442
Sect. ***Ptarmica*** (Mill.) W.D.J. Koch	
7. *A. biserrata* Bieb.	A6 Samsun: Kocadağ, 1200 m, 14.7.2009, T. Akcin, OMUB 6443

The achenes were examined using a stereomicroscope to ensure size, shape, colour and maturity. In order to determine the average achene sizes, 25 mature achenes were measured. For SEM observations, the mature achenes were placed on stubs using double-sided adhesive tape. Following that, they were observed and photographed with a JEOL-Neoscope JCM-5000 scanning electron microscope. The terminology of achene surface patterns adopted is mainly from Barthlott (1981, 1984), Barthlott *et al.* (1998) and Johnson *et al.* (2004).

Results and Discussion

In the present study, achene morphology of seven taxa belonging to the genus *Achillea* were reported for the first time. Within the examined taxa, *A. teretifolia* is endemic to Turkey (Ekim *et al.,* 2000). Achene shapes were oblong-lanceolate in *A. biserrata* and *A. teretifolia*, oblong in *A. millefolium* subsp. *millefolium*, *A. millefolium* subsp. *pannonica, A. coarctata*, *A. biebersteinii* and *A. grandifolia* (Table 2). When achene width and length were considered, different groups were obtained ($P < 0.001$, Table 2). The widest (0.69±0.092 mm) and longest (2.01±0.252 mm) achenes were seen in *A. biserrata* (Table 2). Mature achenes ribbed and glabrous in all studied taxa. Achene colours vary from brown to light brown and yellowish-brown.

The position of carpopodium is almost similar (basal) in all taxa studied. However, carpopodium is with narrow circular ring in *A. teretifolia* (Fig. 1a), *A. millefolium* subsp. *millefolium* (Fig. 1e), *A. millefolium* subsp. *pannonica* (Fig.1h), *A. coarctata* (Fig. 2m), *A. biebersteinii* (Fig. 2j) and *A. grandifolia* (Fig. 2o). On the other hand, *A. biserrata* could be separated from other taxa due to its carpopodium shape (Fig. 1c). Abid and Qaiser (2009) reported that cypselar features may also be utilized for specific delimitation as the species of

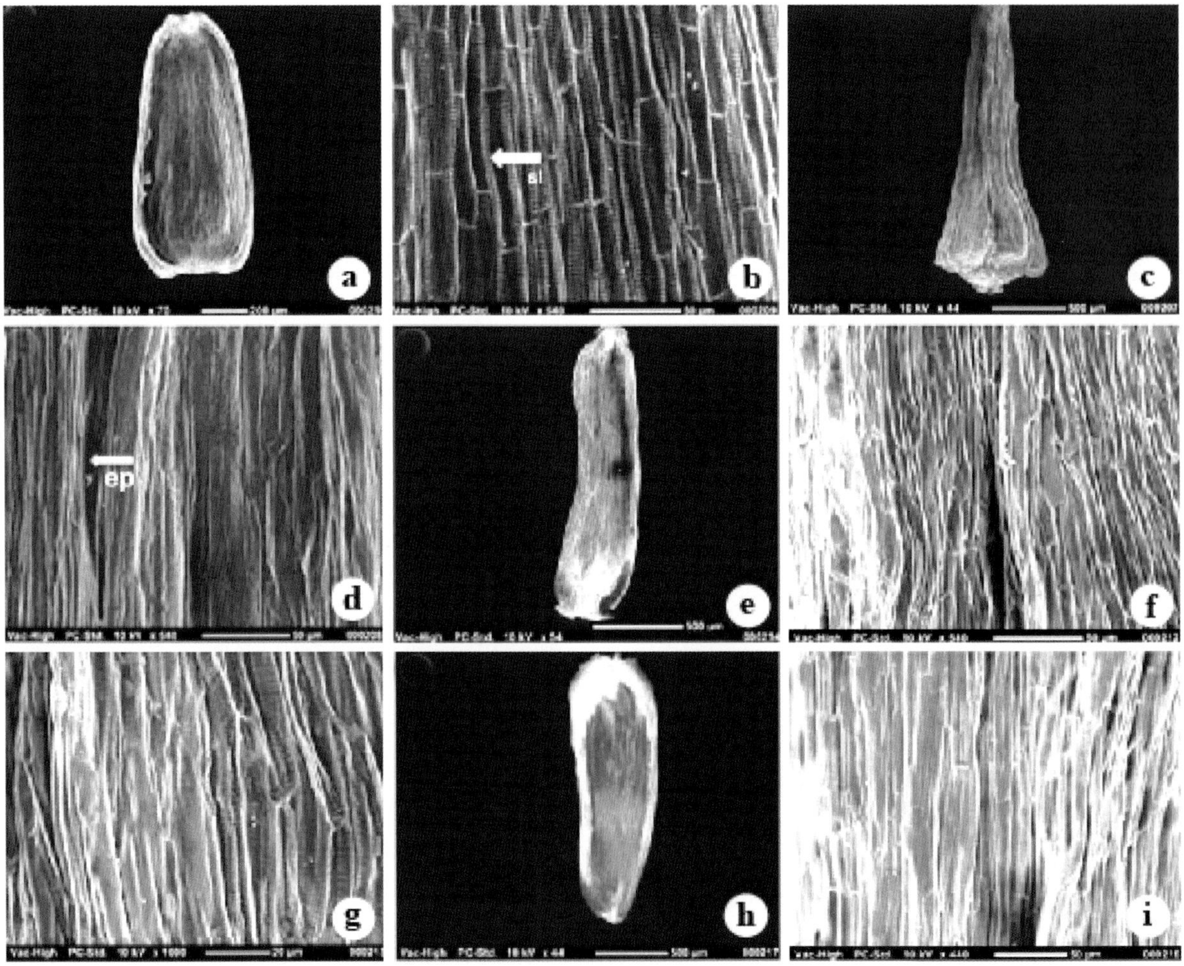

Fig. 1. SEM micrographs of achenes in *Achillea* L. a-b) *A. teretifolia*; c-d) *A. biserrata*; e-g) *A. millefolium* subsp. *millefolium*; h-i) *A. millefolium* subsp. *pannonica*, ep = epidermal cells; sl = slime cells.

Table 2. Achene features of *Achillea* taxa [means in the column the same letter are not significantly different using Tukey's test ($P < 0.001$), data are given as mean ± sd].

Name of taxa	Achene					Carpopodium		
	Shape	Surface	Colour	Width (mm)	Lenght (mm)	Diameter of Carpopodium (µm)	Diameter of Foramen (µm)	Shape
1. *Achillea biebersteinii*	Oblong	Ribbed, glabrous, slime cells forming groups	Brown	0.54 ± 0.096b	1.24 ± 0.092cd	120.68 ± 4.088c	99.04 ± 7.778	With narrow circular ring
2. *A.biserrata*	Oblong	Ribbed, glabrous, slime cells cover the entire surface	Yellowish-brown	0.69 ± 0.092a	2.01 ± 0.252a	215.93 ± 41.751a	108.66 ± 17.605	Ring lobed
3. *A.coarctata*	Oblong	Ribbed, glabrous, slime cells cover the entire surface	Yellowish-brown	0.54 ± 0.048b	1.30 ± 0.038c	139.07 ± 7.742c	133.81 ± 27.195	With narrow circular ring
4. *A.grandifolia*	Oblong-lanceolate	Indefinite ribbed, glabrous, slime cells alternate with the epidermal cells	Light brown	0.30 ±0 .018c	1.12 ± 0.058d	215.91 ± 9.703a	96.40 ± 9.713	With narrow circular ring
5. *A. millefolium* subsp. *millefolium*	Oblong	Ribbed, glabrous, slime cells alternate with the epidermal cells	Light brown	0.33 ± 0.045c	1.63 ± 0.077b	170.13 ± 11.783b	119.54 ± 4.590	With narrow circular ring
6. *A. millefolium* subsp. *pannonica*	Oblong	Ribbed, glabrous, slime cells alternate with the epidermal cells	Light brown	0.53 ± 0.048b	1.74 ± 0.089b	196.35 ±1 8.799a	122.49 ± 24.227	With narrow circular ring
7. *A. teretifolia*	Oblong-lanceolate	Ribbed, glabrous, slime cells alternate with the epidermal cells	Light brown	0.54 ± 0.024b	1.37 ± 0.063c	134.57 ± 12.769b	86.27 ± 11.000	With narrow circular ring

Achillea are grouped on the basis of ribbed and non-ribbed cypsela. *A. millefolium* was observed to have 10-12 ribbed yellowish-brown cypsela, while non-ribbed cypselas are characteristics for *A. wilhemsii* (Abid and Qaiser, 2009). Similarly, *A. millefolium* as indicated by Abid and Qaiser (2009) seperated from *A. wilhemsii* by the presence of circular ring shaped carpopodium, while *A. wilhemsii* has carpopodium with 4-6 lobed ring. The present investigation also support the observation of Abid and Qaiser (2009). The achenes of *A. millefolium* subsp. *millefolium* and *A. millefolium* subsp. *pannonica* have a circular ring shaped carpopodium and the ribbed surface of achenes (Figs 1 e-i). However, carpopodium diameters of the achenes differ in the taxa ($P< 0.001$, Table 2). The widest carpopodium diameter was present in *A. biserrata* (215.93 ± 41.751 µm) followed by *A. grandifolia* (215.91 ± 9.703 µm), while it was narrow in *A. biebersteinii* (120.68 ± 4.088 µm). The diameters of foramen of carpopodium among the studied taxa were not significant ($P < 0.001$, Table 2).

Slime envelope formation is known in several plant families, including the Asteraceae (Kreitschitz and Valles, 2007). Within the *Achillea* genus, slime has been reported previously (Grubert, 1974; Akçin and Akçin, 2010). Our results confirmed the presence of slime in the investigated *Achillea* taxa. Slime cells are usually rectangular and they form ladder-like columns that are elongated in a parallel form to the long axis of the achene (Figs. 1b, 2k,l,n). Slime cells can either cover almost entire surface of the achene as in *A. biserrata* (Fig. 1d) and *A. coarctata* (Fig. 2n), or alternate with the epidermal cells in *A. millefolium* subsp. *millefolium* (Figs 1f-g), and *A. biebersteinii* (Figs 2k-l). A distinct pattern occurs in *A. biebersteinii* achenes (Figs 2k-l), where slime cells form groups scattered over the achene surface. Slime formation on the fruit and/or seed surface is known to be an ecological adaptation to limited availability of water (Huang and Gutterman, 1999; Huang *et al.*, 2000). Several authors showed that the presence of the slime envelope facilitates the adherence of achenes to the soil surface and makes germination easier (Huang *et al.*, 2000; Kreitschitz and Valles, 2007).

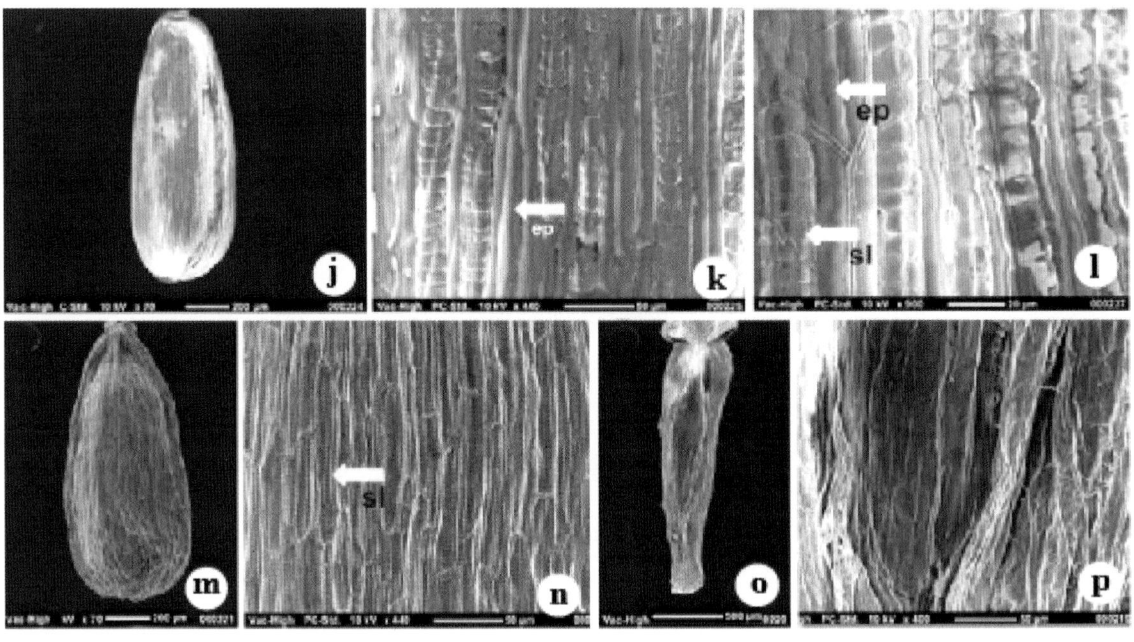

Fig. 2. SEM micrographs of achenes in *Achillea* L. j-l) *A. biebersteinii*; m-n) *A.coarctata*, o-p) *A. grandifolia*, ep = epidermal cells; sl = slime cells.

Taxa such as *A. millefolium* subsp. *millefolium, A. millefolium* subsp. *pannonica*, *A. teretifolia*, *A. coarctata*, *A. biebersteinii* occurring in less fertile and often arid location (Huber-Morath, 1975) are characterized by the presence of numerous strands of slime cells. Slime formation also plays an important role in the control of germination, mostly in plants that grow in the condition of water deficiency in arid and semiarid environments, thus faciliating imbibing and maintenance of the water (Kreitschitz and Valles, 2007).

It was also reported that slime cells in species growing in wet environment such as stream sides and meadows may not mature or function properly (İnceer *et al.*, 2012). However, in *A. biserrata* and *A. grandifolia*, often growing around wet environment such as coniferous forest (Huber-Morath, 1975), the presence of a slime envelope may play an important role in dispersal of this species. Huang *et al.* (2000) reported that the slime can reduce the specific weight of diaspores thus facilitating their transportation with water current (hydrochory). In the present study, there are differences in distribution of the slime cells in the examined taxa. Slime cells in *A. teretifolia* similar to *A. coarctata* are usually ladder-like coloumns elongated parallelly to long axis of the achene (Figs 1b, 2n). However, *A. biebersteinii* has a distinct slime cells forming groups scattered over the achene surface (Figs 2k-l). For this reason, these species could be further seperated by the sculpture of achene surface. *A. grandifolia* is separated from the remaining species by having indefinite individual ribs (Fig. 2p). It was also determined that slime cells alternate with epidermal cells in *A. millefolium* (Figs 1f,g,i), while it covers almost the entire surface of achene in *A. biserrata* (Fig. 1d) and *A. coarctata* (Fig. 2n). These results confirmed that slime cells on the fruit surface is associated with ecological adaptation. Akçin and Akçin (2010) reported that achenes of *A. phrygia* have myxogenic cells (slime-producing) almost throughout the entire surface but slime cells are more restricted in *A. gypsicola*. Such a relationship was reported for Lamiaceae (Mosquero *et al.,* 2004), *Artemisia* (Asteraceae) (Kreitschitz and Valles, 2007) and *Tripleurospermum* (İnceer *et al.*, 2012).

We can conclude that this work helps to charracterize the taxa of *Achillea* L. Studies employing additional taxa of the genus are necessary which might further contribute to utilization of achene micromorphological characters as significant attributes in classification of the species.

Acknowledgements

We would like to thank the staff of Faculty of Science and Literature Directorship of Basic Sciences Research and Applied Center of Balıkesir University (BUTAM) for taking microphotographs of achenes.

References

Abid, R.D. and Qaiser, M. 2002. Cypsela morphology of *Inula* L. (s. str.) and its allied genera (Inuleae: Compositae) from Pakistan and Kashmir. Pak. J. Bot. **34**(3): 207-223.

Abid, R.D. and Qaiser, M. 2007a. Micromorphology of cypsela in the tribe Plucheeae from Pakistan. Pak. J. Bot. **39**(3): 671-677.

Abid, R.D. and Qaiser, M. 2007b. Cypsela morphology of the genus *Anaphalis* DC. (Gnaphalieae-Asteraceae) from Pakistan. Pak. J. Bot. **39**(6): 1897-1906.

Abid, R.D and Qaiser, M. 2009. Taxonomic significance of the cypsela morphology in the tribe Anthemideae (Asteraceae) from Pakistan and Kashmir. Pak. J. Bot. **41**(2): 555-579.

Akcin, T.A and Akcin, A. 2010. Morphological and anatomical characteristics and taxonomical significance of achene micromorphology of *Achillea phrygia* and *A. gypsicola* (Asteraceae), endemic to Turkey. Nordic J. Bot. **28**(1): 65-73.

Akyalcın, H., Arabacı, T. and Yıldız, B. 2011. Pollen morphology of six *Achillea* L. sect. *Achillea* (Asteraceae) species in Turkey. Turkish J. Bot. **35**: 183-201.

Arabacı, T. and Budak, Ü. 2009. *Achillea hamzaoglui* (Asteraceae), a new species from Turkey. Annales Botanici Fennici **46**: 459-463.

Arabacı, T. and Yıldız, B. 2006. Rediscovery of *Achillea boissieri* Hausskn. *ex* Boiss. later 140 years. Feddes Repertorium **117**: 459-463.

Barthlott, W. 1981. Epidermal and seed surface characters of plants: Systematic applicability and some evolutionary aspects. Nordic J. Bot. **1**(3): 345-355.

Barthlott, W. 1984. Microstructural features of seed surfaces. *In*: Heywood, V.H. and Moore, D.M. (Eds), Current Concepts in Plant Taxonomy, Academic Press, London, pp. 95-105.

Barthlott, W., Neinhuis, C., Cutler, D., Ditsch, F., Meusel, I., Theisen, I. and Wilhelmi, H. 1998. Classification and terminology of plant epicuticular waxes. Bot. J. Linn. Soc. **126**(3): 237-260.

Duman, H. 2000. *Achillea* L. *In*: Güner, A., Özhatay, N., Ekim, T. and Başer, K.H.C. (Eds), Flora of Turkey and the East Aegean Islands (Suppl. 2), Vol. **11**, Edinburgh University Press, Edinburgh, pp. 158-159.

Ekim, T., Koyuncu, M., Vural, M., Duman, H., Aytaç, Z. and Adıgüzel, N. 2000. Red Data Book of Turkish Plants, Ankara, No. **18**, pp. 1-245.

Garg, S.K. and Sharma, K.C. 2007. Taxonomical significance of the micromorphological and scanning electron microscopic surface patterns of cypselas in some members of the tribe Heliantheae (Asteraceae). Feddes Repertorium **118**: 165-191.

Grubert, M. 1974. Studies on the distribution of myxospermy among seeds and fruits of Angiospermae and its ecological importance. Acta Biologica Venezuelica **8**: 315-551.

Huang, Z. and Gutterman, Y. 1999. Water absorption by mucilaginous achenes of *Artemisia monosperma*: floating and germination as affected by salt concentrations. Israel J. Plant Sci. **47**: 27-34.

Huang, Z., Gutterman, Y. and Hu, Z. 2000. Structure and function of mucilaginous achenes of *Artemisia monosperma* inhabiting the Negev desert of Israel. Israel J. Plant Sci. **48**(4): 255-266.

Huber-Morath, A. 1975. *Achillea* L. *In*: Davis, P.H. (Ed.), Flora of Turkey and the East Aegean Islands, Vol. **5**. Edinburgh University Press, pp. 224-252.

İnceer, H., Bal, M., Ceter, T. and Pinar, N.M. 2012. Fruit structure of 12 Turkish endemic *Tripleurospermum* Sch. Bip. (Asteraceae) taxa and its taxonomic implications. Plant Syst. Evol. **298**: 845-855.

Johnson, L.A., Huish, K.H. and Portert, J.M. 2004. Seed surface sculpturing and systematic significance in *Gilia* (Polemoniaceae) and segregate genera. International J. Plant Sci. **165**(1): 153-172.

Kreitschitz, A. and Valles, J. 2007. Achene morphology and slime structure in some taxa of *Artemisia* L. and *Neopallasia* L. (Asteraceae). Flora **202**(7): 570-580.

Mosquero, M.A., Juan, R. and Pastor, J. 2004. Observaciones micromorfologicas y anatomicas en nuculas de *Prunella* L. y *Cleonia* L. (Lamiaceae) del suroeste de Espana. Acta Botanica Malacitana **29**: 203-214.

Pandey, A.K. and Kumari, A. 2007. Anatomical patterns of pericarp in Asteraceae. *In*: Chauhan, S.V.S., Rana, A. and Chauhan, S. (Eds), Plant Reproductive Biology and Biotechnology. Aavishkar Publisher, Jaipur, pp. 64-77.

Shekhar, S., Pandey, A.K. and Anderbergh, A.A. 2011. Cypsela morphology and anatomy in some genera formerly placed in *Inula* (Asteraceae: Inuleae-Inulinae). Rheedea **21**(1): 13-22.

Western, T.L., Debra, J.S. and Haughn, G.W. 2000. Differentation of mucilage secretory cells of the *Arabidopsis* seed coat. Plant Physiol. **122**(2): 345-355.

Yakovleva, O.V., Korobkov, A.A. and Boyko, E.V. 2002. Structure of mucilage containing cells in achene pericarp of some species of *Artemisia* (Asteraceae). Botanicheskii Zhurnal **87**: 1-14.

Zhu, S.X., Qin, H.N. and Shih, C. 2006. Achene wall anatomy and surface sculpturing of *Lactuca* L. and related genera (Compositae: Lactuceae) with notes on their systematic significance. J. Integr. Plant Biol. **48**(4): 390-399.

PALYNOLOGICAL FEATURES OF ELEVEN *AETHIONEMA* TAXA FROM TURKEY AND THEIR SYSTEMATIC IMPLICATIONS

MEHMET CENGIZ KARAISMAILOGLU[1]

Department of Botany, Faculty of Science, Istanbul University, 034116 Istanbul, Turkey

Keywords: Aethionema; Brassicaceae; Pollen; Morphology; Turkey; SEM; PCA.

Abstract

Pollen morphology of 11 taxa, including 2 endemic of the genus *Aethionema* W.T. Aiton from Turkey was examined under light and scanning electron microscopes. The pollens of *Aethionema* are mostly isopolar and bilaterally symmetric; spheroidal, prolate, perprolate and subprolate with the polar axes 14.07-26.41 μm and the equatorial axes 7.85-22.02 μm; mostly tricolpate, rarely 2-colpate; surface ornamentation is micro or macro reticulate. The exine thickness varies between 0.66 and 1.91 μm, and in tine thickness ranges from 0.27 to 0.85 μm. It is found that dimension of pollen grains, surface ornamentation, apocolpidium and amb diameter are taxonomically significant.

Introduction

The family Brassicaceae (Cruciferae), one of the largest angiosperm families, consists about 340 genera and 3350 species distributed mostly in temperate Northern Hemisphere (Al-Shehbaz, 1986; Karaismailoglu, 2016). The genus *Aethionema* W.T. Aiton represents with about 45 taxa in Turkey, including 20 endemic taxa (Guner *et al*., 2012). Turkey is one of the biodiversity rich centers of the genus, and its number in outside Anatolia declines gradually (Davis, 1965; Pinar *et al.,* 2007). *Aethionema* having relatively few morphological characters and dimorphism in fruits among individuals of some species poses some taxonomic problems in classification of taxa within the genus (Al-Shehbaz *et al.,* 2007). Besides, some of taxa within genus are of the common convergence in fruits and seeds in the family (Mummenhoff *et al.,* 1997). Using morphological characters in infrageneric delimitation becomes problematic in the genus. Therefore, additional features could make useful contribution to the solution of taxonomical problems in the genus.

The significance of palynological information has been stressed by several workers in the family Cruciferae, *viz.* Inceoglu and Karamustafa (1977), Brochmann (1992), Pinar *et al.* (2009), and Mutlu and Erik (2012). There has been no comprehensive palynological study in the genus *Aethionema*. However, recently Atceken *et al*., (2016) investigated pollen morphology of four species of the genus. Therefore, the present investigation aims to enhance current palynological knowledge of the genus and to evaluate their taxonomic significance as taxonomic characters.

Material and Methods

Plant samples used for investigation were collected from different natural habitats of Turkey. A list of studied specimens is given in Table 1 with collection localities and collection numbers, and specimens were stored in the ISTF (Istanbul University Science Faculty Herbarium).

Specimens for scanning electron microscopy were prepared mounting with silver adhesive on the stub, covered by gold, and examined with a JEOL Neoscope-5000 scanning electron microscope (Karaismailoglu, 2015).

[1]Email: cengiz.karaismailoglu@istanbul.edu.tr

Pollen slides for light microscope were prepared following the technique of Wodehouse (1935). Pollen grains were dyed by safranin, and mounted with a cover slip, examined with Olympus CX21FS1 light microscope, and photographed by Kameram Imaging Software.

Table 1. The examined taxa of *Aethionema* and their locations

Taxa	Locality	Collection no.
Aethionema froedinii Rech. (A1)	Gumushane, Kelkit	Karaismailoglu 213
A. arabicum (L.) Andrz. ex DC. (A2)	Mugla, Koycegiz	Karaismailoglu 194
A. eunomioides (Boiss.) Bornm. (A3)*	Artvin, Yusufeli	Karaismailoglu 169
A. fimbriatum Boiss. (A4)	Nigde, Camardi	Karaismailoglu 275
*A. speciosum*Boiss. et Huet ssp. *speciosum*(A5)	Artvin, Savsat	Karaismailoglu 67
A. speciosum Boiss. et Huetssp. *compactum* Hartvig et Strid (A6)*	Mugla, Koycegiz	Karaismailoglu 260
A. saxatile (L.) R. Br. (A7)	Trabzon, Of	Karaismailoglu 94
A. oppositifolium (Pers.) Hedge (A8)	Agri, Downtown area	Karaismailoglu 164
A. iberideum Boiss. (A9)	Erzurum, Ispir	Karaismailoglu 170
A. armenum Boiss. (A10)	Kahramanmaras, Goksun	Karaismailoglu 206
A. grandiflorum Boiss. et Hohen. (A11)	Mugla, Marmaris	Karaismailoglu 191

*=endemic

Pollen terminology based on Erdtman (1952 and 1969), Brochmann (1992) and Punt *et al.* (1994). Ten palynological characters have been determined to discriminate the 11 taxa of *Aethionema* genus (Table 2), and applied Duncan's multiple range tests for each of them (SPSS, 2006), and dissimilarity matrix was consisted (Table 3). Afterwards, the variations of the determined characters are presented in Figure 3 with the whisker graph. Cluster analysis of taxa was performed in accordance with UPGMA (Fig. 4) (Mohammadi and Prasanna, 2003). Besides, the ordination of taxa is performed with Principal Component Analysis (PCA) (Fig. 5). Computations except Duncan's multiple range tests were made with the MVSP software (Kovach, 2007).

Results and Discussion

The morphological characters of pollen grains of the examined *Aethionema* taxa are presented in Table 2, and photographs are presented in Figures 1 and 2. The pollens are isopolar and vary between prolate (prolate (72.73%), perprolate (9.09 %), and subprolate (9.09%) and spheroidal (9.09%). The polar axes range from 14.07-26.41 µm and the equatorial axes 7.85-22.02 µm. Their dimensions are smaller in *A. arabicum* and *A. speciosum* ssp. *compactum*, but larger in *A. iberideum* and *A. armenum* (Table 2, Figs 1&2).

The most common shape in the examined taxa is prolate (eight taxa), subprolate (one taxon), perprolate (one taxon) and rarely spheroidal (one taxon), respectively (Figures 1-2). The result is compatible with the result of Mutlu and Erik (2012), found in the genus *Arabis* L. of the family Cruciferae.

The aperture and exine characters of taxa can reflect the major criterion for the determination of the phylogenetic relationships (Cronquist, 1968; Ocak *et al.*, 2013). The number of aperture ranges from 2 to 4; and aperture type is mostly tricolpate (basic type), however; dicolpate type is

Table 2. Pollen characteristics of the studied *Aethionema* taxa

Taxa	pollen shape	Polar axes (P)(µm)			Equatorial axes (E)(µm)			(P/E)	Colpus number	Aperture type	Ornamentation
		Min	Max	Mean (1)*	Min	Max	Mean (2)*				
A1	subprolate	14.07	17.11	16.57±0.12g	10.45	13.17	12.49±0.17c	1.32	3	tricolpate	micro reticulate
A2	prolate	15.16	17.29	16.44±0.18g	7.85	9.51	8.76±0.24g	1.87	3	tricolpate	reticulate
A3	prolate	16.58	19.22	17.85±0.24f	10.86	13.25	11.91±0.08d	1.49	3	tricolpate	reticulate
A4	prolate	18.61	20.54	19.13±0.33e	9.77	11.65	10.33±0.27e	1.85	3	tricolpate	coarsely reticulate
A5	prolate	17.53	20.42	18.91±0.24ef	8.96	10.09	9.70±0.11ef	1.94	3	tricolpate	reticulate
A6	prolate	13.59	16.15	14.99±0.18h	8.15	9.91	8.76±0.18g	1.71	2-3	%10 dicolpate, %90 tricolpate	reticulate
A7	prolate	18.21	20.94	19.82±0.12e	9.13	10.86	10.08±0.15e	1.96	2-3	%10 dicolpate, %90 tricolpate	reticulate
A8	prolate	20.16	24.72	22.34±0.16c	13.17	15.90	14.26±0.12b	1.56	3	tricolpate	reticulate
A9	prolate	23.44	25.51	24.05±0.12b	12.87	14.56	13.94±0.08b	1.72	2-3	%5 dicolpate, %95 tricolpate	coarsely reticulate
A10	spheroidal	20.11	22.18	21.45±0.21d	19.56	22.02	21.22±0.16a	1.01	2-3	%5 dicolpate, %95 tricolpate	coarsely reticulate
A11	perprolate	23.19	26.41	25.04±0.15a	8.76	9.46	9.05±0.12f	2.76	3	tricolpate	micro reticulate

Table 2 continued

Taxa	Colpus sizes* Clt(µm) (3)	Clw(µm) (4)	Clt/Clw	Apo* (µm) (5)	Amb* (µm) (6)	Lumina* (µm) (7)	Muri* (µm) (8)	Intine* (µm) (9)	Exine* (µm) (10)
A1	9.81±0.16f	1.08±0.04g	9.08	2.13±0.04d	9.01±0.12f	0.41±0.04d	0.65±0.06a	0.45±0.02c	1.25±0.04cd
A2	14.26±0.27b	2.25±0.08e	6.33	2.25±0.06d	13.05±0.18d	0.66±0.08bc	0.38±0.11cd	0.32±0.06d	0.89±0.08e
A3	9.29±0.44fg	2.86±0.12d	3.24	2.38±0.04d	17.24±0.27ab	0.89±0.06a	0.61±0.8ab	0.68±0.06a	1.47±0.04b
A4	14.15±0.16b	2.14±0.08ef	6.61	4.11±0.12b	10.83±0.16e	0.51±0.08cd	0.71±0.12a	0.44±0.04c	1.08±0.10d
A5	10.86±0.33e	2.51±0.14e	4.32	3.57±0.15c	14.62±0.33c	0.62±0.06c	0.48±0.10abc	0.76±0.12a	1.44±0.08b
A6	13.74±0.27bc	3.53±0.12c	3.89	3.86±0.21bc	14.22±0.15c	0.69±0.04c	0.49±0.06c	0.43±0.04c	1.21±0.06cd
A7	12.55±0.24d	2.62±0.12de	4.79	4.02±0.24bc	13.54±0.18d	0.85±0.08a	0.69±0.05a	0.69±0.06a	1.38±0.08bc
A8	14.77±0.12b	4.96±0.24a	2.97	4.45±0.15b	14.46±0.21c	0.51±0.11cd	0.42±0.08c	0.85±0.06a	1.79±0.04a
A9	9.86±0.18f	4.21±0.16b	2.34	3.41±0.12c	9.41±0.18f	0.28±0.02e	0.45±0.06c	0.27±0.04e	0.66±0.02f
A10	14.83±0.22b	2.05±0.14f	7.23	6.08±0.18a	17.65±0.27a	0.85±0.06a	0.28±0.04d	0.71±0.08a	1.85±0.08a
A11	15.21±0.25a	4.37±0.16b	3.48	4.17±0.24b	15.61±0.21b	0.73±0.04b	0.43±0.06c	0.68±0.10ab	1.91±0.12a

*Different letters are significant at $p = 0.05$ level (Duncan's multiple-range test), colpus length=clt, colpus width=clw, Apocolpidium=Apo, Outline of a pollen grain as viewed from pole=Amb, apocolpium=Apo, Standard Deviation=±, for the taxa abbreviations see Table 1.

also encountered in some taxa e.g. *A. speciosum* ssp. *compactum* (A6), *A. saxatile* (A7), *A. iberideum* (A9) and *A. armenum* (A10). Besides, some taxa showed heteromorphic characters such as 90 % tricolpate and 10 % dicolpate in taxa *A. speciosum* ssp. *compactum* (A6) and *A. saxatile* (A7), 95% tricolpate and 5% dicolpate in taxa *A. iberideum* (A9) and *A. armenum* (A10). These variations in pollen aperture type are declared as heteromorphy in pollens by Inceoglu and Karamustafa (1977) and Ceter *et al.* (2013). Also, the colpus sizes range from 9.29 (*A. eunomioides*) to15.21 (*A. grandiflorum*) µm in length, from 1.08 (*A. froedinii*) to 4.96 (*A. oppositifolium*) µm in width. The colpus membranes are more or less granulate (Table 2, Figures 1&2).

Table 3. Dissimilarity matrix of the examined taxa (for the taxa abbreviations see Table 1).

Taxa	A1	A2	A3	A4	A5	A6	A7	A8	A9	A10	A11
A1	0	-	-	-	-	-	-	-	-	-	-
A2	2.27	0	-	-	-	-	-	-	-	-	-
A3	2.71	2.34	0	-	-	-	-	-	-	-	-
A4	1.96	1.35	2.69	0	-	-	-	-	-	-	-
A5	2.24	1.52	1.30	1.62	0	-	-	-	-	-	-
A6	2.61	0.90	2.23	1.82	1.61	0	-	-	-	-	-
A7	2.25	1.41	1.85	1.04	0.72	1.67	0	-	-	-	-
A8	3.34	2.83	2.69	2.19	2.34	2.97	1.88	0	-	-	-
A9	2.64	3.50	3.29	2.50	2.76	3.85	2.47	2.36	0	-	-
A10	4.66	4.65	3.80	4.19	4.13	4.65	3.92	2.65	4.06	0	-
A11	4.14	3.01	3.21	2.57	2.48	3.26	2.07	1.91	3.07	4.17	0

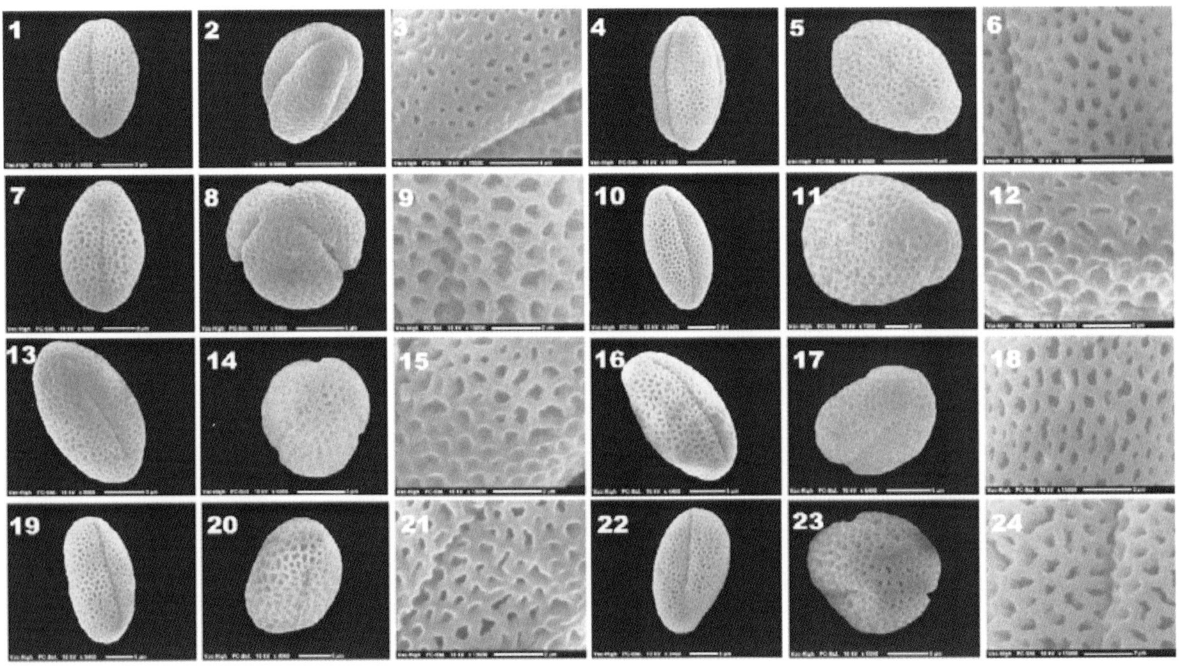

Fig. 1. SEM microphotograph of the studied *Aethionema*: A1: 1-3, A2: 4-6, A3: 7-9, A4: 10-12, A5: 13-15, A6: 16-18, A7: 19-21, A8: 22-24 (For abbreviation taxa see Table 1).

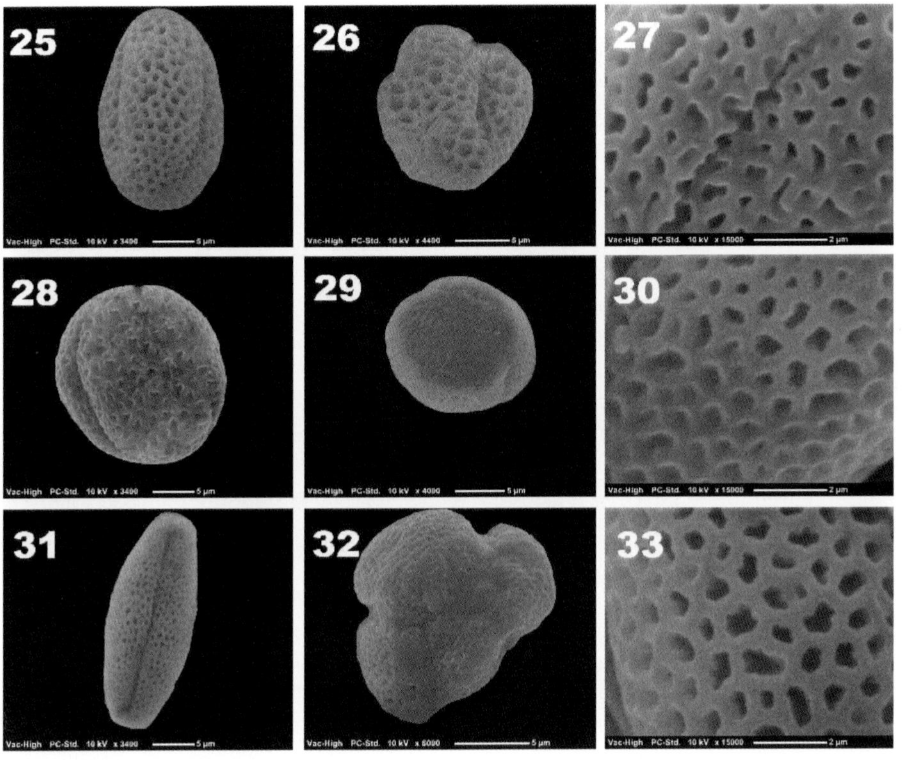

Fig. 1. SEM microphotograph of the studied *Aethionema*: A9: 25-27, A10: 28-30 and A11: 31-33 (For abbreviation taxa see Table 1).

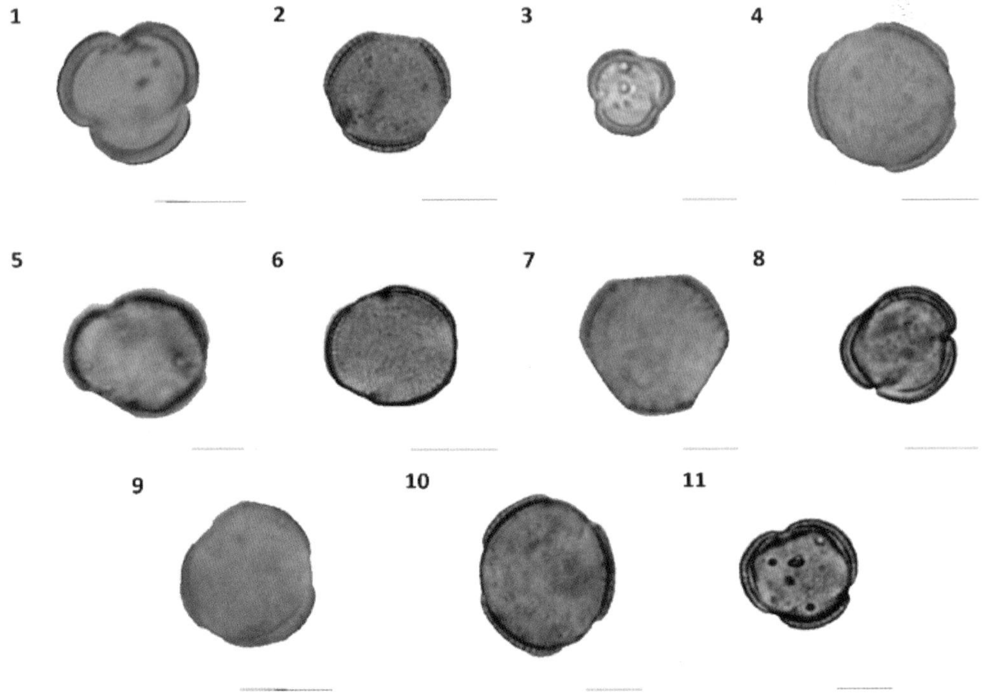

Fig. 2. Light microscope photographs of the studied *Aethionema*: A1:1, A2:2, A3:3, A4:4, A5:5, A6:6, A7:7, A8:8, A9:9, A10:10 and A11:11(Scale bars=5 μm) (For abbreviation taxa see Table 1).

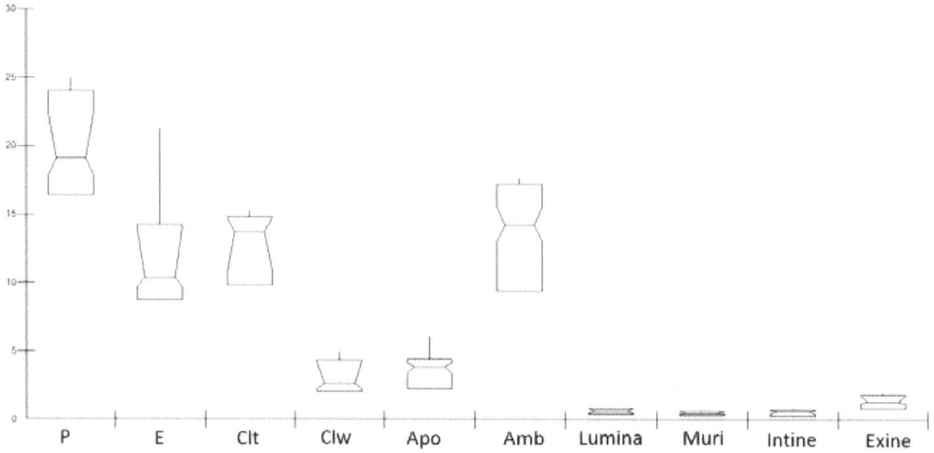

Fig. 3. Correlation among the palynological characters for the studied taxa (for abbreviations of the characters see Table 2).

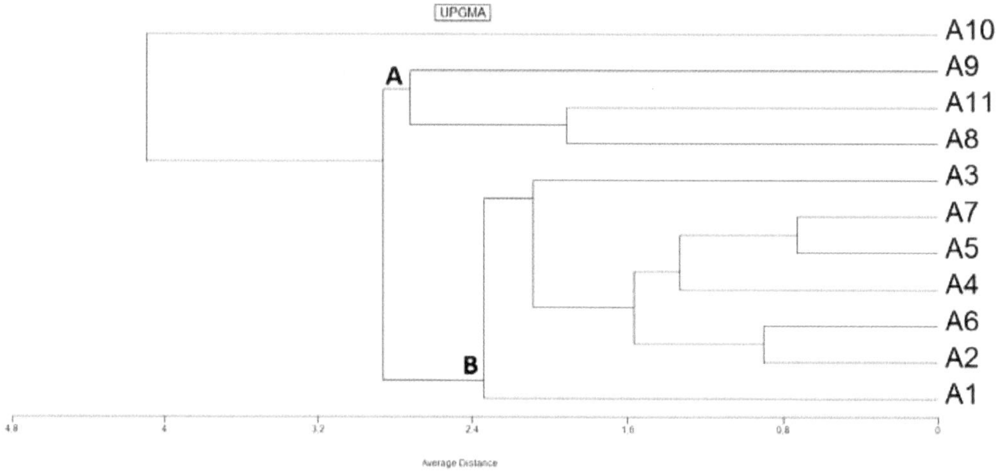

Fig. 4. UPGMA clustering of the examined taxa based on palynological characters (for taxa abbreviations see Table 1).

The exine thickness varies between from 0.66 (*A. iberideum*) and1.91 (*A. grandiflorum*) µm, and it frequently seems thicker in aperture. Besides, the in tine thickness ranges from 0.27 (*A. iberideum*) to 0.85 µm (*A. oppositifolium*). The surface ornamentation in the examined *Aethionema* taxa is reticulate and coarse or micro reticulate with straight or rough muri. Lumina includes 4-7 polygonal or irregular cells. Its diameter varies between 0.28 (*A. iberideum*) and 0.89 (*A. eunomioides*) µm.

The pollen exine ornamentations are of a significant role in delimitation of the some closely related taxa in Cruciferae (Khalik *et al.,* 2002). The present communication reveals that the pollen ornamentations of genus *Aethionema* are reticulate, coarsely reticulate, and micro-reticulate (Table 2). Anchev and Deneva (1997) reported ornamentation of the pollens of Cruciferae family is mostly reticulate and foveolate, which confirms the present result.

Pollen and colpus dimensions and, diameter of the apo and amb figure are remarkably variable to separate taxa, and to be useful for the delimitation of taxa unlike muri, lumina and thickness of intine and exine (Fig. 3). The pollen morphology of *Aethionema* taxa shows a close relationship with other genera of the family, for example; *Arabidopsis* Heynh.(Khan, 2004), *Hesperis* L. (Pinar *et al.,*2009), and *Arabis* (Mutlu and Erik, 2012).

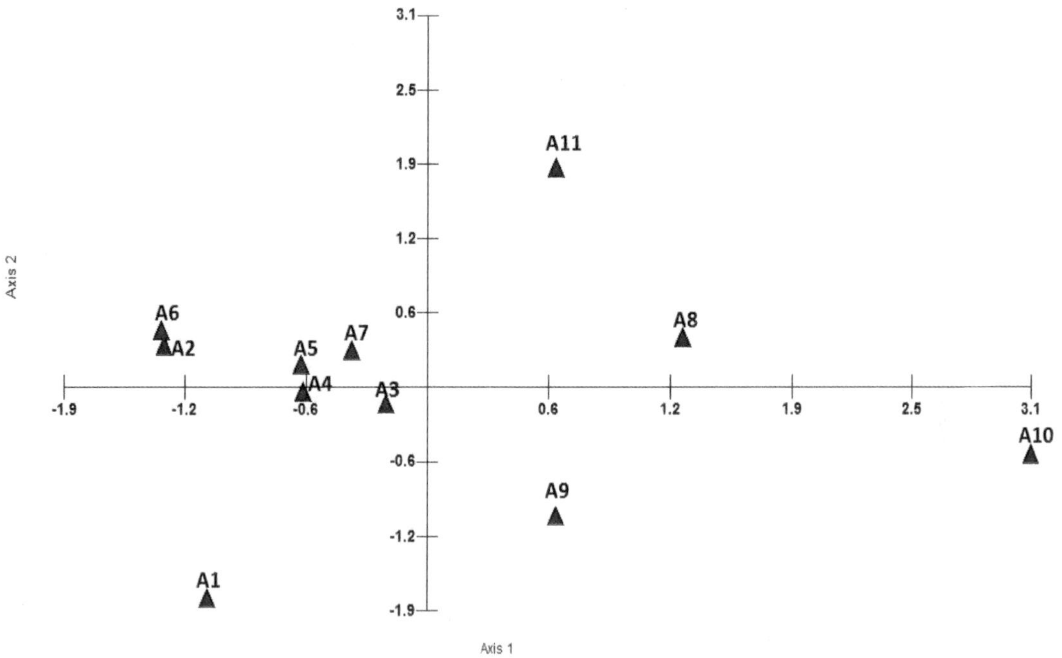

Fig. 5. PCA of the studied taxa based on palynological characters (for taxa abbreviations see Table 1)

A dendrogram of cluster analysis of 11 *Aethionema* taxa has been created based on 10 characters. To define the correlation of the dendrogram and the dissimilarity matrix (Table 3), the co-phenetic correlation coefficient is measured; the higher relationship reflects more preferable in terms of position in the hierarchy. In the cluster analysis of the studied taxa formed two distinctive clades, clade A and clade B. Clade A includes taxa *A. oppositifolium* (A8), *A. iberideum* (A9) and *A. grandiflorum* (A11). Clade B includes taxa *A. froedinii* (A1), *A. arabicum* (A2), *A. eunomioides* (A3), *A. fimbriatum* (A4), *A. speciosum* ssp. *speciosum* (A5), *A. speciosum* ssp. *compactum* (A6) and *A. saxatile* (A7). However, taxon *A. armenum* (A10) remains out the cluster (Fig. 4). It is found that the most closely related taxa are *A. speciosum* ssp. *speciosum* (A5) and *A. saxatile* (A7) (dissimilarity ratio: 0.72), while taxa *A. froedinii* (A1) and *A. armenum* (A10) are the most distantly related (dissimilarity ratio: 4.66) (Fig. 5, Table 3).Clade B consists of most of the taxa studied, which can be attributed as the taxa of this clade having primitive characteristics within *Aethionema.*

The current investigation reveals that palynological data has taxonomic significance, and it offers substantial contribution to the current classification of *Aethionema.*

References

Al-Shehbaz, I.A. 1986. The genera of *Lepidieae* (Cruciferae; Brassicaceae) in the southeastern United States. J. Arnold. Arbor. **67**: 265–311.

Al-Shehbaz, I.A., Mutlu, B. and Donmez, A.A. 2007.The Brassicaceae (Cruciferae) of Turkey, Updated. Turk. J. Bot. **31**: 327–336.

Anchev, M. and Deneva, B. 1997. Pollen morphology of seventeen species from family Brassicaceae (Cruciferae). Phytol. Balcan. **3**:75–82.

Atceken, M.M., Dural, H. and Yilmaz Citak, B. 2016. The morphological, anatomical and palynological investigations on some taxa of genus *Aethionema* A.T. Waiton (Brassicaceae).Biol. Diver. Conserv. **9**: 55–68.

Brochmann, C. 1992. Pollen and seed morphology of Nordic *Draba* (Brassicaceae): phylogenetic and ecological implications. Nord. J. Bot.**1:** 657–673.

Ceter, T., Pinar, N.M., Inceer, H., Hayırlıoglu-Ayaz, S. and Yaprak, A.E. 2013.The comparative pollen morphology of genera *Matricaria* L. and *Tripleurospemum* Sch. Bip. (Asteraceae) in Turkey. Plant Syst. Evol. **299**: 959–977.

Cronquist, A. 1968. The evolution and classification of the flowering plants. Edinburgh: Thomas Nelson Ltd.

Davis, P.H. 1965. Flora of Turkey and the East Aegean Islands, Vol. 1. Edinburgh: Edinburgh Univ. Press.

Guner, A., Aslan, S., Ekim, T., Vural, M. and Babac, M.T. 2012. Türkiyebitkilerilistesi (damarlıbitkiler). [A Checklist of the Flora of Turkey (Vascular Plants)]. Nezahat Gokyigit Botanik Bahcesive Flora Arastırmaları Dernegi Yayını, Istanbul.

Erdtman, G. 1952. Pollen morphology and plant taxonomy Angiosperms. Stockholm: Altham Mass.

Erdtman, G. 1969. Hand book of palynology. Morphology, taxonomy, ecology. An introduction to the study of pollen grains and spores. New York: Hafner Pub.

Inceoglu, O. and Karamustafa, F. 1977. The pollen morphology of plants in Ankara region II. Cruciferac. Comm. Fac. Sci. Univ.Ank. Series C2 **21**: 111–118.

Karaismailoglu, M.C. 2015. Morphological and anatomical features of seeds of Turkish *Romulea* taxa (Iridaceae) and their taxonomic significance. Acta Bot. Croat.**74**: 31–41.

Karaismailoglu, M.C. 2016. Addition to characters of endemic *Aubrieta canescens* subsp. *canescens* Bornm. (Brassicaceae) from Turkey. Bangladesh J. Bot. **45**(3): 509–515.

Khalik, K.A., Maesen, L.J.G., Kopman, W.J.M. and Berg, R.G. 2002. Numerical taxonomic study of some tribes of Brassicaceae from Egypt. Plant Syst. Evol. **233:** 207–221.

Khan, R. 2004. Studies on the pollen morphology of the genus *Arabidopsis* (Brassicaceae) from Pakistan. Pak. J. Bot. **36**: 229–234.

Kovach, W.L. 2007. MVSP - A Multivariate Statistical Package for Windows, ver. 3.1. Kovach Computing Services, Pentraeth, Wales, U.K.

Mohammadi, S.A. and Prasanna, B.M. 2003. Analysis of genetic diversity in crop plants. Salient statistical tools and considerations. Crop. Sci. **43:**1235–1248.

Mummenhoff, K., Franzke, K.A. and Koch, M. 1997. Molecular data reveal convergence in fruit characters used in the classification of *Thlaspi* s.l. (Brassicaceae). Bot. J. Linn. Soc.**125:** 183–199.

Mutlu, B. and Erik, S. 2012. Pollen morphology and its taxonomic significance of the genus *Arabis* (Brassicaceae) in Turkey. Plant Syst. Evol. **298**: 1931–1946.

Ocak, A., PotogluErkara, I., Koyuncu, O., Soydan, K., Yaylacı, O.K., Ozgisi, K. and Kurt, F. 2013. Palynological investigations on some *Hypericum* taxa (Hypericaceae) growing naturally in Turkey. Plant Syst. Evol. **299**: 379–388.

Pinar, N.M., Bayrak, N. and Geven, F. 2007. Seed coat macro-sculpturing in some Turkish *Aethionema* R. Br. (Brassicaceae). Pak. J. Bot. **39**: 1025-1036.

Pinar, N.M., Duran, A., Ceter, T. and Tug, G.N. 2009. Pollen and seed morphology of the genus *Hesperis* L. (Brassicaceae) in Turkey. Turk. J. Bot.**33**: 83–96.

Punt, W., Blackmore, S., Nilsson, S. and Le Thomas, A. 1994. Glossary of pollen and spore terminology. Lab Palaeobot Palynol, Utrect, pp. 71.

SPSS Inc. 2006. SPSS Base 15.0 for windows, Chicago.

Wodehouse, R.P. 1935. Pollen Grains. New York: Mc. Graw Hill.

13

SYSTEMATIC IMPLICATIONS OF ACHENE CHARACTERISTICS IN GENERA *CENTAUREA* L., *CYANUS* MILL., *PSEPHELLUS* CASS. AND *RHAPONTICOIDES* VAILL. (ASTERACEAE)

MEHMET BONA[1]

Department of Botany, Faculty of Science, Istanbul University, Istanbul, Turkey

Keywords: Achene; Asteraceae; Scanning Electron Microscopy; Taxonomy.

Abstract

This study examines the exomorphic achene characteristics of 23 taxa belonging to *Centaurea*, *Cyanus*, *Psephellus* and *Rhaponticoides* in Asteraceae using light microscope and scanning electron microscope (SEM). The exomorphic characteristics studied are shape, size, colour and surface pattern of achene, and hair, length and colour of pappus. The results of the present investigation showed that achene size of the studied taxa is 3–7 × 1–4 mm. Pappus length is 1–14 mm, deciduous for three taxa and inner row is differentiated from the outer, 1–3 mm long, for 10 taxa. SEM studies showed eight different types of achene surface patterns for the studied taxa: Glebulate, glebulate-ruminate, reticulate, ribbed, ruminate, smooth, smooth-glebulate and undulate. Achene characteristics are useful for both intrageneric and intraspecific classification of the studied taxa.

Introduction

Morphological and anatomical studies on fruit and seed structure play an important role in systematics (Kumar *et al.*, 2012). Microstructural details of the seed and fruit coat enable the distinguishing of taxa or the discovery of their affinities. This is especially useful for families in which the identification of particular taxa is complicated (Kumar *et al.*, 2012; Bona, 2013; Piwowarczyk *et al.*, 2014). Furthermore, observations of micromorphological features can also provide us with information about developmental strategies, adaptation to different environmental conditions and evolutionary tendencies within related groups of plants (Kreitschitz and Vallès, 2007; Moazzeni *et al.*, 2010).

The family Asteraceae is one of the largest angiosperm families and comprises about 1,300 genera and 2,500 species distributed over three subfamilies and 17 tribes (Ayad *et al.*, 2012). The genus *Centaurea* L. *s.l.* is one of the largest genera in the family Asteraceae. Among the challenging taxonomic problems persisting in the Asteraceae is the delimitation of the genus *Centaurea* (Bancheva and Raimondo, 2013; Ranjbar *et al.*, 2013). *Centaurea* has recently been divided into four genera, namely *Centaurea*, *Rhaponticoides*, *Psephellus* and *Cyanus* (Wagenitz and Hellwig, 2000; Greuter, 2003a, b). *Cyanus*, however, is not widely accepted by different authors (Susanna and Garcia-Jacas, 2007).

Though previous studies support the use of achene surface patterns as diagnostic characters at species and subspecies levels for the genus *Centaurea s.l.*, understanding the importance of these characters at generic and subgeneric levels requires further studies (Uysal *et al.*, 2005; Çelik *et al.*, 2005a, b; Aksoy *et al.*, 2010; Okay and Demir, 2010; Shabestari *et al.*, 2013; Bona, 2014; Candan *et al.*, 2015). With this in mind, this study examines the exomorphic achene characters of 23 taxa belonging to the genera *Centaurea*, *Cyanus*, *Psephellus* and *Rhaponticoides* in Asteraceae by using light microscope (LM) and scanning electron microscope (SEM).

---

[1]Email: mehmetbona@gmail.com

Materials and Methods

The materials of this study are the mature achenes of 23 Asteraceae taxa belonging to the genera *Centaurea*, *Cyanus*, *Psephellus* and *Rhaponticoides*. The collected specimens were kept at the Istanbul University, Faculty of Pharmacy, Department of Pharmaceutical Botany Herbarium (ISTE). The exomorphic characteristics of the achene are achene colour, size, shape and surface pattern, and pappus colour, hair, and length (Table 1). All these characteristics are described, illustrated and compared. Up to 50 (at least 10) mature achenes for each taxon were measured and observed under the light microscope. During scanning electron microscopy, two mature achenes were selected and mounted onto stubs with double-sided adhesive tape, and were then coated with gold. The achene surfaces were examined from the lateral sides. For each sample, photographs of testa were taken using the JEOL JSM-5600 at a magnification of 22×–50×, 1000×, and 3000×. The terminology of achene characteristics in this work is based on the descriptions used by Barthlott (1981), Stearn (1992), Koul *et al.* (2000), and Bojňanský and Fargašová (2007).

Results

The results of the present investigation showed that the achene size of the studied taxa is 3–7 × 1–4 mm. The pappus length is changing from 1 to 14 mm, deciduous for three taxa and the inner row is differentiated (1–3 mm long) for 10 taxa. Testa cells of all the studied taxa appear regularly arranged and elongated-parallel with the seed surface. The results of the studied taxa are distinguished below.

1. ***Centaurea antiochia*** Boiss. var. ***antiochia***: Achene straw-coloured, oblong, 5–6 × 2–3 mm, pappus scabrous, 5–6 mm long, inner row differentiated from outer row, c. 2 mm long. Seed surface pattern smooth. Cell boundaries thin and cell centres at ± equal levels with the boundaries (Fig. 1a–c).

2. ***C. arifolia*** Boiss.: Achene dark brown, oblong, 4–5 × 3 mm, pappus scabrous, 7–8 mm long, inner row differentiated from outer row, c. 1 mm. Seed surface pattern glebulate-ruminate. Cell boundaries very thin and cell centres raised above boundaries (Fig. 1d–f).

3. ***C. carduiformis*** DC. subsp. ***carduiformis*** var. ***thrinciifolia*** (DC.) Wagenitz: Achene blackish green, oblong, 5–6 × 3 mm, pappus dark silver, scabrous, 8–9 mm long, inner row differentiated from outer row, c. 2 mm long. Seed surface pattern undulate. Cell boundaries thin and cell centres at ± equal levels with the boundaries. Testa cells are apparently imbricate; this condition shows itself as waved layers on the edge of the achene (Fig. 1g–i).

4. ***C. cassia*** Boiss.: Achene black, oblong, 3 × 1.4–1.5 mm, pappus white, scabrous, c. 1 mm long. Seed surface pattern ruminate. Cell boundaries very thin and cell centres raised above the boundaries (Fig. 1j–l).

5. ***C. cheirolopha*** (Fenzl) Wagenitz: Achene light brown-straw-coloured, oblong, 4–5 × 2 mm, pappus absent or very short, scabrous, c. 1 mm long. Seed surface pattern glebulate. The cell boundaries are broad and appear raised above the cell centres (Fig. 1m–o).

6. ***C. drabifolia*** subsp. ***cappadocica*** (DC.) Wagenitz: Achene greyish, oblong-elliptic, 4–5 × 2 mm, pappus white-straw-coloured, plumose, 6–7 mm long, deciduous. Seed surface pattern glebulate-ruminate. The cell boundaries are thin and have a smooth structure and the cell centres are placed at ± equal levels with the boundaries (Fig. 1p–s).

Table 1. Pappus and achene characteristics of the studied taxa.

Taxa	Pappus					Achene			
	Colour	D/P	Hair	Length	Length (inner)	Colour	Shape	Size	Surface
Centaurea antiochia var. *antiochia*	Straw	P	Sc	5–6	2	Straw	Oblong	5–6 × 2–3	Smooth
C. arifolia	Purple to dark brown	P	Sc	7–8	1	Dark brown	Oblong	4–5 × 3	Glebulate-ruminate
C. carduiformis subsp. *carduiformis* var. *thrinciifolia*	Dark silver	P	Sc	8–9	2	Blackish green	Oblong	5–6 × 3	Undulate
C. cassia	White	P	Sc	1	N	Black	Oblong	3 × 1.5	Ruminate
C. cheirolopha	Straw	P	Sc	1	N	Light brown to straw	Oblong	4–5 × 2	Glebulate
C. drabifolia subsp. *cappadocica*	White-straw	D	Pl	6–7	N	Greyish	Oblong-elliptic	4×5 × 2	Glebulate-ruminate
C. drabifolia subsp. *floccosa*	Straw	P	Pl	8–10	N	Yellow to green	Oblong	4–5 × 2	Smooth
C. foliosa	Dark purple	P	Sc	5–6	N	Dark silver to brown	Oblong	5 × 3	Glebulate-ruminate
C. lycopifolia	Straw	P	Sc	1	N	Dark green to brown	Oblong	4–6 × 2	Glebulate-ruminate
C. pseudoscabiosa subsp. *araratica*	Silver brown	P	Sc	6	2	Dark brown	Oblong	5–6 × 3	Reticulate
C. ptosimopappa	Straw to purple	D	Sc	5–6	N	Black	Oblong	5–6 × 2.0–2.5	Glebulate-ruminate

Table 1. Contd.

Taxa	Pappus					Achene			
	Colour	D/P	Hair	Length	Length (inner)	Colour	Shape	Size	Surface
C. virgata	White	D	Sc	2	N	Brown to grey	Oblong	3–4 × 1–2	Glebulate-ruminate
Cyanus bourgaei	Straw	P	Sc	1	N	Straw to brown	Oblong	5 × 2	Smooth
Cy. cheiranthifolius subsp. *purpurascens*	Straw	P	Sc	2	N	Straw to dark green	Oblong	6–7 × 3	Smooth-glebulate
Cy. depressus	Straw	P	Sc	5–6	2	Dark green	Oblong	5–6 × 2–2.5	Smooth
Cy. segetum	Straw to red	P	Sc	2	N	Dark green	Narrowly oblong	3 × 1	Smooth-glebulate
Psephellus appendicigerus	Straw	P	Pl	13–14	N	Straw	Narrowly ovate	6–7 × 3	Ribbed
P. bornmuelleri	Purple	P	Sc	8	3	Yellow	Lanceolate	6–7 × 3	Smooth
P. brevifimbriatus	Straw to brown	P	Sc	7	3	Straw	Oblong	6 × 3	Undulate
P. mucronifer	Straw	P	Sc	4	1	Straw to greyish	Narrowly ovate	6 × 3	Smooth
P. pulcherrimus	Straw	P	Pl	10–12	N	Dark green to brown	Lanceolate	5–6 × 2	Ribbed
P. pyrrhoblepharus	Straw to yellow	P	Sc	2–4	1	Straw to green	Narrowly oblong	5–7 × 2	Ruminate
Rhaponticoides wagenitziana	Brown	P	Sc	7–8	2	Dark brown	Broadly oblong	6–7 × 4	Reticulate

Note: D, deciduous; N, not different; P, persistent; Pl, plumose; Sc, scabrous.

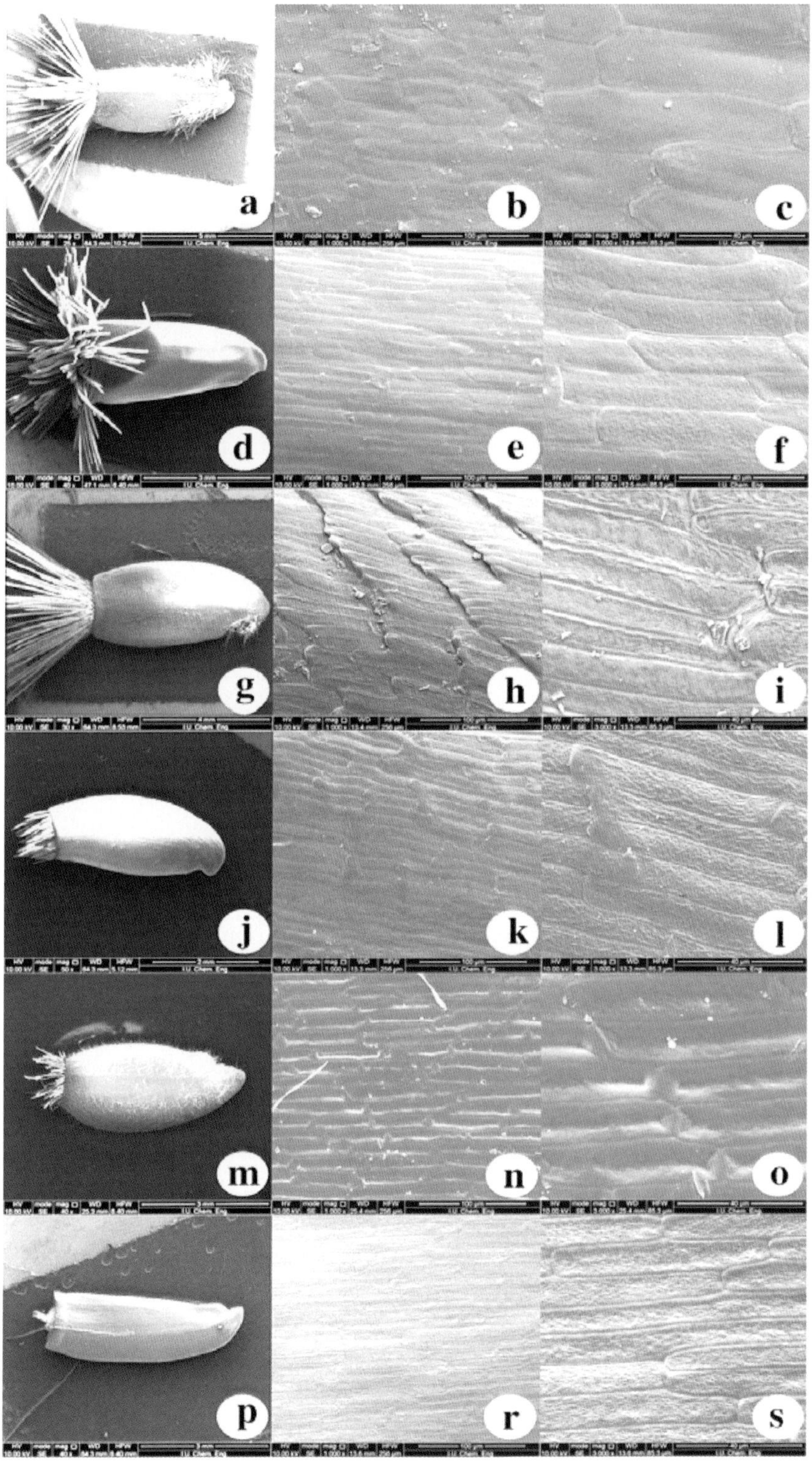

Fig. 1. SEM micrographs of achenes of *Centaurea antiochia* var. *antiochia* (a–c); *C. arifolia* (d–f); *C. carduiformis* subsp. *carduiformis* var. *thrinciifolia* (g–i); *C. cassia* (j–l), *C. cheirolopha* (m–o); *C. drabifolia* subsp. *cappadocica* (p–s).

7. ***C. drabifolia*** subsp. ***floccosa*** (Boiss.) Wagenitz & Greuter: Achene yellow-green, oblong, 4–5 × 2 mm, pappus straw-coloured, plumose, 8–10 mm long. Seed surface pattern smooth. Cell boundaries are thin and the boundaries appear raised above the cell centres (Fig. 2a–c).

8. ***C. foliosa*** Boiss. & Kotschy *ex* Boiss.: Achene dark silver-brown, oblong, c. 5 × 3 mm, pappus dark purple, scabrous, 5–6 mm long. Seed surface pattern glebulate-ruminate. Cell boundaries are thin and smooth and appear raised above the cell centres (Fig. 2d–f).

9. ***C. lycopifolia*** Boiss. & Kotschy *ex* Boiss.: Achene dark green-brown, oblong, 4–6 × 2 mm, pappus straw-coloured, scabrous, c.1 mm long. Seed surface pattern glebulate-ruminate. Cell boundaries are very thin and the cell centres are raised above the boundaries (Fig. 2g–i).

10. ***C. pseudoscabiosa*** subsp. ***araratica*** (Azn.) Wagenitz: Achene dark brown, oblong, 5–6 × 3 mm, pappus silvery brown, scabrous, 6 mm long, inner row differentiated from outer row, c. 2 mm long. Seed surface pattern reticulate. Cell boundaries are broad and have a smooth structure and the boundaries seem distinctly raised above the cell centres (Fig. 2j–l).

11. ***C. ptosimopappa*** Hayek: Achene straw-coloured to green when young and blackish when mature, oblong, 5–6 × 2.0–2.5 mm, pappus straw-coloured or sometimes purplish, scabrous, 5–6 mm long, deciduous. Seed surface pattern glebulate-ruminate. Cell boundaries are broad and have a smooth structure and the boundaries seem distinctly raised above the cell centres (Fig. 2m–o).

12. ***C. virgata*** Lam.: Achene brown-grey, oblong, 3–4 × 1–2 mm, pappus white, scabrous, c. 2 mm long, deciduous. Seed surface pattern glebulate-ruminate. Cell boundaries are thin and smooth and appear raised above the cell centres (Fig. 2p–s).

13. ***Cyanus bourgaei*** (Boiss.)Wagenitz & Greuter: Achene straw-coloured to brown, oblong, c. 5 × 2 mm, pappus straw-coloured, scabrous, c.1 mm long. Seed surface pattern smooth. Cell boundaries are thin and the cell centers are ± equal with the boundaries. Testa cells are sulcate at the centres of the cells (Fig. 3a–c).

14. ***Cy. cheiranthifolius*** (Willd.) Soják subsp. ***purpurascens*** (DC.) Wagenitz: Achene straw-coloured to dark green, oblong, 6–7 × 3 mm, pappus straw-coloured, scabrous, c. 2 mm long. Seed surface pattern smooth-glebulate. The cell boundaries are thin and the cell centres are ± equal with the boundaries (Fig. 3d–f).

15. ***Cy. depressus*** (M. Bieb.) Soják: Achene dark green, oblong, 5–6 × 2–2.5 mm, pappus straw-coloured, scabrous, 5–6 mm long, inner row differentiated from outer row, c. 2 mm long. Seed surface pattern smooth. Cell boundaries are very thin and the cell centres are raised above the boundaries (Fig. 3g–i).

16. ***Cy. segetum*** Hill: Achene dark green, narrowly oblong, c. 3 × 1 mm, pappus straw-coloured to red, scabrous, c. 2 mm long. Seed surface pattern smooth-glebulate. The cell boundaries are thin and have a smooth structure and the cell centres are ± equal with the boundaries (Fig. 3j–l).

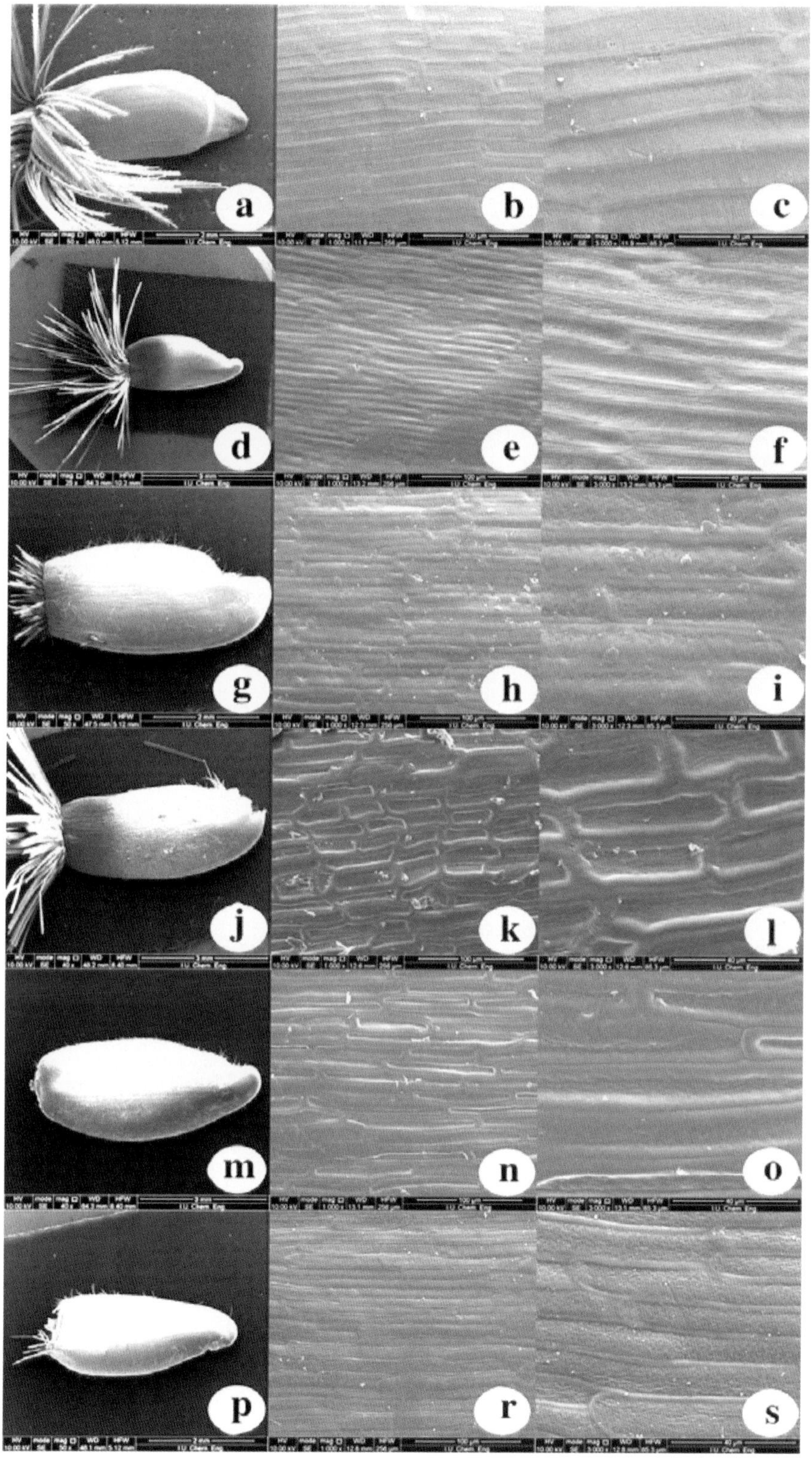

Fig. 2. SEM micrographs of achenes of *Centaurea drabifolia* subsp. *floccosa* (a–c); *C. foliosa* (d–f); *C. lycopifolia* (g–i); *C. pseudoscabiosa* subsp. *araratica* (j–l); *C. ptosimopappa* (m–o); *C. virgata* (p–s).

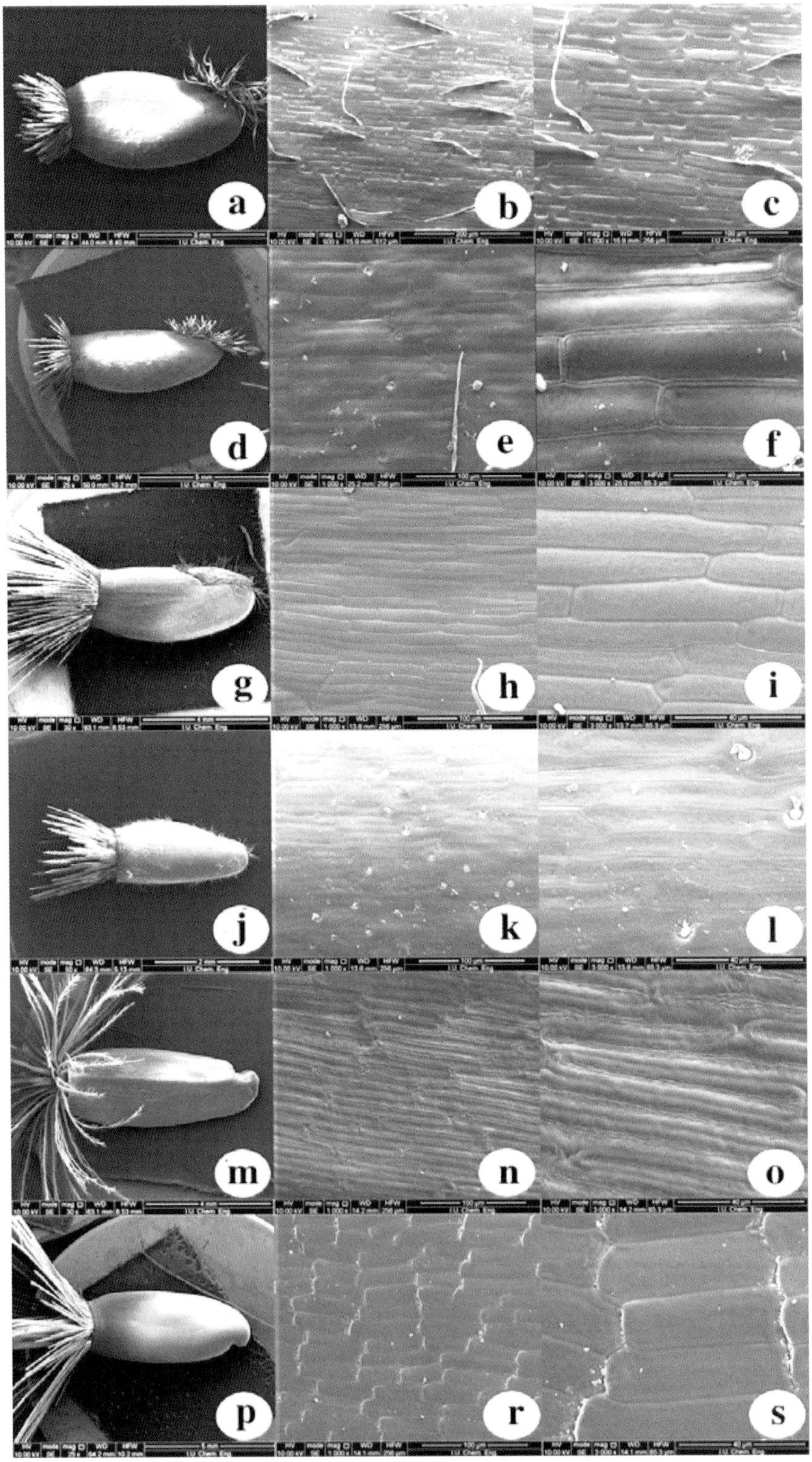

Fig. 3. SEM micrographs of achenes of *Cyanus bourgaei* (a–c); *Cy. cheiranthifolius* subsp. *purpurascens* (d–f); *Cy. depressus* (g–i); *Cy. segetum* (j–l); *Psephellus appendicigerus* (m–o); *P. bornmuelleri* (p–s).

17. ***Psephellus appendicigerus*** (K. Koch) Wagenitz**:** Achene is straw-coloured, narrowly ovate, 6–7 × 3 mm, pappus light straw-coloured, plumose, 13–14 mm long. Seed surface pattern ribbed. The cell boundaries are broad and have smooth structure and appear raised above the cell centres. Cell centres are sulcate (Fig. 3m–o).

18. ***P. bornmuelleri*** (Hausskn. *ex* Bornm.)Wagenitz**:** Achene yellowish, lanceolate, 6–7 × 3 mm, pappus purple, scabrous, 8 mm long, and its inner row is differentiated from the outer row, c. 3 mm long. Seed surface pattern smooth. The cell boundaries are thin and the cell centres are ± equal with the boundaries. Testa cells are apparently imbricate; this condition shows itself as waved layers (Fig. 3p–s).

19. ***P. brevifimbriatus*** (Hub.-Mor.) Wagenitz: Achene straw-coloured, oblong, c. 6 × 3 mm, pappus straw-coloured to light brown, scabrous, 7 mm long, and its inner row is differentiated from the outer row, c. 3 mm long. Seed surface pattern undulate. The cell boundaries are thin and the cell centres are ± equal with the boundaries. Cell centres are slightly sulcate (Fig. 4a–c).

20. ***P. mucronifer*** (DC.) Wagenitz: Achene straw-coloured to greyish, narrowly ovate, c. 6 × 3 mm, pappus straw-coloured, scabrous, c. 4 mm long, and its inner row differentiated from the outer row, c. 1 mm long. Seed surface pattern smooth. The cell boundaries are thin and have smooth structure and the centres of the cells are placed at ± equal levels with the boundaries (Fig. 4d–f).

21. ***P. pulcherrimus*** (Willd.) Wagenitz: Achene dark green to brown, lanceolate, 5–6 × 2 mm, pappus straw-coloured, plumose, 10–12 mm long. Seed surface pattern ribbed. The cell boundaries are very thin and the centres of the cells are raised above the boundaries. Cell centers are ribbed (Fig. 4g–i).

22. ***P. pyrrhoblepharus*** (Boiss.) Wagenitz: Achene straw-coloured to green, narrowly oblong, 5–7 × 2 mm, pappus straw-coloured to yellow, scabrous, 2–4 mm long, and its inner row different, c. 1 mm long. Seed surface pattern ruminate. The cell boundaries are broad and have smooth structure and the boundaries appear raised above the cell centres (Fig. 4j–l).

23. ***Rhaponticoides wagenitziana*** (Bancheva & Kit Tan) M.V. Agab. & Greuter: Achene dark brown above, yellow at base, broadly oblong, 6–7 × 4 mm, pappus brown, scabrous, 7–8 mm long, inner row differentiated from outer row, c. 2 mm long. Seed surface pattern reticulate. Cell boundaries are distinctly broad and appear seem distinctly raised above the cell centres. Achenes have waved layers which are independent from cells and cell boundaries (Fig. 4m–o).

Discussion

Rhaponticoides wagenitziana is the only species of the studied taxa which belongs to genus *Rhaponticoides*. The achenes are dark brown above, yellow at base and have waved layers which are independent from the cells and cell boundaries. These two characteristics seem specific to *R. wagenitziana*. More work is required to confirm the use of characteristics for delimitation of the genus *Rhaponticoides*.

Psephellus coruhensis, *P. turcicus* and *P. psephelloides* have reticulate seed surface patterns (Duran and Hamzaoğlu, 2005; Duran *et al.*, 2009). Our study shows that pappus length, colour and achene colour characteristics show differences at the specific level. Additionally, the achene

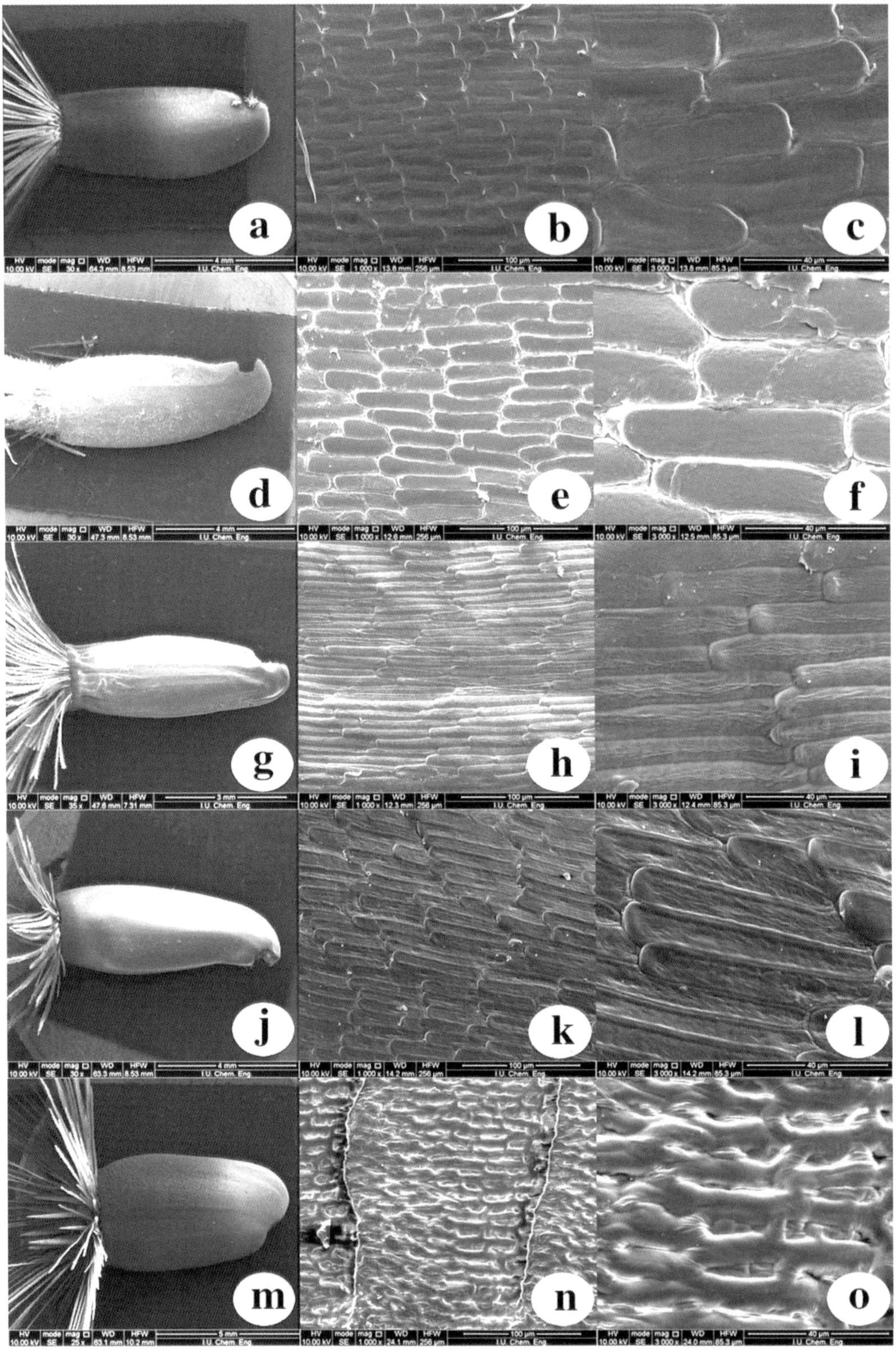

Fig. 4. SEM micrographs of achenes of *Psephellus brevifimbriatus* (a–c); *P. mucronifer* (d–f); *P. pulcherrimus* (g–i); *P. pyrrhoblepharus* (j–l); *Rhaponticoides wagenitziana* (m–o).

surface pattern can be ribbed, smooth or undulate for the genus *Psephellus*. Because of the overlap, achene characteristics are not useful for determining generic limits in *Psephellus*. Despite that, achene characteristics provide strong support in the delimitation of the studied *Psephellus* taxa at the specific level. *P. appendicigerus* and *P. pulcherrimus* were placed into Section *Aetheopappus* (Wagenitz, 1975). These two taxa are easily separated from other *Psephellus* taxa by their ribbed achene surface pattern and their pappus, which do not have differentiated inner rows. Other studied *Psephellus* taxa belong to section Psepheloideae (Wagenitz, 1975) in which their inner pappus row is differentiated. These findings show that achene characteristics could be useful for sectional classification of the genus *Psephellus*.

There is no specific pappus length, pappus colour, achene size or colour for the genus *Cyanus,* but all the studied *Cyanus* taxa have smooth or smooth-glebulate achene surface pattern.

This study supports the use of achene surface patterns as diagnostic characters at both specific and infraspecific levels in the genus *Centaurea.* There are other achene characteristics, including achene length and colour, pappus length and colour, that might be helpful to distinguish more taxa. However, though achene characteristics provide strong support in the delimitation of the studied taxa at specific level, understanding the importance of these characteristics for intrageneric classification of the genus *Centaurea* requires further study. *Centaurea drabifolia* subsp. *cappadocica* and *C. drabifolia* subsp. *floccosa* are separated from each other based on stem length in the Flora of Turkey (Wagenitz, 1975). This study shows that achene characteristics are useful in the separation of these taxa, because pappus length, achene colour and achene surface patterns of these taxa are different from each other. The SEM study of achene surfaces showed that the cell wall in *C. kurdica* is thicker than that of *C. sclerolepis*, especially on the margins and could be a useful character in the delimitation of species (Uysal *et al.*, 2005).

In conclusion, the achene characteristics of 23 taxa belonging to the genera *Centaurea*, *Cyanus*, *Psephellus* and *Rhaponticoides* were evaluated in this study. Results support the use of achene characteristics as diagnostic characters for intrageneric and intraspecific classification of these taxa. Achene characteristics provide strong support for the delimitation of the studied taxa at specific level, but understanding the importance of these characteristics at generic and intrageneric levels needs further studies. Hopefully, this investigation will encourage additional studies about achene morphology of this complex family to further elucidate the complex taxonomy of the Asteraceae.

Acknowledgements

The author thanks the Scientific Investigation Project Coordinator of Istanbul University (project no: 42493) and Edinburgh Botanic Garden (Sibbald) Trust for their financial support. Thanks are also due to Dr. Sabina Knees for her help in improving the manuscript.

References

Aksoy, N., Ataşlar, E., Efe, A. and Güneş, N. 2010. *Centaurea yaltirikii* subsp. *dumanii* subsp. nov. (*C.* sect. Pseudoseridia, Asteraceae) in Marmara region of Turkey. J. Food Agric. Environ. **8**: 1212−1215.

Ayad, R., Ababsa, Z.E.A., Belfadel, F.Z., Akkal, S., León, F., Brouard, I. and Medjroubi, K. 2012. Phytochemical and biological activity of Algerian *Centaurea melitensis*. International Journal of Medicinal and Aromatic Plants **2**: 151–154.

Bancheva, S. and Raimondo, F.M. 2013. A new *Centaurea* species (Asteraceae) from Mt Sakar, South-Eastern Bulgaria. Plant Biosyst. **147**: 800–805.

Barthlott, W. 1981. Epidermal and seed surface characters of plants: systematic applicability and some evolutionary aspects. Nord. J. Bot. **1**: 345–355.

Bojňanský, V. and Fargašová, A. 2007. Atlas of seeds and fruits of Central and East-European flora: The Carpathian Mountains region. Springer, Dordrecht.

Bona, M. 2013. Seed-coat microsculpturing of Turkish *Lepidium* (Brassicaceae) and its systematic application. Turk. J. Bot. **37**: 662–668.

Bona, M. 2014. Achene characteristics of Turkish *Centaurea* (Asteraceae) and their systematic application. Bangladesh J. Botany **43**: 163–168.

Candan, F., Uysal, T., Tugay, O., Bozkurt, M. and Demirelma, H. 2015. The examinations of achene ultrastructural features of section *Acrolophus* (*Centaurea*/Asteraceae) via scanning electron microscopy (SEM). Turk. J. Bot. DOI: 10.3906/bot-1503-51.

Çelik, S., Uysal, I. and Menemen, Y. 2005a. *Centaurea* species in Turkey (A): *Centaurea odyssei* Wagenitz (Asteraceae) in Kazdagı (Mt. Ida) National Park. International Journal of Biodiversity Science and Management **1**:113–120.

Çelik, S., Uysal, I., Menemen, Y. and Karabacak, E. 2005b. Morphology, anatomy, ecology, pollen and achene structure of *Centaurea consanguinea* DC. (Sect. Acrolophus). International Journal of Botany **1**: 85–89.

Duran, A. and Hamzaoğlu, E. 2005. *Psephellus turcicus* sp. nov. (Asteraceae), a new chasmophyte species from central Anatolia, Turkey. Bot. J. Linn. Soc. **148**: 495–500.

Duran, A., Öztürk, M., and Doğan, B. 2009. A new species of the genus *Psephellus* (Asteraceae) from North-East Anatolia, Turkey. Ozean Journal of Applied Sciences **2**: 103–111.

Greuter, W. 2003a. The Euro+Med treatment of *Cardueae* (Compositae) -generic concepts and required new names. Willdenowia. **33**: 49–61.

Greuter, W. 2003b. The Euro+Med treatment Senecioneae and the minor Compositae tribes-generic concepts and required new names, with an addentum to *Cardueae*. Willdenowia. **33**: 245–250.

Koul, K.K., Nagpal, R. and Raina, S.N. 2000. Seed coat microsculpturing in *Brassica* and allied genera (subtribe Brassicinae, Raphaninae, Moricandiinae). Ann. Bot. **86**: 385–397.

Kreitschitz, A. and Vallès, J. 2007. Achene morphology and slime structure in some taxa of *Artemisia* L. and *Neopallasia* L. (Asteraceae). Flora **202**: 570–580.

Kumar, V., Kodandaramaiah, J. and Rajan, M.V. 2012. Leaf and anatomical traits in relation to physiological characteristics in mulberry (*Morus* sp.) cultivars. Turk. J. Bot. **36**: 683–689.

Moazzeni, H., Zarre, S., Al-Shehbaz, I.A. and Mummenhof, K. 2010. Phylogeny of *Isatis* (Brassicaceae) and allied genera based on ITS sequences of nuclear ribosomal DNA and morphological characters.Flora **205**: 337–343.

Okay, Y. and Demir, K. 2010. Critically endangered endemic *Centaurea tchihatcheffii* Fisch. And Mey. and its propagation possibilities. Afr. J. Agric. Res. **5**: 3536–3541.

Piwowarczyk, R., Halamski, A.T. and Durska, E. 2014. Seed and pollen morphology in the *Orobanche alsatica* complex (Orobanchaceae) from central Europe and its taxonomic significance. Aust. Syst. Bot. **27**: 145–157.

Ranjbar, M., Negaresh, K. and Karamian, R. 2013. *Centaurea shahuensis* and *C. ravansarensis* spp. Nov. (Asteraceae, Carduae) from west Iran. Nord. J. Bot. **30**: 1–7.

Shabestari, E.S.B., Attar, F., Riahi, H. and Sheidai, M. 2013. Seed morphology of the *Centaurea* species (Asteraceae) in Iran. Phytol. Balcan. **19**: 209–214.

Stearn, W.T. 1992. Botanical Latin. David & Charles, London.

Susanna, A. and Garcia-Jacas, N. 2007. Tribe Cardueae. *In*: Kadereit, J.W. and Jeffrey, C. (Eds) The Families and Genera of Vascular Plants. Vol. **8**. Springer Verlag, Berlin, Heidelberg & New York. pp. 123–146.

Uysal, I., Çelik, S. and Menemen, Y. 2005. Morphology, anatomy, ecology, pollen and achene features of *Centaurea polyclada* DC. (Sect. Acrolophus) in Turkey. J. Biol. Sci. **5**: 176–180.

Wagenitz, G. 1975. *Centaurea* L. *In*: Davis, P.H. (Ed.), Flora of Turkey and the East Aegean Islands, Vol. **5**. Edinburgh University Press, Edinburgh, pp. 465–585.

Wagenitz, G. and Hellwig, F.H. 2000. *Psephellus* Cass. (Compositae, Cardueae) revisited with a broadened concept. Willdenowia **30**: 29–44.

14

MORPHOMETRICS OF THE TRIBE *MERREMIEAE* AUSTIN (CONVOLVULACAE) FROM INDIA

S. A. DESHMUKH AND VINOD B. SHIMPALE[1]

Department of Botany, The New College, Kolhapur- 416 012, (MS), India

Keywords: Convolvulaceae; *Merremieae*; Numerical Taxonomy; Principal component analysis.

Abstract

Sixteen species of the tribe *Merremieae* Austin (Convolvulaceae) from India were morphometrically analyzed with the help of Principal component analysis (PCA) and cluster analysis to explain the relationship between them. PCA showed that quantitative characters like corolla breadth, fruit breadth and length of paracot leaf play important role in bringing together all the species in the same tribe while the characters like leaf length, leaf breadth, petiole length, pedicel length, calyx length and calyx breadth play vital role in the delimitation of taxa within the tribe *Merremieae*. Cluster analysis and dendrogram revealed that, the genus *Operculina* S. Manso is segregated from *Merremia* Denns. *ex* Endlich.; genus *Hewittia* Wight. & Arn. is very close to *Merremia,* probably originated from *M. aegyptia* (L.) Urban and *M. dissecta* (Jacq.) Hall. f. cluster and genus *Xenostegia* Austin & Staples is isolated in the separate cluster.

Introduction

The Convolvulaceae is one of the largest families of angiosperms consist of 52 genera and 1650 species (Mabberley, 2008) and mainly distributed in the tropical and subtropical regions of the world. Some members of the family Convolvulaceae are medicinally as well as economically important (Austin, 1982; Chopra *et al*., 1995; Kamalutheen *et al*., 2009; Sahu and Gupta, 2014).

The tribe *Merremieae* Austin is a diverse tribe in the family Convolvulaceae. Due to many overlapping characters in the members of tribe *Merremieae*, their proper identification is confusing, which may be due to the lack of sound morphological characters to define it and increase in the number of published species in this tribe. The tribe possesses four genera, *viz*., *Hewittia* R. Wight & Arnott, *Merremia* Dennst. *ex* Endl., *Operculina* Silva Manso and *Xenostegia* Austin & Staples. The genus *Hewittia* is represented by a single species, *Merremia* by 70 species, *Operculina* by 15 species and *Xenostegia* by only two species in the world (Mabberley, 2008). The present study reveals that, in India the genus *Hewittia* is represented by a single species, while the genera *Operculina* and *Xenostegia* consist of two species each, and the genus *Merremia* is represented by 11 species.

Numerical taxonomy plays important role in segregation of taxa. Cluster analysis (CA) and Principal component analysis (PCA) techniques are commonly used in determining the phylogenetic relationships among different taxa. Applying various methods of numerical taxonomy many authors, e.g. Sneath and Sokal (1973), Chiapella (2000), Gomez-Campo *et al.* (2001), Sonibare *et al.* (2004), Henderson (2006), Soladoye *et al.* (2010), Rahman *et al.* (2013) and many others have interpreted interrelationship among different genera and families of flowering plants. According to Sonibare *et al.* (2004) cluster analysis provides a hierarchical classification of entities (taxa) based on the similarity matrix, while PCA is a second method used for reducing the dimensions of the original data which allows visual interpretation of the relationships.

[1]Corresponding author. Email: shimpale@yahoo.com

Using macromorphological data and seedling character variations the present study aims to delimit the following 16 species of the tribe *Merremieae* and to evaluate systematic relationships within the tribe.

Materials and Methods

Plant specimens:

Sixteen species of the tribe *Merremieae* are listed in Table 1. Freshly collected as well as herbarium specimens deposited in BSI, BAMU, PBL, SUK and herbarium of the The New College, Kolhapur were used in this study. Fresh specimens were collected from different localities of India during 2005 to 2011. Some morphological characters are based on herbarium specimens (Table 1). The species were identified and authenticated with the help of relevant literature (Hooker, 1885; Cooke, 1958; Austin and Staples, 1980; Johari, 1983; Gamble, 1986; Bhandari, 1995; Biju, 1997; Almeida and Almeida, 2001; Singh *et al.*, 2001; Bhagat *et al.*, 2009). All the specimens are deposited in the herbarium of The New College, Kolhapur (NCK), India.

Table 1. List of species of the tribe *Merremieae* along with voucher specimens.

No.	Species	Specimens examined
1	*Hewittia malabarica* (L.) Suresh	Maharashtra: Ratnagiri, Bhatia beach, 2.11.2011, Shimpale 9835 (NCK)
2	*Merremia aegyptia* (L.) Urban	Maharashtra: Sangali, Miraj, 8.11.2006, Shimpale 3998 (NCK)
3	*M. dissecta* (Jacq.) Hall. f.	Maharashtra: Osmanabad, Kasabe Tadawale, 25.1.2009, Shimpale 4098 (NCK)
4	*M. gangetica* (L.) Cuford.	Maharashtra: Kolhapur, Rankala, 12.12.2012, Shimpale 5620 (NCK)
5	*M. hederacea* (Burm. f.) Hall. f.	Maharashtra: Pune, Baramati, 12.11.2008, Shimpale 2209 (NCK)
6	*M. peltata* (L.) Merr.	Great Nicobar Island: Dwivedi 17024 without date (PBL)
7	*M. quinquefolia* (L.) Hall. f.	Maharashtra: Kolhapur, Sarnobatwadi, 9.2.2010, Shimpale 5629 (NCK); Karnataka Belgaum district, Nipani, 23.3.2010, VBS 63 (SUK)
8	*M. rajsthanensis* Bhandari	Rajsthan: Jodhpur, Sardarsamand, 29.8.1975, Bhandari 1976 (CAL)
9	*M. rhyncorhiza* (Dalz.) Hall. f.	Karnataka: Belgaum, Kankumbi, 17.8.2008, Shimpale 3342 (NCK)
10	*M. tuberosa* (L.) Rendle	Maharashtra: Kolhapur, Tarabai Park, 29.12.2009, Shimpale 2629 (NCK)
11	*M. umbellata* (L.) Hall. f.	Maharashtra: Kolhapur, Dajipur, 22.3.2006, Shimpale 269 (NCK)
12	*M. vitifolia* (Burm.f.) Hall. f.	Maharashtra: Kolhapur, Dajipur, 22.3.2006 Shimpale 276 (NCK)
13	*Operculina tansaensis* Sant. & Patel	Maharashtra: Thane, Thansa lake, 28.11.2010, Shimpale 3229 (NCK)
14	*O. turpethum* (L.) S. Manso	Maharashtra: Ratnagiri, Dapoli, 16.1.2008, Shimpale 998 (NCK)
15	*Xenostegia filiformis* (Thunb.) Almeida	Karnataka: Khanapur, 2.1.2001, Shimpale 794 (NCK)
16	*X. tridentata* (L.) Hall. f.	Maharashtra: Sindhudurg, Malvan, 12.11.2009, Shimpale 754 (NCK); Nanded District, Dharmabad, 12.12.1993 Madhukar 6716 (BAMU)

Morphometric studies:

Morphometric studies were carried out on freshly collected as well as herbarium specimens that are deposited in BSI, BAMU, PBL, SUK and The New College, Kolhapur (NCK). Measurements were taken for 19 selected quantitative characters e.g. leaf length, leaf breadth, petiole length, pedicel length, calyx length, calyx breadth, corolla length, corolla breadth, stamen length, filament length, anther length, ovary length, style length, fruit length, fruit breadth, seed length, seed breadth, apical notch length, paracot length, and recorded on record sheets using as many as numbers of specimens were available for each taxonomic operational unit (OTU). The mean and standard deviation values for all the 19 quantitative characters were calculated and processed for PCA and CA (Kovach, 1999) by keeping data standardized at similarity matrix and tolerance of Eigen analysis set at 1E-010.

Results and Discussion

Nineteen parameters of 16 species of the tribe *Merremieae* from India were examined using numerical methods. The morphological features employed for delimitation of the 16 species with their means and standard deviations are shown in Table 2.

Similarity matrix based on correlation of *Merremieae* species (Table 3) shows that close resemblance of species could be observed when certain characters are employed. It is observed that, there is significant correlation between leaf length and corolla breadth, leaf length and corolla length, leaf breadth and petiole length, leaf breadth and calyx length, petiole length and calyx length, petiole length and corolla breadth, pedicel length and paracot length, calyx length and calyx breadth, calyx breadth and fruit breadth, corolla length and corolla breadth, corolla length and style length, corolla breadth and style length, stamen length and paracot length, filament length and paracot length, style length and fruit breadth, style length and seed length, fruit length and seed length, fruit length and fruit breadth.

Cluster analysis shows that *Xenostegia tridentata* distinctly differs from *Hewittia malabarica* and *Merremia aegyptia* while both the species of the genus *Xenostegia*, distinctly differs from *M. dissecta*, *M. hederacea*, *M. peltata*, *M. quenquifolia*, *M. rajsthanensis*, *M. rhyncorhiza*, *M. tuberosa*, *M. vitifolia*, *Operculina tansaensis* and *O. turpethum* (Table 4). *Merremia tuberosa* distinctly differs with *H. malabarica*, *M. aegyptia*, *M. gangetica*, *M. quenquifolia*, *M. rajsthanensis* and *M. umbellata*. It was also determined that *Operculina turpethum* is closely related with the *O. tansaensis*, *M. vitifolia*, *M. rhyncorhiza*, *M. dissecta* and *M. aegyptia,* while *M. vitifolia* closely relate with *M. umbellata*, *M. rajsthanensis*, *M. quenquifolia*, *M. dissecta* and *H. malabarica*. The relationships among the species of the tribe *Merremieae* is shown in Figure 1.

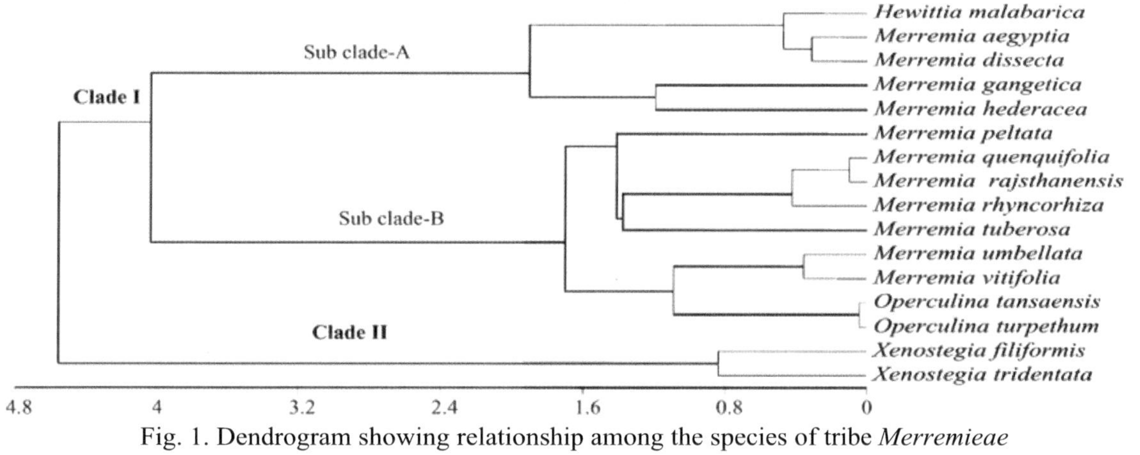

Fig. 1. Dendrogram showing relationship among the species of tribe *Merremieae*

Table 2. Quantitative characters of Tribe *Merremieae* species with Mean and Standard Deviation

Characters	H1	M1	M2	M3	M4	M5	M6	M7	M8	M9	M10	M11	O1	O2	X1	X2
Leaf length	7.8± 0.57	8±1.2	8±2.25	1.3±0.69	5.5±1.3	18.2±6.4	3.8±1.1	3.7±1.15	10±4.77	13.9±5.49	7±2.5	7±2.8	11±3.4	9±3.7	6±2.4	1.2±0.66
Leaf breadth	3.3± 0.91	3.2±0.9	10±3.4	1.8±0.78	4.8±0.98	17.4±5.2	6.5±2.4	6.7±2.2	4.5±2.1	12.9±4.15	3±0.55	9±3.5	8±2.7	8.5±2.9	0.9±0.1	0.35±0.11
Petiole length	2.8± 0.12	6± 2	4.5±1.8	1.8±0.57	4.7±1.25	11±3.4	4.5±1.7	8.9±3.77	9±2.75	11.4±1.9	1±0.2	3±1.15	10±2.9	7±2.5	0.2±0.01	0.2±0.05
Pedicel length	0.8± 0.05	4± 0.75	1.8±0.9	0.2±0.07	0.45±0.21	2±1.1	0.7±0.2	1.1±0.22	0.5±0.1	1.7±0.69	0.9±0.31	0.9±0.21	1.8±0.88	1.7±0.61	7±2.15	0.6±0.2
Calyx length	1.0± 0.65	2±0.18	2.5±0.18	0.5±0.05	5.5±1.3	1.9±0.9	0.7±0.18	1.1±0.18	1±0.3	2.2±0.9	1±0.5	1.4±0.28	1.8±0.75	1.7±0.82	0.9±0.2	0.5±0.11
Calyx breadth	0.6± 0.12	0.6±0.02	1.2±0.88	0.3±0.04	3.2±0.95	1.3±0.88	0.6±0.22	0.7±0.1	0.6±0.25	2.1±0.87	0.8±0.34	0.7±0.08	1.2±0.3	1.3±0.4	0.4±0.03	0.35±0.07
Corolla length	2.5± 0.18	2.6±0.16	3±0.92	0.6±0.045	0.8±0.05	5.5±1.57	1.8±0.99	1.4±0.33	4±0.15	5.5±2.2	2.4±1.75	2.7±1.23	4.4±0.6	4±1	2±0.07	0.8±0.06
Corolla breadth	3.8± 0.2	2.4±0.16	3.2±0.90	0.4±0.035	1.1±0.7	10±4.11	1.8±0.87	2±0.8	4.5±1.27	5.5±1.85	2.5±0.99	4±1.8	5.1±1.1	4.2±0.92	1±0.04	1±0.05
Stamen length	0.8± 0.06	1.9±0.69	1.7±0.69	0.5±0.02	1.1±0.09	2±0.69	1.2±0.72	1.3±0.06	1.2±0.8	1.1±0.3	1.5±0.07	1.5±0.2	2.3±0.4	2.2±0.7	0.9±0.08	0.9±0.07
Filament Length	0.6± 0.15	1.5±0.57	1±0.4	0.3±0.023	0.9±0.08	1.7±0.77	0.9±0.32	1±0.6	0.4±0.16	0.8±0.13	1.3±0.09	0.9±0.1	0.8±0.03	0.7±0.05	0.6±0.07	0.6±0.07
Anther length	0.2± 0.16	0.4±0.01	0.7±0.02	0.2±0.017	0.2±0.01	0.3±0.07	0.3±0.02	0.3±0.04	0.8±0.17	0.3±0.02	0.3±0.03	0.6±0.05	1.6±0.04	1.5±0.07	0.3±0.02	0.3±0.01
Ovary length	1.0± 0.7	0.2±0.01	0.3±0.01	0.1±0.09	0.1±0.01	0.2±0.09	0.2±0.018	0.2±0.03	0.2±0.01	0.2±0.04	0.2±0.011	0.1±0.02	0.2±0.01	0.1±0.01	0.2±0.01	0.2±0.02
Style length	0.9± 0.28	1±0.08	1.4±0.09	0.2±0.08	0.5±0.04	2.3±0.15	1±0.09	1.1±0.6	1.3±0.67	2.8±1.2	1.2±0.05	1.1±0.08	2.1±0.09	1.9±0.2	1±0.02	1.1±0.05
Fruit length	0.9±0.35	1±0.08	0.9±0.08	0.2±0.07	0.6±0.07	1.5±0.95	0.8±0.08	0.6±0.2	1±0.4	3.6±0.85	1.3±0.88	1.4±0.08	0.8±0.03	0.75±0.01	0.6±0.03	0.5±0.02
Fruit breadth	0.6± 0.14	1±0.07	1.2±0.08	0.4±0.04	0.7±0.065	1.7±0.67	1±0.22	0.7±0.15	0.9±0.38	3.2±1.05	0.8±0.44	1.5±0.05	1.7±0.04	1.5±0.52	0.6±0.04	0.3±0.01
Seed length	0.4± 0.1	0.4±0.008	0.6±0.07	0.2±0.015	0.26±0.017	0.6±0.05	0.4±0.1	0.5±0.2	0.5±0.19	1.6±0.3	0.58±0.17	0.66±0.07	0.75±0.04	0.6±0.02	0.28±0.02	0.22±0.01
Seed breadth	0.3± 0.7	0.4±0.011	0.4±0.01	0.2±0.011	0.22±0.01	0.4±0.22	0.3±0.05	0.4±0.06	0.4±0.19	1.2±0.24	0.39±0.06	0.51±0.04	0.73±0.06	0.64±0.02	0.3±0.01	0.2±0.02
Apical notch length	0.7± 0.4	0.64±0.02	0.7±0.02	0.17±0.01	0.36±0.07	0.51±0.18	0.22±0.05	0.18±0.01	0.1±0.04	0.11±0.02	0.29±0.04	0.47±0.05	0.8±0.02	0.9±0.01	0.94±0.04	0.8±0.04
Paracot length	1.3± 0.88	1.7±0.47	1.6±0.51	0.46±0.025	0.7±0.67	1.37±0.6	0.97±0.14	0.9±0.3	0.8±0.22	1.1±0.92	0.88±0.03	1.86±0.03	1.6±0.09	1.7±0.08	1.1±0.52	0.9±0.03

H1 = *Hewittia malabarica*, M1 = *Merremia aegyptia*, M2 = *M. dissecta*, M3 = *M. gangetica*, M4 = *M. hederacea*, M5 = *M. peltata*, M6 = *M. quinquefolia*, M7 = *M. rajsthanensis*, M8 = *M. rhyncorhiza*, M9 = *M. tuberosa*, M10 = *M. umbellata*, M11 = *M. vitifolia*, O1 = *Operculina tansaensis*, O2 = *O. turpethum*, X1 = *Xenostegia filiformis*, X2 = *X. tridentata*

Table 3. Principal component analysis (PCA) in Tribe *Merremieae*

Similarity matrix	Leaf length	Leaf breadth	Petiole length	Pedicel length	Calyx length	Calyx breadth	Corolla length	Corolla breadth	Stamen length	Filament length	Anther length	Ovary length	Style length	Fruit length	Fruit breadth	Seed length	Seed breadth	Apical notch length	Paracot length
Leaf length	1																		
Leaf breadth	0.714	1																	
Petiole length	0.634	0.839	1																
Pedicel length	0.422	0.011	-0.028	1															
Calyx length	0.516	0.517	0.486	0.145	1														
Calyx breadth	0.514	0.6	0.531	-0.036	0.905	1													
Corolla length	0.917	0.692	0.617	0.392	0.237	0.299	1												
Corolla breadth	0.881	0.78	0.685	0.169	0.284	0.355	0.929	1											
Stamen length	0.643	0.614	0.546	0.373	0.426	0.348	0.653	0.66	1										
Filament length	0.467	0.463	0.3	0.358	0.423	0.318	0.333	0.425	0.647	1									
Anther length	0.376	0.347	0.392	0.151	0.143	0.137	0.521	0.437	0.693	-0.097	1								
Ovary length	0.147	-0.089	-0.073	-0.011	-0.155	-0.179	0.113	0.193	-0.228	-0.082	-0.214	1							
Style length	0.741	0.623	0.534	0.372	0.209	0.364	0.879	0.831	0.648	0.37	0.484	-0.007	1						
Fruit length	0.703	0.589	0.435	0.189	0.297	0.435	0.728	0.694	0.284	0.393	-0.006	0.06	0.723	1					
Fruit breadth	0.773	0.826	0.691	0.262	0.441	0.565	0.832	0.773	0.575	0.356	0.409	-0.16	0.82	0.828	1				
Seed length	0.677	0.706	0.589	0.16	0.293	0.473	0.774	0.71	0.408	0.247	0.307	-0.02	0.813	0.889	0.922	1			
Seed breadth	0.354	0.34	0.369	0.158	0.186	0.31	0.478	0.307	0.251	-0.127	0.471	-0.083	0.526	0.567	0.669	0.789	1		
Apical notch length	0.085	-0.219	-0.32	0.551	0.071	-0.097	0.091	0.042	0.342	0.09	0.379	0.216	0.147	-0.265	-0.102	-0.231	-0.103	1	
Paracot length	0.58	0.464	0.287	0.533	0.271	0.11	0.627	0.613	0.741	0.448	0.556	0.146	0.557	0.346	0.539	0.397	0.327	0.611	

Table 4. Cluster Analysis revealing the relationship in Tribe *Merremieae*

Distance matrix	*H. malabarica*	*M. aegyptia*	*M. dissecta*	*M. gangetica*	*M. hederacea*	*M. peltata*	*M. quenquifolia*	*M. rajsthanensis*	*M. rhyncorhiza*	*M. tuberosa*	*M. umbellata*	*M. vitifolia*	*O. tansaensis*	*O. turpethum*	*X. filiformis*	*X. tridentata*
H. malabarica	0															
M. aegyptia	0.472	0														
M. dissecta	0.449	0.314	0													
M. gangetica	1.084	1.689	1.906	0												
M. hederacea	0.848	0.882	0.699	1.194	0											
M. peltata	1.359	1.08	0.601	4.04	1.983	0										
M. quenquifolia	0.323	0.516	0.393	0.739	0.638	1.413	0									
M. rajsthanensis	0.499	0.452	0.402	1.006	0.594	1.28	0.1	0								
M. rhyncorhiza	0.384	0.57	0.441	1.715	1.037	0.817	0.422	0.41	0							
M. tuberosa	1.452	1.185	0.708	3.821	1.836	0.448	1.38	1.273	0.825	0						
M. umbellata	0.237	0.562	0.565	0.986	0.861	1.702	0.373	0.68	0.686	1.658	0					
M. vitifolia	0.326	0.467	0.129	1.586	0.86	0.801	0.268	0.425	0.387	0.825	0.357	0				
O. tansaensis	0.817	0.513	0.243	2.757	1.313	0.423	0.807	0.689	0.361	0.513	1.092	0.426	0			
O. turpethum	0.65	0.43	0.126	2.327	1.079	0.526	0.596	0.553	0.351	0.622	0.84	0.265	0.04	0		
X.filiformis	1.022	0.972	1.568	1.351	1.866	3.351	1.364	1.698	1.991	3.325	0.732	1.519	2.296	1.936	0	
X. tridentata	1.202	1.845	2.256	0.418	1.794	4.556	1.259	1.689	2.251	4.398	0.89	1.876	3.142	2.664	0.837	0

UPGMA dendrogram based on cluster analysis, mean character difference and constrained clustering strategy reveals that, greater affinity exists in between *Operculina tansaensis* and *O. turpethum* (Fig. 1). It is also pointed out that, in dendrogram, 16 species clearly forms distinct two clades i.e. I and II. Clade I is divided into subclade A and subclade B. Subclade A comprises *Hewittia malabarica, M. aegyptia, M. dissecta, M. gangetica* and *M. hederacea,* while subclade B comprises *M. peltata, M. quenquifolia, M. rajsthanensis, M. rhyncorhiza, M. tuberosa*, *M. umbellata, M. vitifolia*, *O. tansaensis* and *O. turpethum*. Clade II comprises *Xenostegia filiformis* and *X. tridentata.* On the basis of dendrogram it is clear that both the *Xenostegia* species distantly differs with the remaining species of tribe *Merremieae*, which supports Austin and Staples (1980) segregation of the genus *Xenostegia* from *Merremia*. The present study is the first report depicting correlation between the members of the tribe *Merremieae* from India and also predicts the phylogenetic relationship within the members of the tribe.

Acknowledgements

Authors are thankful to the Head of the Botany department and Principal, The New College, Kolhapur for laboratory facilities. Thanks are due to the following persons and Institutes for permitting to consult their herbaria. The Director, Botanical Survey of India, Calcutta; Dr. P. G. Diwakar Ex-Deputy Director, BSI, Pune; Prof. S. R. Yadav Head of the Botany department, Shivaji University, Kolhapur; Dr. A. S. Dhabe, BAMU, Aurangabad and Dr. M. Y. Kamble, Scientist, BSI, Port Blair.

References

Almeida, M.R. and Almeida, S.M. 2001. Convolvulaceae *In:* Flora of Maharashtra. St. Xavier College, Mumbai, Vol. **III B**, pp. 303-350.

Austin, D.F. 1982. *Operculina turpethum* (Convolvulaceae) as a medicinal plant in Asia. Economic Botany, **36(3)**: 265-269.

Austin, D.F. and Staples, G.W. 1980. *Xenostegia*, a new genus of Convolvulaceae. Brittonia **32**(4): 533-536.

Bhagat, R.B., Shimpale, V.B. and Deshmukh, R.B. 2009. Convolvulaceae *In:* Flora of Baramati. Bhagat Publications, Pune, pp. 212- 220.

Bhandari, M.M. 1995. Convolvulaceae *In:* Flora of the Indian Desert. MPS Reports, Jodhpur, pp. 216-237.

Biju, S.D. 1997. Taxonomic and morphologic studies in family Convolvulaceae of Southern Peninsular India. Unpublished Ph. D. thesis, University of Calicut, India.

Chiapella, J. 2000. The *Deschampsia cespitosa* complex in central and Northern Europe. A morphological analysis. Bot. J. Linn. Soc. **134**: 495-512.

Chopra, R.N., Nayar, S.L. and Chopra, I.C. 1995. Glossary of Indian Medicinal Plants. Council of Scientific and Industrial Research, New Delhi, p. 330.

Cooke, T. 1958 (Repr. ed.). Convolvulaceae *In:* The Flora of the Presidency of Bombay, Vol. **2**. Botanical Survey of India, Calcutta, pp. 290-331.

Gamble, J.S. 1986. Convolvulaceae *In:* Flora of the Presidency of Madras. Bishen Singh Mahendra Pal Singh, Dehradun, India, pp. 901-930.

Gomez-Campo, C., Herranz-Sanz, J.M. and Montero-Riquelme, F. 2001. The genus *Coincya* Rouy (Cruciferae) in South-Central Spain revisited: A morphometric analysis of population structure. Bot. J. Linn. Soc. **135**: 125-135.

Henderson, F.M. 2006. Morphology and anatomy of palm seedlings. Bot. Rev. **72**: 273-329.

Hooker, J.D. 1885. Flora of British India, Vol. **4**. L. Reeve & Co. Ltd., London, pp. 179-228.

Johari, S.C. 1983. Genus *Ipomoea* L. in India, Ph. D. thesis (Unpublished). University of Rajasthan, Jaipur, India.

Kamalutheen, M., Gopalakrishnan, S. and Syed Ismail, T. 2009. Anti-inflammatory and anti-arthritic activities of *Merremia tridentata* (L.) Hall.f. E- Journal of Chemistry **4**: 943-948.

Kovach, W.L. 1999. MVSP-Multivariate statistical package for windows, version 3.1, Pentraeth Wales, UK. Kovach computing services.

Mabberley, D.J. 2008. Mabberly's Plant Book – a portable dictionary of plants, their classification and uses (Third Edition). Cambridge University Press, Cambridge, UK, pp. 1-1021.

Rahman, M.O., Rahman, M.Z. and Begum, A. 2013. Numerical taxonomy of the genus *Senna* Mill. from Bangladesh. Bangladesh J. Plant Taxon. **20**(1): 77-83.

Sahu, P.K. and Gupta, S. 2014. Medicinal plants of Morning glory: Convolvulaceae Juss. of Central India (Madhya Pradesh and Chhattishgarh). Biolife **2**(2): 463-469.

Singh, N.P., Lakshminarasimhan, P., Karthikeyan, S. and Prasanna, P.V. (Eds). 2001. Convolvulaceae *In:* Flora of Maharashtra State **2**: 454 - 473. Botanical Survey of India, Calcutta.

Sneath, P.H.A. and Sokal, R.R. 1973. Numerical Taxonomy: The Principles and Practice of Numerical Classification (2nd Edition), San Francisco: Freeman, pp. 1-573.

Soladoye, M.O., Sonibare, M.A. and Chukwuma, E.C. 2010. Morphometric study of the genus *Indigofera* Linn. (Leguminosae-Papilionoideae) in South-Western Nigeria. Int. J. Bot. **6**(3): 343-350.

Sonibare, M.A., Jayeola, A.A. and Egunyomi, A. 2004. A morphometric analysis of the genus *Ficus* Linn. (Moraceae). Afr. J. Biotechnol. **3**: 229-235.

15

MORPHOMETRIC ANALYSIS OF THE GENUS *MONOCHORIA* (PONTEDERIACEAE) IN SRI LANKA

SAVITHRI SAROJANI UDAGE AND DEEPTHI YAKANDAWALA[1]

Department of Botany, University of Peradeniya, Sri Lanka

Keywords: Monochoria vaginalis; Morphological data; Sub-palisade cavities; Principal coordinate anaysis.

Abstract

The genus *Monochoria* is represented in Sri Lanka by only two species, *M. hastata* and *M. vaginalis*. Of the two species *M. vaginalis* exhibits a high morphological diversity. A morphometric analysis was performed on the Sri Lankan members of the genus *Monochoria* to evaluate the morphological diversity exhibited by comparing 34 vegetative, anatomical and reproductive characters. Both cluster and principal coordinate analyses resulted in four clusters of which one corresponded to *M. hastata,* while *M. vaginalis* was divided into three phenetic groups indicating that the Sri Lankan *M. vaginalis* is a species complex involving more than one taxon. The length ratio of the lower spathe petiole to inflorescence stalk was identified as an informative character in delimiting the phenetic groups, a character that has not been considered before. Further, the occurrence of sub-palisade cavities that are filled with a red colored liquid in the leaves of two phenetic groups was a novel character for the genus as well as the family Pontederiaceae.

Introduction

The genus *Monochoria* C. Presl belongs to the family Pontederiaceae is distributed widely in the tropical and warm temperate regions of the Old world, Asia and Australia. Sri Lanka harbors two native members, *Monochoria vaginalis* (Burm. f.) C. Presl *ex* Kunth and *M. hastata* (L.) Solms (Dassanayake, 2000). The plants often inhabit wetlands and ditches, while *M. vaginalis* is more often found in paddy fields where it is considered a weed.

The taxonomy of the genus has been a point of debate and addressed by many taxonomists. Verdcourt (1961) in his taxonomic treatment to African *Monochoria,* recognized *M. vaginalis* as a species and raised the varieties then recognized under *M. vaginalis* to species rank, *M. africana* N. E. Br. and *M. brevipetiolata* Verdc. Both are endemic to Africa. During the worldwide revision of the genus, Cook (1989) recognized eight species under the genus, *M. korsakowii* Regel & Maack, *M. vaginalis* (Burm. f.) C. Presl *ex* Kunth, *M. hastata* (L.) Solms, *M. elata* Ridl., *M. africana* N.E.Br., *M. brevipetiolata* Verdc., *M. cyanea* (F. Muell.) F. Muell. and *M. australasica* Ridl. A comprehensive taxonomic treatment on Asian *Monochoria* by Wang *et al.* (2004) recognizes five species, *M. korsakowii* Regel & Maack, *M. hastata* (L.) Solms, *M. elata* Ridl., *M. valida* G. X. Wang & Nagam. and *M. vaginalis* (Burm.f.) Kunth. They also recognized two varieties of *M. vaginalis*, viz. *M. vaginalis* var. *angustifolia* G. X. Wang and *M. vaginalis* var. *vaginalis* (Wang *et al.*, 2004). The study however did not include any specimens from Sri Lanka.

According to Verdcourt (1961), the Asian, East African and West African populations of *M. vaginalis* belong to three distinct taxa. Wang *et al.* (2004) is of the opinion that the greater variability observed in the vegetative morphology of *Monochoria* due to adaptation to different

[1]Corresponding author. Email: deepthiy@pdn.ac.lk

habitats may have led the early taxonomists to recognize 'species' and 'varieties'. A more recent study on Thai *M. vaginalis* also corroborates that the species exhibit a great variation in their morphology (Tungmunnithum, 2015). The studies using morphological and molecular data have indicated that *M. vaginalis* in Thailand consists of more than one taxon. The most recent taxonomic treatment on Sri Lankan *Monochoria* in the Revised Handbook to the Flora of Ceylon (Dassanayake, 2000) recognizes only two species: *M. vaginalis* and *M. hastata*. However, field studies recently carried out in Sri Lanka encountered *M. vaginalis* populations with distinct morphological characters and the results of a preliminary morphometric analysis by Udage and Yakandawala (2011) supported these observations. Therefore, the present study was carried out with the aim of re-evaluating the patterns of morphological variation within *M. vaginalis* and to understand the species limits of the genus *Monochoria* in Sri Lanka.

Materials and Methods

Sample collection:

Seventy-four live plant samples of *Monochoria* bearing flowers/fruits were collected, from populations of all possible locations in Sri Lanka, covering the three major climatic zones, wet, dry and intermediate zones. The specimens were then studied in detail. A reference collection of living plants was also maintained at the Department of Botany, University of Peradeniya, Sri Lanka during the study period (2010-2015).

Data collection:

The morphological features of the samples were studied in detail and recorded along with the characters traditionally used to identify and distinguish taxa within the genus. The morphological characters were examined and recorded from at least three specimens of a population. A minimum of five measurements were taken from an individual specimen for a particular character, and the mean measurement was taken as the particular character value. Macroscopic parts were observed under a dissecting microscope and a stereomicroscope (LEICA L2). The leaf anatomy was studied either with free hand sections or microtome sections. The microtome sections were stained in safranin. All character measurements were taken using a ruler (smallest measurement 1 mm) or an eye piece graticule (smallest measurement 0.1 mm) where appropriate. Following a preliminary examination of 88 characters, 34 characters (23 - quantitative and 11 - qualitative) were scored for the analysis based on the significance of variation (Table 1). The data were entered into an Excel spreadsheet (Microsoft Excel version 2007) and was later transformed into a file format suitable for morphometric analysis.

Statistical analysis:

The Principal Coordinate Analysis (PCoA) and the Hierarchical Cluster Analysis (CA) were carried out using the statistical software PAST (Version 2.15) (Hammer *et al.*, 2001). The cluster solution was selected from the best suitable algorithm where, Gower Distance was used to calculate the similarity measures with the 'paired group' (UPGMA) option and the Single Linkage Algorithm with the highest Cophenetic Correlation Value. The ordination analysis was performed with Gower Distance (transformation exponent C=2) to generate a distance matrix for the use in the PCoA. Since the PCoA gives the distance between the OTU's rather than the correlation between characters, this is suitable for a mixed character data, as it will not be distorted by binary characters as with Principal Components Analysis (PCA) (Cupido, 2003). From the results of these analyses, each major, consistently recovered clusters were identified.

Results and Discussion

The UPGMA dendrogram (cophenetic correlation coefficient = 0.8587) resolved four discrete clusters of OTUs (hereafter referred to as Phenetic Groups 1-4), which separated respectively at approximately 0.425, 0.35, and 0.225 distance units (Fig.1). The first four (principal) eigenvalues recovered from the PCoA (2.4495, 0.4264, 0.2390 and 0.1286) accounted for 74.456% of the total

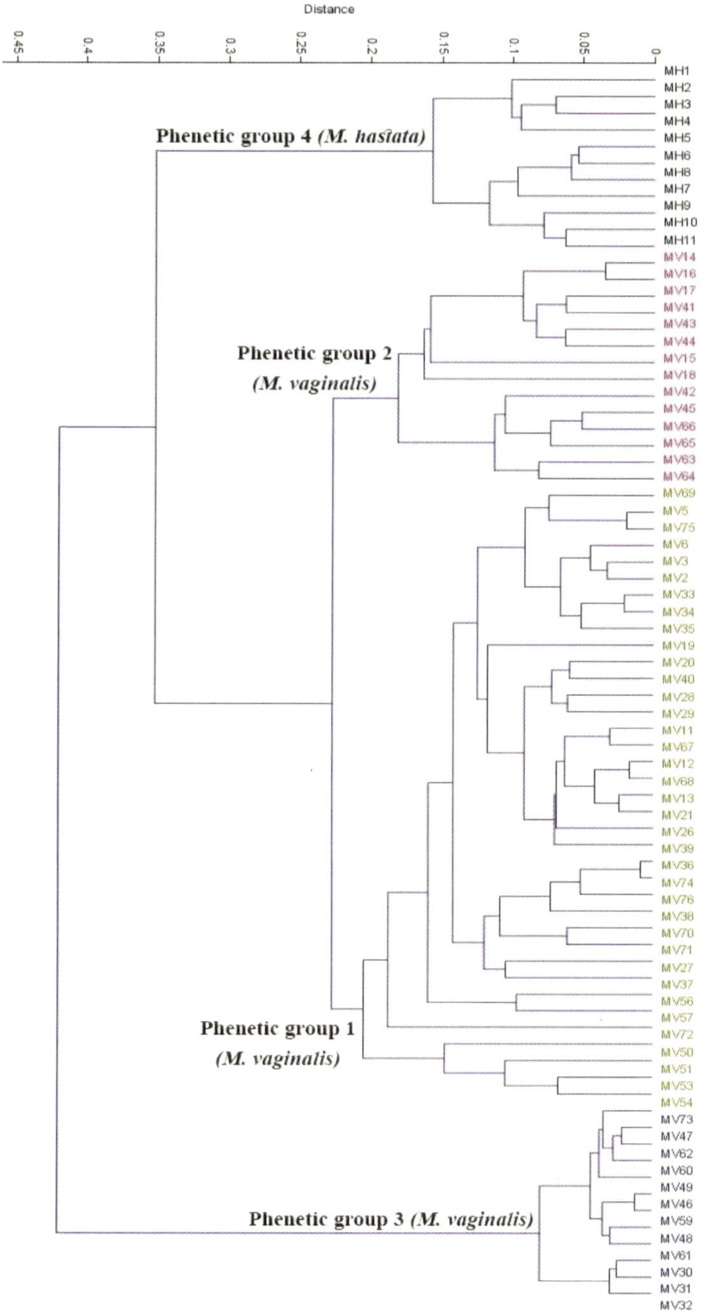

Fig. 1. Dendrogram resulted from Cluster analysis of OTUs of *Monochoria* in the morphometric analysis. The OTUs corresponding to the three clusters of *M. vaginalis* are marked as *M. vaginalis* Phenetic Groups 1, 2 and 3.

variance (56%, 9.79%, 5.49% and 2.954% respectively). A plot of the first and second coordinates (which provided the greatest separation of OTUs) resulted in a separation similar to that obtained by the CA. Here the PCoA also resolved four discrete clusters (Fig.2) with each corresponding exactly to one of the clusters indicated by the UPGMA dendrogram (Fig.1). However, few members of the Phenetic Groups 1 and 2 overlapped along both first and second coordinate axes (Fig.2).

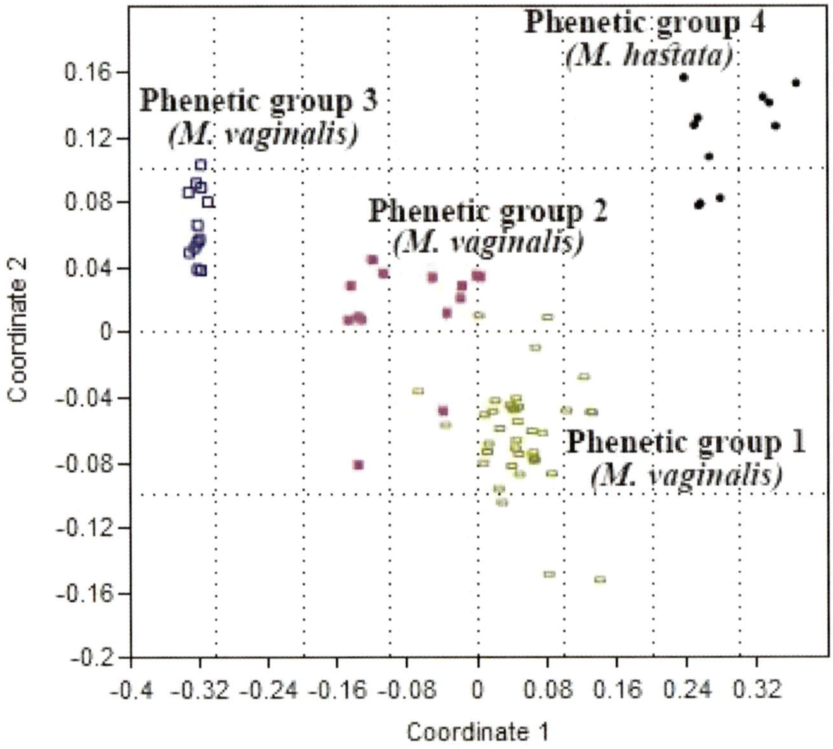

Fig. 2. Scatter plot resulted from PCoA showing the four distinct phenetic groups of *Monochoria* OTUs generated from the evaluated study material.

The Cluster Analysis and the PCoA recognized *M. hastata* as a distinct group while dividing the OTU's of *M. vaginalis* into three distinct clusters, indicating the involvement of more than one phenetic group within *M. vaginalis* in Sri Lanka. The significant differences in leaf lobe length and inner and outer tepal widths separate the three phenetic groups of *M. vaginalis* from each other. The variations in leaf width and the position where the inflorescence arose (length ratio of the lower spathe petiole: inflorescence stalk) clearly separated the Phenetic Group 1 from the other two groups. The differences in flower diameter, inner and outer tepal lengths and inner and outer tepal widths contributed to the separation of the Phenetic Group 3 from the other two groups. The ranges of other characters overlapped among the three phenetic groups (Table 2).

The Phenetic or multivariate methods in taxonomic studies provide insights on how to interpret phenetic results to make valid taxonomic decisions. In the absence of a universally acceptable species concept, it is up to the individual taxonomist to define species level taxa (Davis and Goldman, 1993; Cupido, 2003). The phenetic species concept of Sneath and Sokal (1973) is an empirical method that considers distinct phenetic clusters produced by overall similarity as species, without making assumptions about speciation. Clusters of overall similarity are formed between objects as a function of their individual similarities in each of the many characters in which they are being compared. These phenetic clusters possess a group of partially correlated characters rather than a fixed character and therefore pattern related. The pattern related species concepts are feasible for practicing taxonomists because they use criteria that are based on observed patterns of character variation in delineating species (Cupido, 2003).

Table 1. List of characters used in the multivariate analysis.

	Quantitative characters		Qualitative characters
1	Plant height up to leaf base	1	Leaf shape
2	Leaf width	2	Leaf base
3	Leaf length - lamina	3	Lamina texture
4	Leaf length - lobe	4	Presence of sub palisade cavities filled with a red liquid filled with a
5	Peduncle length	5	Arrangement of flowers in the inflorescence
6	Peduncle width	6	Bract shape
7	Length ratio of the lower spathe petiole: inflorescence stalk	7	Bract texture
8	Number of flowers per inflorescence	8	Flower color
9	Bract width	9	Persistent perianth surrounding the capsule
10	Bract length	10	Persistent tepal texture in capsule
11	Pedicel length	11	Capsule shape
12	Pedicel width		
13	Flower diameter		
14	Outer tepal length		
15	Outer tepal width		
16	Inner tepal length		
17	Inner tepal width		
18	Style length		
19	Ovary length		
20	Anther length of large stamen		
21	Filament length of large stamen		
22	Anther length of small stamen		
23	Filament length of small stamen		

The results of the morphometric analyses, CA and PCoA, further suggest that the Sri Lankan *M. vaginalis sens. lat.* may be a species complex with three different phenetic groups. These were recognized from the analyses as *M. vaginalis* Phenetic Group 1, *M. vaginalis* Phenetic Group 2 and *M. vaginalis* Phenetic Group 3 supporting the field observations made during the past years. The comparison of morphological features with those in the published literature, tentatively identified the Phenetic Group 1 as *M. vaginalis* but would still need confirmation by scoring type specimens of *M. vaginalis* and conducting phylogenetic analyses. None of the phenetic groups exhibited features of *M. vaginalis* var. *angustifolia* recognized by Wang *et al.* (2004) on the basis of having narrow lanceolate mature leaves, $3\text{-}7 \times 0.3\text{-}2.0$ cm with the basal lobes up to 0.2 cm long, apex acute or acuminate and 3-7 flowered raceme.

Although the vegetative characters of *Monochoria* are highly variable, even within a single species, many of them have proven to be of taxonomic significance. Wang *et al.* (2004) discuss the taxonomic utility of some of these morphological features in detail. The leaf characters such as length, width and the basal lobe length were taxonomically useful characters in the study of Asian *Monochoria* (Wang *et al.,* 2004). Similarly, they were useful in the present study in delimiting the already recognized *M. hastata* and the putative phenetic groups within *M. vaginalis*. The Phenetic

Group 1 which is tentatively identified as *M. vaginalis*, exhibited the shortest basal leaf lobe with a shallow sinus compared to the other two phenetic groups.

Table 2. Comparison of distinct morphological characters among the three phenetic groups of *M. vaginalis* and *M. hastata.*

Character	*M. vaginalis* phenotype 1	*M. vaginalis* phenotype 2	*M. vaginalis* phenotype 3	*M. hastate*
Leaf length (cm) (Excluding the lobes)	3.8-9.3	5.6-14.9	5.1-7.6	5.9-12.9
Leaf lobe length (cm)	0.2- 0.8	1.8- 2.6	0.9- 1.7	2.2-3.5
Leaf width (cm) (at the broadest point)	2.4-6.6	5.4-7.5	5.1-8.1	4.3-6.7
Leaf shape	Ovate - lance ovate	Widely ovate	Widely ovate	Ovate - lance ovate
Leaf base	Cordate with a shallow sinus	Cordate with a deep sinus	Cordate with a large sinus	Sagittate
Lamina texture	Smooth	Relatively thick	Rough upper surface	Smooth
Sub palisade cavities filled with a red liquid	Absent	Occasionally present	Present	Absent
The length ratio of the lower spathe petiole: inflorescence stalk	(1:2)-(1:3)	1:1	1:1	1:4
Number of flowers per inflorescence	5-11	15-19	10-19	12-51
Flower arrangement	Lax	Dense	Dense	Dense (helically arranged)
Peduncle length (cm)	1.5- 3.0	1.0- 1.4	1.0- 1.4	2.0-2.5
Flower diameter (cm)	1.6- 2.7	1.6- 2.7	0.8- 1.5	1.9-2.9
Pedicel length (cm)	0.5 - 2.0	0.5 - 2.0	0.1- 0.4	1.2-3.5
Arrangement of the persistent perianth around the capsule	Irregularly twisted around ripe fruit	Irregularly twisted around ripe fruit	Not twisted, but partially covers the capsule	Twisted and screwed on the ripe fruit in a regular manner
Capsule shape	Widely elliptic	Widely elliptic	Ovate	Widely elliptic
seed shape	Oblong	Widely elliptic	Widely elliptic	Oblong

In addition, the sub-palisade cavities that are filled with a red coloured liquid were observed in the leaves of the Phenetic Groups 3 and 2 (to a lesser extent in Group 2). This character is a novel character for the genus *Monochoria* and the family Pontederiaceae. The leaf surface of the phenetic group 3 was rough or blister like; the blisters are visible to the naked eye as well as under the stereomicroscope (Fig. 4). Therefore, the presence of sub-palisade cavities filled with a red coloured liquid seems to be a taxonomically important character in delimiting the entities within *M. vaginalis sens. lat.* Although the sub-palisade cavities were recorded in our study, a recent study on comparative anatomical characteristics of *M. vaginalis* and *M. hastata* conducted in India (Narayanan and Kaliappan, 2014) did not record any cavities/glands in the sub-palisade layer. The populations of the Phenetic Group 3 were mostly distributed in the dry zone of Sri Lanka but several co-occurring populations of Phenetic Groups 1 and 3 were also encountered. The populations of the Phenetic Group 2 were relatively restricted in their distribution.

The floral features including inflorescence characters were also of taxonomic importance in the genus *Monochoria*. *Monochoria* occurring in Asia possess sub-umbellate inflorescences, racemes or branched panicles (Cook, 1989; Wang *et al*., 2004). Of the two species occurring in Sri Lanka, *M. hastata* possesses a sub-umbellate inflorescence, while *M. vaginalis* possesses a raceme type inflorescence. According to Wang *et al.* (2004), the flowering sequence of *M. hastata* is definite while it is indefinite blooming in *M. vaginalis.* However, according to the field observations, inflorescences of all specimens encountered under *M. vaginalis* in Sri Lanka show a definite sequence in the blooming pattern. This observation corroborates with few other literature. Although Trimen (1898) identified the inflorescence type of *M. vaginalis* as a sub-sessile raceme, he states that the terminal flower is the first to open. Further, Soerjani *et al.* (1987) also state that the inflorescence is a raceme and flowers open either simultaneously or from top to bottom within a few days. Most literature does not mention the flower opening sequence of *Monochoria*. The inflorescence of *M. vaginalis* Phenetic Group 1 was lax, whereas it was dense in the other two phenetic groups. The flower diameter was larger (1.6- 2.7 cm) in the *M. vaginalis* Phenetic Groups 1 and 2, while that of the Phenetic Group 3 was relatively smaller (0.8- 1.5 cm).

Fig. 3. Comparison of the morphological characters among the three phenetic groups: leaves, proportion of the lower spathe petiole: inflorescence stalk, inflorescence, fruits and seeds of Phenetic Group 1 (A, D, G, J, M) Phenetic Group 2 (B, E, H, K, N) and Phenetic Group 3 (C, F, I, L, O) respectively.

During the present study, the position from where the inflorescence arose was taxonomically informative i.e. the length ratio of the lower spathe petiole: inflorescence stalk. This character was not taken into account in previous taxonomic treatments. The inflorescence of *Monochoria* is terminal where it is subtended by two spathes, upper and lower. In both *M. vaginalis* and *M. hastata*, the upper spathe is reduced. In *M. vaginalis* (Phenetic Group 1) this proportion was between 1:2 - 1:3 where the inflorescence arose closer to the lower spathe. In the other two Phenetic groups (2 and 3) it was 1:1 or even more towards the base of the plant. Further, during this study we did not encounter specimens with the petiole length of the lower spathe was as short as 0.5 cm as recorded by Cook (1989).

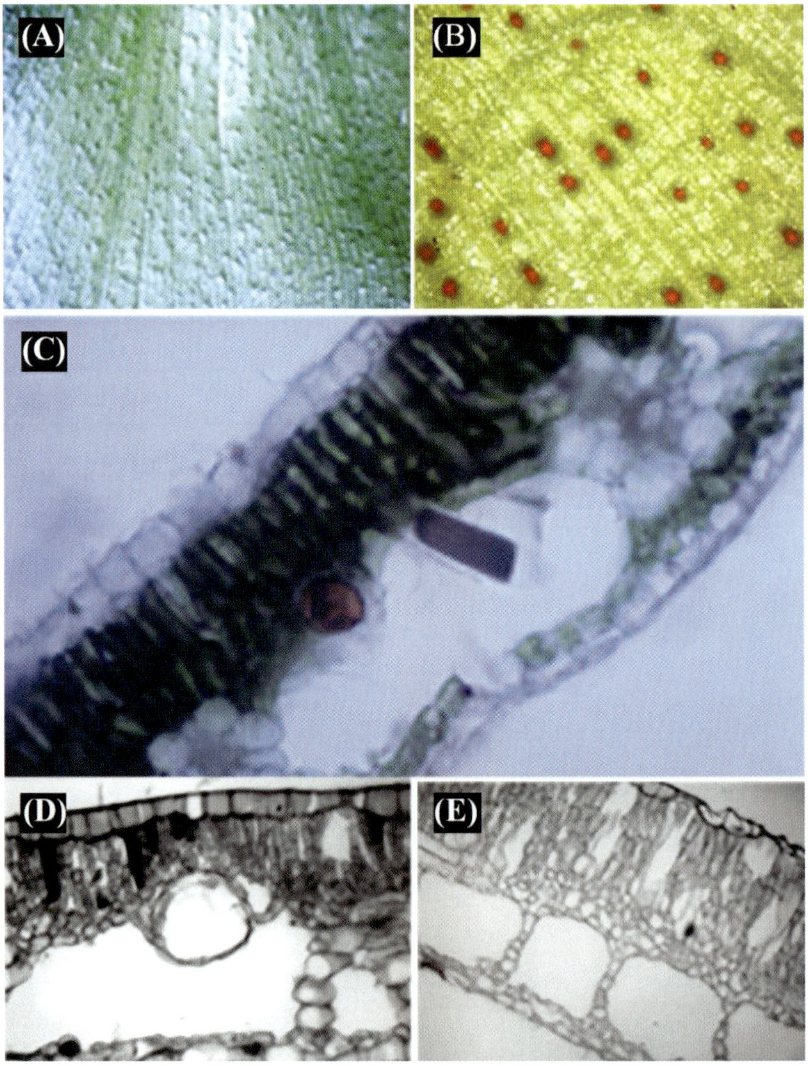

Fig. 4. Sub-epidermal glands in the Phenetic group 3. (A): Leaf upper surface, the glands are visible as blisters; (B): Leaf upper surface under a stereomicroscope, the glands are visible as red dots; (C): Leaf section under the light microscope (x10x10); (D): Leaf sections (Microtome) under the light microscope (x10) and (E): Leaf sections (Microtome) of Phenetic group 1 with no glands under the light microscope (x10).

Fruit and seed characters were also of taxonomic importance in the genus *Monochoria*. The capsule of *Monochoria* is usually covered by a persistent perianth. The persistent tepals/perianth of *M. hastata* is screwed, covering the fruit, while in *M. vaginalis* Phenetic Group 1 and Phenetic Group 2, the perianth parts are arranged spirally but not neatly screwed. In Phenetic Group 3, the perianth parts do not totally cover the fruit. Even though not included in the analysis, the examination of seeds of a larger subset of individuals indicates that the seed shapes are also taxonomically useful in recognition of the three phenetic groups within *M. vaginalis* in Sri Lanka (Fig. 3). The three *M. vaginalis* phenetic groups showed variation in the seed size and shape; seeds of the Phenetic Group 1 were the largest (0.834-0.869 x 0.594-0.617 mm) while Group 3 seeds were the smallest (0.393-0.399 x 0.280-0.301 mm). The chalazal end of *M. vaginalis* Phenetic Group 1 was depressed while that in the other two groups was obtuse.

The results of the multivariate analyses of morphological data suggest that the Sri Lankan *M. vaginalis sens. lat.* is a species complex that includes two distinct phenetic groups as well as the true *M. vaginalis*. These phenetic groups have also been supported by molecular studies using RAPD data (Jayewardene *et al.* (2013 and Gunaratne *et al.* (2014) where the polymorphic banding patterns were different among the three *M. vaginalis* phenetic groups. Moreover, a phytochemical investigation by Ileperuma *et al.* (2014) reported different chemical profiles for the three phenetic groups of *M. vaginalis*. The Phenetic Group 3 with the sub-palisade cavities expressed a chemical profile with the highest brine shrimp lethality/toxicity and haemolytic activity which could possibly be abbreviated to the chemical substances deposited. Further, the results of the present study were similar to the results of the recent study conducted in Thailand (Tungmunnithum, 2015), in which a taxonomic revision was carried out to evaluate the species limits of *M. vaginalis* and the results indicated the involvement of more than one taxon within the species.

Resolving the species circumscriptions is important in terms of evaluating the biodiversity and its conservation. It is feared that many species will become extinct even before they are described due to clearing and destruction happening around the world. In order to conserve the species and their diversity, scientists should be able to correctly identify the species, locate their populations and estimate the remaining extents of those populations. To support the correct taxonomic identification, species complexes such as *Monochoria vaginalis sens. lat.* should be studied in detail and taxonomic species boundaries should be defined. The present study contributes to understanding the diversity and the putative number of entities within *Monochoria vaginalis sens. lat. M. vaginalis* is an important medicinal plant and the identification of the correct species with the expected medicinal value is a requisite. Further studies of Sri Lankan *M. vaginalis* using molecular data are currently underway and expectantly will shed additional merit in the identification of the *M. vaginalis* and the description of species corresponding with the phenetic groups. In addition, once the species limits are resolved, revised taxonomic keys will be developed to correctly identify the species.

References

Cook, C.D.K. 1989. Taxonomic revision of *Monochoria* (Pontederiaceae). *In*: Tan, K (Ed). The Davis and Hedge Festschrift: Plant taxonomy, phytogeography and related subjects. Edinburgh, University Press. pp. 149–184.

Cupido, C.N. 2003. Systematic studies in the genus *Merciera* (Campanulaceae): A re-assessment of species boundaries. Adansonia **25**(1): 33–44.

Dassanayake, M. D. 2000. Pontederiaceae. *In:* Dassanayake, M.D. and Clayton W.D. (Eds.), *A Revised Handbook to the Flora of Ceylon* 14. Oxford & IBH Publishing Co. Pvt. Ltd., New Delhi, pp. 259–264.

Davis, J.I. and Goldman, D.H. 1993. Isozyme variation and species delimitation among diploid populations of the *Puccinellia nuttalliana* complex (Poaceae): character fixation and the discovery of phylogenetic species. Taxon **42**: 585–599.

Gunaratne, U.S., Samaraweera, P. and Yakandawala, D.M.D. 2014. Genetic diversity among populations of *Monochoria vaginalis* detected by random amplified polymorphic DNA (RAPD) validates three phenetic groups found in Sri Lanka. Proceedings of The Peradeniya University International Research Sessions, iPURSE – 2014.Peradeniya, Sri Lanka **18.** p. 610 (Abs.)

Hammer, Ø., Harper, D.A.T. and Ryan, P.D. 2001. PAST: Paleontological statistics software package for education and data analysis. *Palaeontol. Electronica* **4**: 1–9.

Ileperuma, C.V.K., Jayasinghe, U.L.B., Yakandawala, D.M.D. and Kumar, N.S. 2014. A re-assessment of species boundaries of the genus *Monochoria* (Pontederiaceae) in Sri Lanka using phytochemical data. Proceedings of the Peradeniya University International Research Sessions, iPURSE – 2014, Peradeniya, Sri Lanka **18**, p 589 (Abs.)

Jayawardhana, B.J.G., Samaraweera P. and Yakandawala, D.M.D. 2013. Genetic Diversity of *Monochoria vaginalis* in Sri Lanka by using Random Amplified Polymorphic DNA (RAPD) Markers. *Proceedings and abstracts of the Peradeniya University Research Sessions*, PURSE – 2012, *Peradeniya, Sri Lanka* **17**(1), P 247 (Abs.)

Narayanan, K. B. and Kaliappan, I. 2014. Comparative Anatomical Characteristics of Emergent Aquatic Herbs- Monochoria vaginalis (Burm. F.) Presl. and Monochoria hastata Solms. (Pontederiaceae). International Journal of Botany, 10: 13-23. DOI: 10.3923/ijb.2014.13.23.

Sneath, P.H.A. and Sokal, R.R. 1973. Numerical taxonomy: the principles and practice of numerical classification. W.H. Freeman & Co., San Francisco.

Soerjani, M., Kostermans, A.J.G.H. and Tjitrosoepomo, G. 1987. Weeds of rice in Indonesia. Balai Pustaka, Jakarta, Indonesia. pp. 484 – 489.

Trimen, H. 1898. A Handbook to the Flora of Ceylon Part 4, Dulau & Co. London, 295 -296.

Tungmunnithum, Duangjai 2015. Morphological and Molecular Evidences of the Greatest Variable *Monochoria* Species in Thailand. BOTANY 2015, July 25 - 29, 2015, The Shaw Conference Centre Edmonton, Alberta, Canada.

Udage, S. and Yakandawala, D. 2011. Morphological variation and species boundaries of the genus *Monochoria* (Pontederiaceae) in Sri Lanka. Proceedings of the Peradeniya University Research Sessions, PURSE – 2011, Peradeniya, Sri Lanka **16** (1): 189.

Verdcourt, B. 1961. The genus *Monochoria* Presl (Pontederiaceae) in Africa. *Kirkia* **1**: 80-83.

Wang, G-X., Wei, L, Xiao-Chun, W. and Itoh, K. 2004. Taxonomy of the genus *Monochoria* (Pontederiaceae) in Asia. Current Topics in Plant Biology **5**: 39–52.

16

POLLEN MORPHOLOGY OF THE TRIBE PHYLLODOCEAE (ERICOIDEAE, ERICACEAE) AND ITS TAXONOMIC SIGNIFICANCE[1]

A. K. M. GOLAM SARWAR[2] AND HIDEKI TAKAHASHI

Laboratory of Systematic Botany, Graduate School of Agriculture, Hokkaido University, Japan

Keywords: Pollen morphology; Exine sculpture; Phyllodoceae; Taxonomic significance.

Abstract

Pollen morphology of 13 taxa belonging to 5 genera of the tribe Phyllodoceae (Ericaceae) was examined by means of light and scanning electron microscopy (LM and SEM, respectively), or SEM alone. In Phyllodoceae, 3-colpor(oid)ate, minute to medium, oblate pollen grains are united usually in tetrahedral tetrads. Pollen tetrads are generally characterized by the presence of viscin threads except *Elliottia pyroliflora*, *Epigaea asiatica* and *Phyllodoce caerulea*. The absence of viscin threads might indicate to a secondary loss, since these are present at least in some species within all the genera of Phyllodoceae. The pollen morphological data confirm the infra- and inter-generic relationships as identified by molecular phylogeny of Phyllodoceae and/or vice-versa. Although various palynological characters were found to be taxonomically important at different taxonomic levels, the apocolpial exine sculpture is emerged as one of the most important palynological features of systematic importance. The rugulate apocolpial exine with striate secondary sculpture and a reduced colpus might be apomorphic palynological character states for this tribe and Ericaceae as well. Hitherto, it is the first SEM study of *Rhodothamnus* pollen.

Introduction

The Phyllodoceae Drude is one of the most morphologically heterogeneous group that apparently has no morphological synapomorphy (Kron, 1997). Moreover, the inclusion of *Epigaea* in the tribe Phyllodoceae might increase the morphological heterogeneity of this group and the monophyly of this group is not well supported by morphology. Nearly all descriptions of the tribe have diagnosed the group based upon suites of characters that are homoplasious within the broader Ericoideae, rather than recognizing any particular potentially synapomorphic character (Kron *et al.*, 2002). Gillespie and Kron (2010) used molecular data to clarify tribe-level relationships within the Ericoideae and to propose a new classification that includes five tribes, namely Rhodoreae, Empetreae, Ericeae, Bryantheae and Phyllodoceae; the Phyllodoceae is sister to a clade comprised of the other four tribes. Most members of Phyllodoceae are native to temperate-boreal regions of the Northern Hemisphere. The flowers are variable in shape, size and colour. Some species have pollen dispersed by curved stamens that spring out from the flower centre, explosively. Many species in this group are also poisonous, e.g., *Kalmia*, both to livestock and people. The molecular analyses indicated two strongly supported clades: *Kalmia s.l.* (including *Leiophyllum* and *Loiseleuria*) and a *Phyllodoce* clade (including *Epigaea*, *Kalmiopsis* and *Rhodothamnus*), and *Elliottia* is sister to *Kalmia* + *Phyllodoce* clade (Kron *et al.*, 2002; Gillespie and Kron, 2013). In addition, at least one taxon from all currently recognized tribes within the Ericoideae have at some point been classified within the Phyllodoceae, illustrating the difficulty in determining relationships of these taxa based on morphological or anatomical evidence alone (Gillespie and Kron, 2013).

---

[1]A part of Ph.D. Thesis

[2]Present address: Department of Crop Botany, Bangladesh Agricultural University, Mymensingh 2202, Bangladesh. E-mail: drsarwar@bau.edu.bd

Pollen of the family Ericaceae is very diverse – monads vs. tetrads, or as polyads of indefinite number of tetrads; and its phylogenetic importance have been well documented (Kron *et al.*, 2002; Sarwar, 2007). Among the genera of Phyllodoceae, the pollen morphology of *Kalmia* has already been reported (Sarwar and Takahashi, 2012), but fragmentary palynological information is only available for other genera of this tribe (Sladkov, 1954; Oldfield, 1959; Stevens, 1971; Bohm *et al.*, 1978; Comtois and Larouche, 1981; Clements, 1995; Kron *et al.*, 2002; Radcliffe *et al.*, 2010). Light microscopy (LM) was mainly employed in these studies; and scanning electron microscopy (SEM) was employed in a few cases. The aims of the present study are to provide the palynological data of the genera of Phyllodoceae in addition to *Kalmia*, by using both LM and SEM, and to discuss its systematic significance in light of the recent molecular phylogenetic relationship of this tribe.

Materials and Methods

Pollen morphology of 13 taxa belonging to 5 (out of 7) genera of the tribe Phyllodoceae, was examined by means of light microscopy (LM) and scanning electron microscopy (SEM), or SEM alone (Table 1). Pollen morphology of the genus *Kalmia* has been studied in details (Sarwar and Takahashi, 2012) and the data has been included in discussion. The remaining genus *Kalmiopsis* is endemic to Oregon State, USA and could not be included due to unavailability of pollen samples. Polliniferous materials used in this investigation were taken from the dried specimens of the herbaria GB, S and SAPS. Abbreviations of names of the herbaria are according to the Index Herbariorum (Holmgren *et al.*, 1990).

Table 1. List of Phyllodoceae taxa used in this study along with their voucher specimens.

No.	Name of the taxa	Voucher specimens
1.	*Bejaria aestuans* Mutis *ex* L.	Venezuela: Edo, Tachira Hwy. 9, 37 km south of Delicias, 28.01.1978, J.L. Luteyn *et al.* 5296 (S)
2.	*B. racemosa* Vent.	USA: Florida, Dade Co., west of Fulford, 10.02.1930, H.N. Moldenke 601 (S)
3.	*B. resinosa* Mutis *ex* L. f.	Colombia: District Cauca, El Tambo, 08.11.1936, K. von Sneidern 1069 (S)
4.	*B. subsessilis* Benth.	Ecuador: Prov. Loja-Zamora road, 17.06.1979, B. Løjtnant & U. Molau 15010 (GB)
5.	*Elliottia bracteata* (Maxim.) Benth. *et* Hook. f.	Japan: Hokkaido, Prov. Teshio, Masike-gun, Masike-cho, Mt. Shokanbetsu, 28.07.1983, H. Takahashi 4500 (SAPS)
6.	*E. paniculata* (Sieb. *et* Zucc.) Benth. *et* Hook. f.	Japan: Hokkaido, Hiyama-sicho, Kudoo-gun, Taisei-cho, Mt. Ohta-san, 06.08.1987, H. Takahashi 7802 (SAPS)
7.	*E. pyroliflora* (Bong.) Brim & Stevens	USA: Alaska, Juneau Quadrangle, Mt. Roberts behind, August 1967, L.A. Viereck 8624 (S)
8.	*Epigaea asiatica* Maxim.	Japan: Hokkaido, Hiyama-shicho, Kaminokuni-cho, 18.04.1982, M. Hara 5212 (SAPS)
9.	*E. repens* L.	USA: New York, Tomplins Co., slopes on east side of valley of Cayuga Intel, 13.04.1935, R.I. Clausen 19207 (S)
10.	*Phyllodoce aleutica* (Spreng.) A. Heller	Japan: Hokkaido, Jyozankei, Mt. Yoici-dake, 02.09.1982, H. Takahashi *et al.* 3666 (SAPS)
11.	*P. caerulea* (L.) Bab.	Japan: Hokkaido, the Hidaka range, Mt. Poroshiri – Mt. Tottabetsu, 01.08.1983, H. Takahashi 4569 (SAPS)
12.	*P. nipponica* Makino var. *oblong-ovata* (Tatew.) Toyokuni	Japan: Hokkaido, Hidaka range, Mt. Poroshiri – Mt. Tottabetsu, 01.08.1983, H. Takahashi 4568 (SAPS)
13.	*Rhodothamnus chamaecistus* (L.) Rchb.	Austria: Kärnten, Loibl-pass, 28.05.1960, I. Segelberg *s.n.* (S)

The preparation of pollen grains for LM and SEM, and pollen parameter studied follow Sarwar and Takahashi (2012). Pollen slides of all collection are deposited at the Hokkaido University Museum, Sapporo, Japan. Pollen size and shape classes were made following Erdtman (1986) and descriptive terminology follows Punt *et al.* (2007). The infrageneric classifications of *Bejaria* and *Phyllodoce* follow Clements (1995) and Good (1926 cf. Gillespie and Kron, 2013), respectively.

Results and Discussion

Pollen grains are united usually in tetrahedral tetrads, sometimes compact tetrahedral tetrads and/or with other configurations (Table 2). Viscin threads are usually present, sometime swelling at base in *Bejaria aestuans*, or absent in *Elliottia pyroliflora*, *Epigaea asiatica*, and *Phyllodoce caerulea*. The shape of the pollen grains varies from oblate to suboblate. Costae are usually distinct, but indistinct in *Bejaria*, 3-colpor(oid)ate (Fig. 1B), rarely 4-colporate in *Rhodothamnus chamaecistus* (Fig. 2K), ectoaperture (colpus) margins are distinct in all species. Endoapertures are usually distinct, but indistinct or absent in *Bejaria subsessilis* and *Phyllodoce aleutica*, lalongate, sometimes H-shaped in *Epigaea repens*; and endocracks are usually indistinct or absent in *Phyllodoce* and *Rhodothamnus*. Septum perforations are only observed in *Phyllodoce nipponica* var. *oblong-ovata*. The Phyllodoceae pollen ranges in mean values: D 30.0-53.3 μm, P 15.8-26.9 μm, E 22.0-38.7 μm, D/d 1.31-1.39, P/E 0.66-0.76, 2f 11.5-21.5 μm, W 0.9-2.3 μm, 2f/D 0.26-0.56, endoaperture length 0.8-2.0 μm, width 7.4-13.4 μm, apocolpial exine 1.7-3.0 μm thick, and septum 0.9-2.8 μm thick; tectate, apocolpial exine sculpture from verrucate to rugulate (Table 2).

In SEM, apocolpial exine sculptures mainly constitute two distinct groups, viz. i) primary exine sculpture indistinct, secondary sculpture gemmate-pilate (Figs 1C, E, G–H, L) or intermediate type (Fig. 1M); ii) primary exine sculpture moderately to coarsely rugulate or rugulate-psilate or intermediate types (Figs 2B, C, D, G-I, L); colpus membrane largely granulate or granuloid.

The genera of the Phyllodoceae employed in the present study reveal variation in SEM. In *Bejaria*, pollen surface is somewhat flat, primary apocolpial exine sculpture indistinct, secondary sculpture finely (diam. < 0.5 μm) gemmate-pilate (Type FG; Figs 1C, E, G, H); colpus membrane largely granulate or granuloid.

In *Elliottia*, pollen surface is uneven to somewhat flat, primary apocolpial exine sculpture indistinct, secondary sculpture unit moderately (diam. > 0.5 μm) gemmate-pilate (Type MG; Fig. 1L), or intermediate type (MG/R; Fig. 1M); colpus sometimes narrow and elongate, membrane granulate. The pollen surface of *Epigaea* is somewhat flat, apocolpial exine sculpture coarsely rugulate, the rugulae transversely striate and intermediate type (RS/R; Fig. 2B), or coarsely rugulate-psilate and intermediate type (R/P; Fig. 2C); colpus membrane granulate. Pollen surface in *Phyllodoce* is uneven and rugged, apocolpial exine sculpture intermediate type (R/FG; Fig. 2G); or the surface is somewhat flat, exine sculpture moderately to coarsely rugulate (Type R; Figs 2H-I); colpus membrane granulate or smooth. In *Rhodothamnus*, pollen surface is somewhat flat, apocolpial exine sculpture coarsely rugulate-psilate, the rugulae loosely arranged and clearly striate (Type RS; Fig. 2L); colpus membrane granuloid.

The present study reveals that in Phyllodoceae, both medium or minute and oblate pollen grains are united usually in tetrahedral tetrads (Table 2; Sarwar and Takahashi, 2012). Pollen tetrads of this tribe are generally characterized by the presence of viscin threads except in some species (Table 2; Sarwar and Takahashi, 2012). Results of the present study, in general, support previous palynological observations, however, in several instances, marked differences among the observations are apparent e.g., viscin threads on pollen tetrads of *Epigaea*. We did not observe any

Table 2. Variation in pollen characters of the tribe Phyllodoceae showing mean value in µm and standard deviation. D tetrad diameter, P polar length, d(E) equatorial diameter, Apo. apocolpium, n.d. not discerned.

Name of Species	Config-uration[1]	D	P	d	D/d	P/E	Ectoaperture			2f/D	Endoaperture		Apo. exine thickness	Septum thickness	Ornamentation[2]	Remarks[3]
							Length (2f)	Width (W)	2f/W		Length	Width				
Bejaria																
Section *Bejaria*																
B. aestuans	T(+V)	39.7±2.3	21.2±1.5	29.0±1.7	1.37	0.73	15.4±1.6	1.4±0.8	11.00	0.39	1.4±1.0	13.2±3.3	3.0±0.3	2.8±0.8	FG	1
B. resinosa	T(+V)	45.3±2.1	25.0±1.3	32.7±1.2	1.39	0.76	11.8±1.3	1.7±1.0	6.94	0.26	1.0±0.4	12.9±3.2	2.4±0.5	1.1±0.5	FG	1
B. subsessilis	T(+V)	48.4±2.4	25.0±0.7	35.7±1.7	1.36	0.70	13.4±1.2	1.5±0.4	8.93	0.28	1.3±0.7	13.4±3.5	2.6±0.5	1.7±0.6	FG	1, 2
Section *Racemosae*																
B. racemosa	T(+V)	40.4±2.1	20.9±1.5	30.8±1.2	1.31	0.68	15.0±1.6	1.3±0.6	11.54	0.37	1.2±1.0	12.8±4.4	2.3±0.3	1.6±0.3	FG	1, 3
Elliottia bracteata	T(+V)	52.1±3.1	26.9±1.8	37.5±2.1	1.39	0.72	20.0±3.4	1.0±0.4	20.00	0.38	2.0±1.1	11.6±2.0	1.8±0.2	1.2±0.4	MG	
E. paniculata	T(+V)	45.7±1.4	24.4±1.6	33.7±1.4	1.36	0.72	19.2±3.1	1.7±0.2	11.29	0.42	1.6±0.5	8.6±1.0	1.8±0.2	1.2±0.4	MG/R	4
E. pyroliflora	(T)	53.3±1.6	26.7±2.1	38.7±1.8	1.38	0.69	20.7±2.8	2.3±0.8	9.00	0.39	1.7±1.1	10.8±1.6	2.1±0.2	1.5±0.3	-	
Epigaea asiatica	T	n.d.	n.d.	n.d.	n.d.	n.d.	n.d.	n.d.	n.d.	n.d.	n.d.	n.d.	n.d.	n.d	RS/R	
E. repens	CT(+V)	38.5±1.5	19.8±1.2	29.0±1.0	1.33	0.68	21.5±1.9	2.1±0.3	10.24	0.56	0.8±0.4	8.1±2.3	2.5±0.4	1.4±0.1	R/P	
Phyllodoce																
Subgen. *Eu-Phyllodoce*																
P. aleutica	T(+V)	33.0±1.7	15.8±2.0	24.1±1.1	1.37	0.66	n.d.	n.d.	n.d.	n.d.	n.d.	n.d.	1.7±0.1	1.0±0.3	R/FG	4
P. caerulea	(T)	30.0±1.7	16.1±1.2	22.0±1.3	1.36	0.73	15.8±1.2	0.9±0.3	17.56	0.53	1.3±0.6	7.6±2.9	1.9±0.3	0.9±0.3	R	
Subgen.*Parabryanthus*																
P. nipponica var. *oblong-ovata*	(T)(+V)	33.7±2.7	17.2±1.6	25.2±1.0	1.34	0.68	11.5±1.4	2.3±0.7	5.00	0.34	0.9±0.5	7.4±1.7	1.7±0.1	1.1±0.3	R	2, 4, P
Rhodothamnus chamaecistus	T(+V)	42.4±1.8	22.3±1.1	31.5±1.3	1.35	0.70	17.7±1.6	1.6±0.4	11.06	0.42	1.4±0.8	10.4±1.2	1.8±0.3	1.5±0.6	RS	4, 5

[1] T: Tetrahedral tetrads, (+V): Viscin threads present, (T): Tetrahedral tetrads with other configurations, CT: Compact tetrahedral tetrads

[2] Exine ornamentation type by SEM. FG: Finely gemmate-pilate, MG: Moderately gemmate-pilate, R: Moderately to coarsely rugulate, RS: Coarsely rugulate-psilate with clearly striate rugulae, MG/R or RS/R or R/P or R/FG: Intermediates. For details please consult with the text.

[3] 1: Costae indistinct, 2: Endoaperture indistinct, 3: Apocolpial region small, 4: Endocracks absent/indistinct, 5: Rarely 4-aperturate, P: Perforated septum

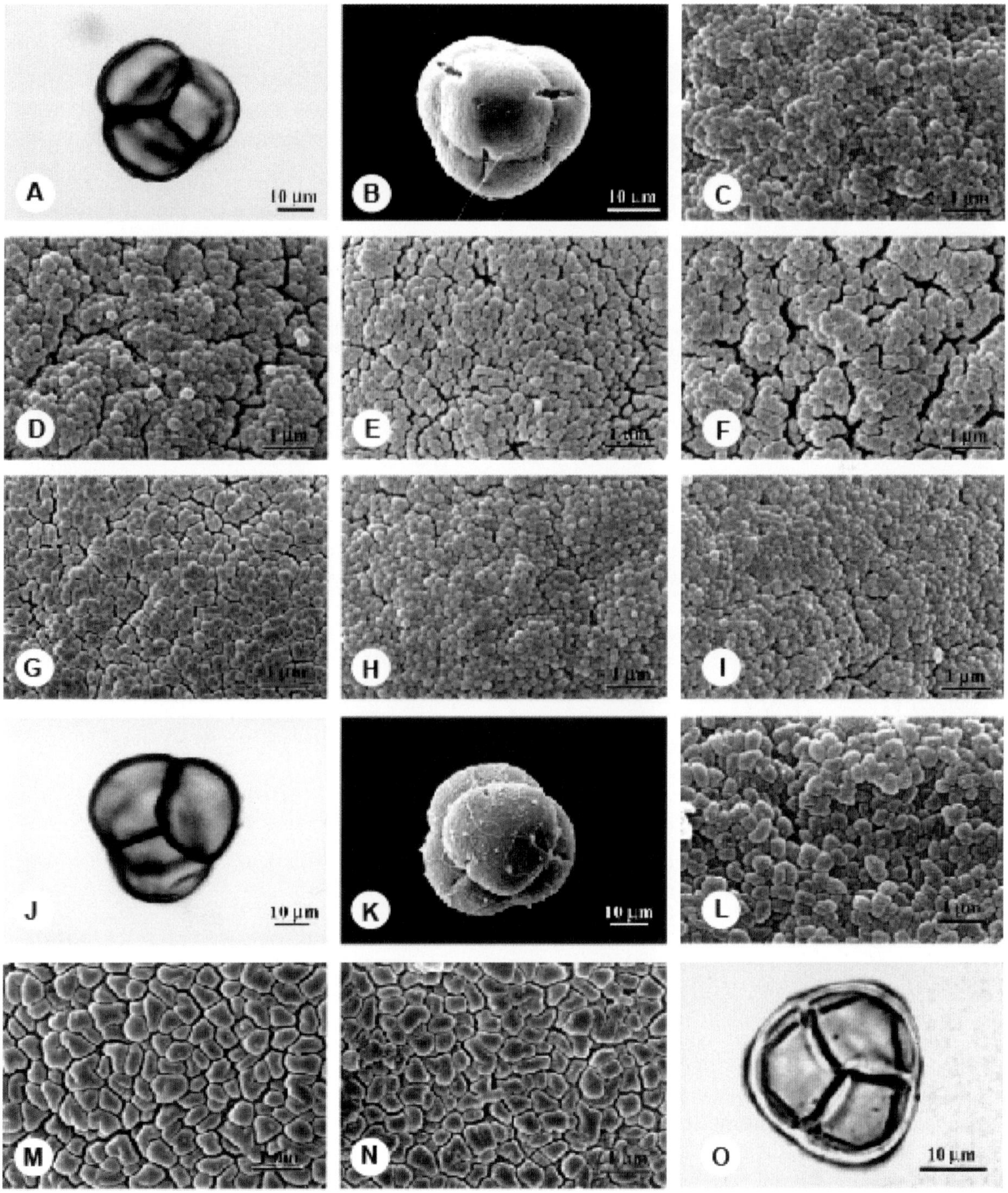

Fig. 1. LM and SEM pollen micrographs. A. *Bejaria aestuans* (Lutyen *et al.* 5296); B. *B. subsessilis* (Løjtnant & Molau 15010); C–D. *B. aestuans* (Lutyen *et al.* 5296); E–F. *B. resinosa* (Sneidern 1069); G. *B. subsessilis* (Løjtnant & Molau 15010); H–I. *B. racemosa* (Moldenke 601); J–K. *Elliottia paniculata* (Takahashi 7802); L. *E. bracteata* (Takahashi 4500); M–N. *E. paniculata* (Takahashi 7802); O. *Epigaea repens* (Clausen 19207). Pollen tetrads at polar view (A–B, J–K, O); pollen tetrad with viscin threads (B, K); micrographs with apocolpial exine sculpture details (C, E, G, H, L, M); micrographs with mesocolpial exine sculpture details (D, F, I, N). Scale bars = 1 μm (C–I, L – N), 10 μm (A, B, J, K, O).

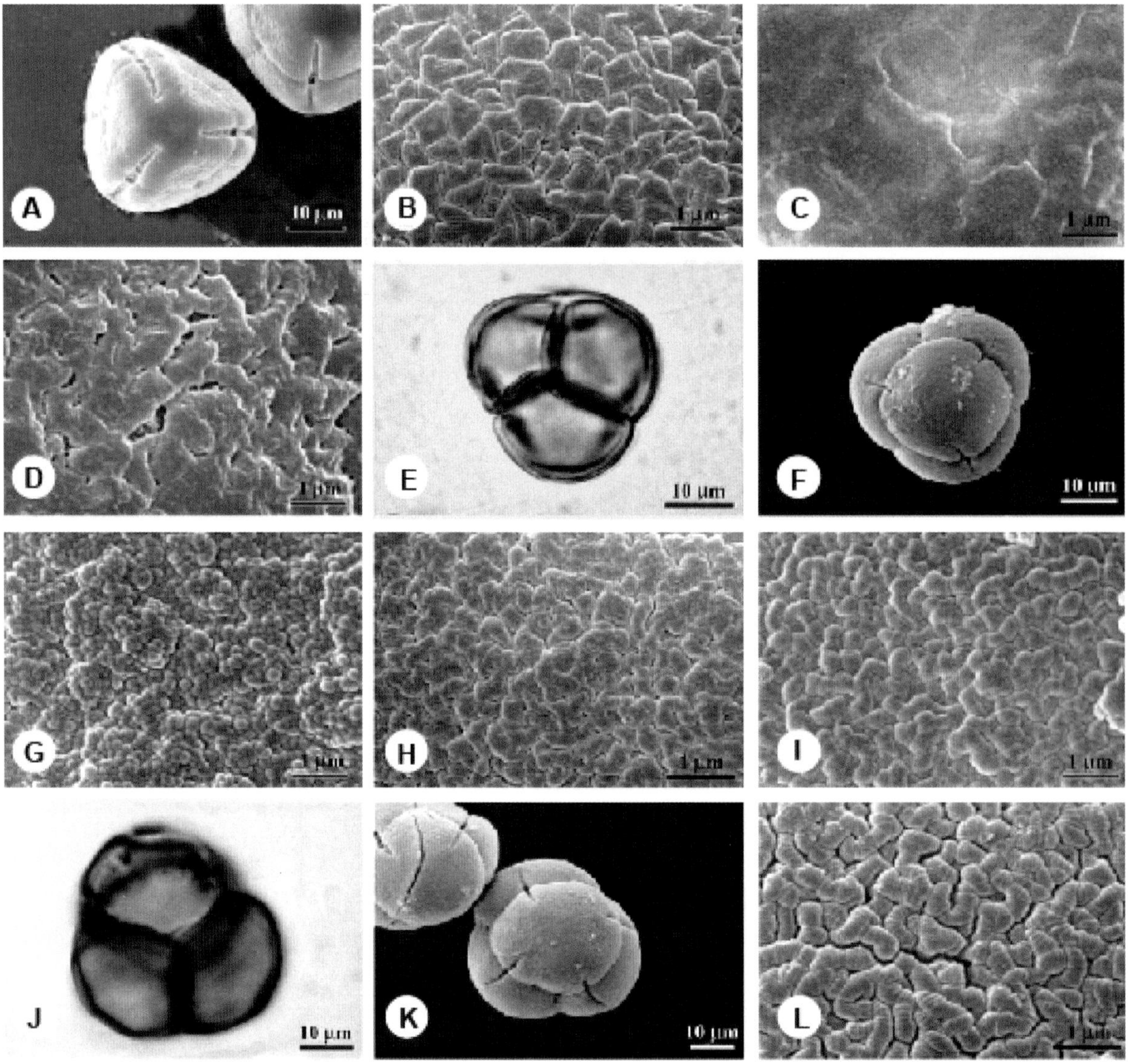

Fig. 2. LM and SEM pollen micrographs. A. *Epigaea repens* (Clausen 19207); B. *E. asiatica* (Hara 5212); C–D. *E. repens* (Clausen 19207); E. *Phyllodoce caerulea* (Takahashi 4569); F–G. *P. aleutica* (Takahashi *et al.* 3666); H. *P. caerulea* (Takahashi 4569); I. *P. nipponica* var. *oblong-ovata* (Takahashi 4568); J–L. *Rhodothamnus chamaecistus* (Segelberg s.n.). Pollen tetrads at polar view (A, E–F, J–K); tetrad with viscin threads (A, F, K); micrographs with apocolpial exine sculpture details (B–C, G–I, L); micrographs with mesocolpial exine sculpture details (D). Scale bars–1 μm (B–D, G–I, L), 10 μm (A, E, F, J, K).

viscin threads on the pollen tetrads of *E. asiatica* (Table 2), although it was reported that all three species of *Epigaea* have viscin threads with their pollen (Stevens, 1971).

Pollen tetrads with viscin threads are often regarded as indicating to entomophilous mode of pollination in the Ericaceae (Waha, 1984; Gillespie and Kron, 2013); however, there are many entomophilous taxa in Ericaceae without viscin threads on their pollen tetrads (Buchmann, 1983). High seed-set success was reported for *Phyllodoce aleutica* whose pollen tetrads have viscin threads, on the other hand, co-occurring *P. caerulea* whose pollen tetrads do not have viscin threads shown the lower seed-set success (Kasagi and Kudo, 2003). The presence or absence of viscin threads may possibly affect the crossing and selfing ability of these two *Phyllodoce* species. Viscin threads might be anticipated in all species with open-campanulate inflorescence, since these are associated with insect pollination. The absence of viscin threads in *P. caerulea* might indicate to a secondary loss (Table 2), since viscin threads are present at least in some species within some genera of the Phyllodoceae (Gillespie and Kron, 2013).

The genera *Epigaea*, *Rhodothamnus*, *Phyllodoce* and *Kalmiopsis*, are consistently made a clade in all molecular analyses (Kron *et al*., 2002; Gillespie and Kron, 2010, 2013); the close relationship within these four genera also represented in their palynological features. For example, all these genera possess apocolpial exine sculptures rugulate and/or its derivatives (Table 2; Figs 2B, C, G–I, L). The exine sculpture has also been proven as a useful taxonomic tool at the infrageneric level (Sarwar and Takahashi, 2006, 2012), although various palynological characters have been found to be important at different taxonomic levels. For example, *Elliottia bracteata* and *E. paniculata* are sister taxa in all molecular analyses (Gillespie and Kron, 2013) and are characterised by the similar value (0.72) of P/E ratio (Table 2).

The exine sculpture of *Rhodothamnus chamaecistus*, with clearly striate secondary sculpture on the rugulae (Type RS; Fig. 2L), is significantly different than that of other members of this tribe except *Epigaea asiatica* (RS/R; Fig. 2B), and very much similar to exine sculpture of members of the tribe Vaccinieae of subfamily Vaccinioideae (Sarwar and Takahashi, 2007). Hitherto, it is the first SEM study of *Rhodothamnus* pollen. In combined analyses of morphological and molecular data, *Epigaea repens* is strongly supported as sister to a clade of both *Rhodothamnus* species (Gillespie and Kron, 2013). The similarity in exine sculpture may be one of the indications of closeness of these two genera (Sarwar, 2007). The apocolpial exine with striate secondary sculpture may be an apomorphic palynological character state for this tribe, and Ericaceae as well (Kron *et al*., 2002; Sarwar and Takahashi, 2007).

The apocolpial exine sculpture types vary often within a single genus. For example, the exine sculptures in *Epigaea asiatica* and *E. repens*, are distinctly different (Type RS/R vs. R/P; Figs 2B, D). The other (external) morphological features of *E. asiatica* also differ considerably from those of *E. repens* (Stevens, 1969), which might be due to disjunct geographic distribution. The infrageneric variation in palynological features due to geographic distribution has also been reported for some other genera of Ericaceae e.g., *Pyrola* (Takahashi, 1986) and *Enkianthus* (Sarwar and Takahashi, 2006).

The generic circumscription of the tribe Bejarieae (Kron *et al*., 2002) and the systematic position of *Bejaria* was a matter of debate for long time (Gillespie and Kron, 2010). Based on palynological, both quantitative and qualitative features it was reported that the monotypic genus *Bryanthus* is clearly distinguished from other members of the tribe Bejarieae - *Bejaria* and *Ledothamnus* (Sarwar, 2007). The molecular phylogenetic studies later confirmed the above finding and transferred the genus *Bejaria* to the tribe Phyllodoceae and construct a new tribe Bryantheae (includes *Bryanthus* and *Ledothamnus*) (Bush and Kron, 2008; Gillespie and Kron, 2010, 2013). Between two sections, *Bejaria* sec. *Bejaria* is characterized by larger values of P, D/d, P/E and thicker apocolpial exine compared to those of *Bejaria* sec. *Racemosae* (Table 2). The close relationship between *B. aestuans* and *B. subsessilis* is supported by values of D/d, ecto- and endo-aperture length and apocolpial exine thickness (Table 2; Bush and Kron, 2008). Moreover, the monophyly of *Bejaria* may be supported by apocolpial exine sculpture (Table 2; Figs 1C, E, G–H). The pollen tetrads of *Bejaria* are similar to those of some *Rhododendron* species of the tribe Rhodoreae. But these similarities may be due to homoplasy. Indistinct costae and lower 2f/D in *Bejaria* are stable and distinct palynological characters within the Ericoideae except for some species of *Rhododendron* and *Therorhodion* (Sarwar, 2007). These palynological features might imply an evolutionary tendency to a reduced (smaller) ectoaperture (Warner and Chinnappa,

1986), and might be an apomorphic pollen character state for this genus as well as the family Ericaceae (Sarwar, 2007). Among the other genera of the tribe Phyllodoceae, the systematic significances of pollen morphological features of *Kalmia* have previously been discussed in details (Sarwar and Takahashi, 2012).

The pollen morphological data confirm the infra- and inter-generic relationships as identified by molecular phylogeny of Phyllodoceae (Gillespie and Kron, 2013) and/or vice-versa. Although various palynological characters were found to be taxonomically important at different taxonomic levels, the apocolpial exine sculpture is emerged as one of the most important palynological features of systematic importance. The rugulate apocolpial exine with striate secondary sculpture, and a reduced colpus might be apomorphic palynological character states for the tribe Phyllodoceae and the family Ericaceae.

Acknowledgements

We wish to express our sincere thanks to the Directors and Curators of the following herbaria: GB, S and SAPS, for allowing us to examine and/or for sending specimens on loan to sample polliniferous materials. The first author is grateful to the MEXT (Japanese Ministry of Education, Culture, Sports, Science and Technology) Scholarship during the period of this study.

References

Bohm, B.A., Brim, S.W., Hebda, R.J. and Stevens, P.F. 1978. Generic limits of the tribe Cladothamneae (Ericaceae), and its position in the Rhododendroideae. J. Arnold Arbor. **59**: 311-341.

Buchmann, S.L. 1983. Buzz pollination in angiosperms. *In*: Jones, C.E. and Little, R.J. (Eds), Handbook of Experimental Pollination Biology. Sci. Acad. Eds., New York, pp.73-113.

Bush, C.M. and Kron, K.A. 2008. A phylogeny of *Bejaria* (Ericaceae: Ericoideae) based on molecular data. J. Bot. Res. Inst. Texas **2**: 1193-1205.

Clements, S.E. 1995. *Bejaria*. *In*: Luteyn, J.L. (Ed.), Ericaceae - Part II: The superior-ovaried genera. Fl. Neotrop. Monogr. **66**, New York Bot. Gard., Bronx, pp. 54-106.

Comtois, P. and Larouche, A. 1981. Morphologie pollinique des Éricales du Québec. Natur. Can. **108**: 245-262 (in French).

Erdtman, G. 1986. Pollen Morphology and Plant Taxonomy – Angiosperms. E. J. Brill, Leiden, 553 pp.

Gillespie, E.L. and Kron, K.A. 2010. Molecular phylogenetic relationships and a revised classification of the subfamily Ericoideae (Ericaceae). Mol. Phyl. Evol. **56**: 343-354.

Gillespie, E.L. and Kron, K.A. 2013. Molecular phylogenetic relationships and molecular evolution within the tribe Phyllodoceae (Ericoideae, Ericaceae). Syst. Bot. **38**: 752-763.

Good, R.D.O. 1926. The genera *Phyllodoce* and *Cassiope*. Le J. de Bot. **64**: 1-10.

Holmgren, P.K., Holmgren, N.H. and Barnett, L.C. (Eds). 1990. Index Herbariorum, Part **I**: The Herbaria of the World. 8th ed., New York Bot. Gard., Bronx, 704 pp.

Kasagi, T. and Kudo, G. 2003. Variation in bumble bee preference and pollen limitation among neighboring populations: comparisons between *Phyllodoce caerulea* and *Phyllodoce aleutica* (Ericaceae) along snowmelt gradients. Am. J. Bot. **90**: 1321-1327.

Kron, K.A. 1997. Phylogenetic relationships of Rhododendroideae (Ericaceae). Am. J. Bot. **84**: 973-980.

Kron, K.A., Judd, W.S., Stevens, P.F., Crayn, D.M., Anderberg, A.A., Gadek, P.A., Quinn, C.J. and Luteyn, J.L. 2002. Phylogenetic classification of Ericaceae: molecular and morphological evidence. Bot. Rev. **68**: 335-423.

Oldfield, F. 1959. The pollen morphology of some of the West European Ericales - Preliminary descriptions and a tentative key to their identification. Pollen *et* Spores **1**: 19-48.

Punt, W., Hoen, P.P., Blackmore, S., Nilsson, S. and Le Thomas, A. 2007. Glossary of Pollen and Spore Terminology. Rev. Palaeob. Palynol. **143**: 1-81.

Radcliffe, C.A., Affolter, J.M. and Wetzstein, H.Y. 2010. Floral morphology and development in Georgia Plume, *Elliottia racemosa* (Ericaceae), a rare coastal plain endemic. J. Amer. Soc. Hort. Sci. **135**: 487-493.

Sarwar, A.K.M. Golam. 2007. Pollen morphology and its systematic significance in the Ericaceae. Ph.D. dissert., Hokkaido University, Japan. http://eprints.lib.hokudai.ac.jp/dspace/handle/2115/46925.

Sarwar, A.K.M. Golam and Takahashi, H. 2006. Pollen morphology of *Enkianthus* (Ericaceae) and its taxonomic significance. Grana **45**: 161-174.

Sarwar, A.K.M. Golam and Takahashi, H. 2007. An overview of pollen morphology and its systematic significance within the blueberry tribe Vaccinieae (Vaccinioideae; Ericaceae). Jpn. J. Palynol. **53**: 87-104.

Sarwar, A.K.M. Golam and Takahashi, H. 2012. Pollen morphology of *Kalmia* (Phyllodoceae, Ericaceae) and its taxonomic significance. Bangladesh J. Plant Taxon. **19**: 123-133.

Sladkov, A.N. 1954. Morphological description of the pollen of the Pyrolaceae, Monotropaceae, Ericaceae, Vacciniaceae and Empetraceae of the European part of USSR. Works Inst. Geo. USSR Acad. Sci. **61**: 119-156 (in Russian).

Stevens, P.F. 1969. Taxonomic studies in the Ericaceae. Ph. D. Thesis, Univ. Edinburgh, UK, 678 pp.

Stevens, P.F. 1971. A classification of the Ericaceae: subfamilies and tribes. Bot. J. Linn. Soc. **64**: 1-53.

Takahashi, H. 1986. Pollen morphology of *Pyrola* and its taxonomic significance. Bot. Mag. (Tokyo) **99**: 137-154.

Waha, M. 1984. Ultrastructure and function of pollen connecting threads in Ericaceae and other Angiosperm families. Plant Syst. Evol. **147**: 189-203.

Warner, B.G. and Chinnappa, C.C. 1986. Taxonomic implications and evolutionary trends in Canadian Ericales. Can. J. Bot. **64**: 3113-3126.

POLLEN MORPHOLOGY OF *TEUCRIUM* L. (LAMIACEAE, AJUGOIDEAE) IN LIBYA

REAM I. MARZOUK[1], SALAMA M. EL-DARIER AND ABDEL BASET M. ASKAR[2]

University of Alexandria, Faculty of Science,
Botany and Microbiology Department, 21511, MoharamBek; Alexandria, Egypt.

Keywords: Lamiaceae; Libya; Pollen exine sculpture; SEM; *Teucrium*.

Abstract

Pollen grains of 11 taxa of *Teucrium* from Libya were examined using light microscopy (LM) and scanning electron microscopy (SEM) in order to provide better insight on the evaluation of palynological attributes for species characterization of *Teucrium* with special emphasis on the five Libyan endemic. Two main pollen shapes were documented; subprolate and prolate or perprolate in *T. fruticans*. The exine sculpture inspected at surface, operculum and pole were mostly verrucate, perforate or scabrate. *Teucrium fruticans* attained the phenomenon of pollen dimorphism with two distinct shapes and specifics sculpture for each form. The results validated the taxonomic significance of pollen grains for the discrimination among *Teucrium* species in Libya.

Introduction

Teucrium L. is the second-largest genus of subfamily Ajugoideae with cosmopolitan distribution over represented in the Mediterranean area (Navarro and EL Oualidi, 2000a; Radulović *et al.*, 2012; Yasaman *et al.*, 2016). The Mediterranean region is its main center of diversity, being represented by around 90% of the total *Teucrium* species in the world (Blanca *et al.*, 2017).

The pollen attributes were potentially useful for both species identification and phylogenetic implication (Abdel Khalik, 2016). These characters endorsed to be stable and of taxonomic significance for generic and specific delimitation of *Teucrium*. The size, exine sculpture and density of supratectal elements seemed to be of systematic value (Abu-Assab and Cantino, 1992; Dinç and Ozturk, 2008; Oybak-Dönmez and Inceoğlu, 1988; Oybak-Dönmez *et al.*, 1999; Navarro *et al.*, 2004).

In Libya, *Teucrium* is represented by eleven taxa of which *T. apollinis* Maire & Weiller, *T. barbeyanum* Asch. & Taub. *ex* E.A. Durand & Barratte, *T. davaeanum* Coss., *T. lini-vaccarii* Pam. and *T. zanonii* Pam. are believed to be endemic to Libya (Marzouk *et al.*, 2016).

The present study was performed to assess the taxonomic value of pollen grains for species characterization of *Teucrium*. It was also achieved to appraise these characters to the Libyan endemic species for the opportunity of their genetic resources conservation.

Material and Methods

Eleven taxa of *Teucrium* were collected from eleven locations in Libya, starting from Al-Gabal Al-Akhdar in the east to Gabal Naffusah (Garian) in the west, along 4500 sq. km. during two flowering seasons: 2009 and 2010. The studied species were sorted under three sections: Chamaedrys, Polium and Teucrium (Table 1). The voucher specimens are kept at the herbaria of Alexandria University (ALEX) and Omar EL-Mukhtar University (Libya).

[1]Corresponding author: Email: reammarzouk@yahoo.com
[2]Omar EL-Mukhtar University; Faculty of Science; El-Beyda; Libya.

The samples were acetolyzed following Erdtman's technique (Erdtman, 1952). 1-3 specimens for each taxon were subjected in this work; at least 30 pollen grains per taxon were examined by using Zeiss light microscope with a micrometer eye-piece. For SEM, the anthers were transferred directly on a stub with double-sided tape, coated for 5 minutes with gold in a polaron JEC-1100E coating unit, and then photographed with JEOL JSM-5300 SEM (Faculty of Science, Alexandria University). The applied terminology based on Punt *et al.* (2007).

Table 1. *Teucrium* specimens used in the present study (sections after Siddiqi, 1985).

Section	Specimen number	Species	Coordinates
Chamaedrys	1-3	**T. barbeyanum* Asch. & Taub. ex E. A. Durand & Barratte	Shahhat Susah Coordinate pair: N32° 50°30.42 E21° 51°7.2
Polium	4-6	**T. apollinis* Maire & Weiller	
	7-8	*T. capitatum* L.	Tarhonah Coordinate pair: N32° 29°48.78 E13° 37°37.08
	9-11	**T. davaeanum* Coss.	Wadi El Quttarh Coordinate pair: N32° 01°35.82 E20° 24°45.48
	12-14	**T. lini-vaccarii* Pamp.	Quasser–El Quaar Coordinate pair: N32° 35°20.28 E13° 50°18.18
	15-17	*T. polium* L.	Sirut Coordinate pair: N31° 08°56.1 E16° 34°35.16
	18-19	*T. polium* subsp. *flavovirens* Batt.	Al Hameida escarpment Coordinate pair: N32° 24°52.98 E20° 32°17.88
	20-22	**T. zanonii* Pamp.	Dryannah Coordinate pair: N32° 19°42.12 E20° 16°34.86
Teucrium	23-25	*T. brevifolium* Schreb.	Lathroun–Ras El Hellal Coordinate pair: N32° 52 305'°0 E22° 15°6.12
	26-28	*T. campanulatum* L.	Wadi Errieg Coordinate pair: N32° 32 230°0 E20° 42°56.82
	29-30	*T. fruticans* L.	El Rabtta–El Assbeh Coordinate pair: N32° 07°12.6 E21° 52°16.14

*Endemic species.

Results and Discussion

The taxonomic complexity of *Teucrium* is reflected in the changes in its classification using different characteristics as pollen morphology (Díez *et al.*, 1993), karyology (Valdés-Bermejo and Sánchez-Crespo, 1978), indumentum characteristics (Manzanares *et al.*, 1983; El Oualidi and Puech, 1993; Navarro and EL Oualidi, 2000b), phytochemistry (Harborne *et al.*, 1986; Velasco-Negueruela and Pérez-Alonso, 1990; Bukhari *et al.*, 2014) and amino acid composition (Juani *et al.*, 2004). Meanwhile, there is an imminent danger for genetic erosion of Libyan wild species due to the increment in drought and anthropogenic activities, that has resulted in habitat loss and fragmentation leading to the diminishing of the germplasm reservoir (Al-Idrissi *et al.*, 1996). The

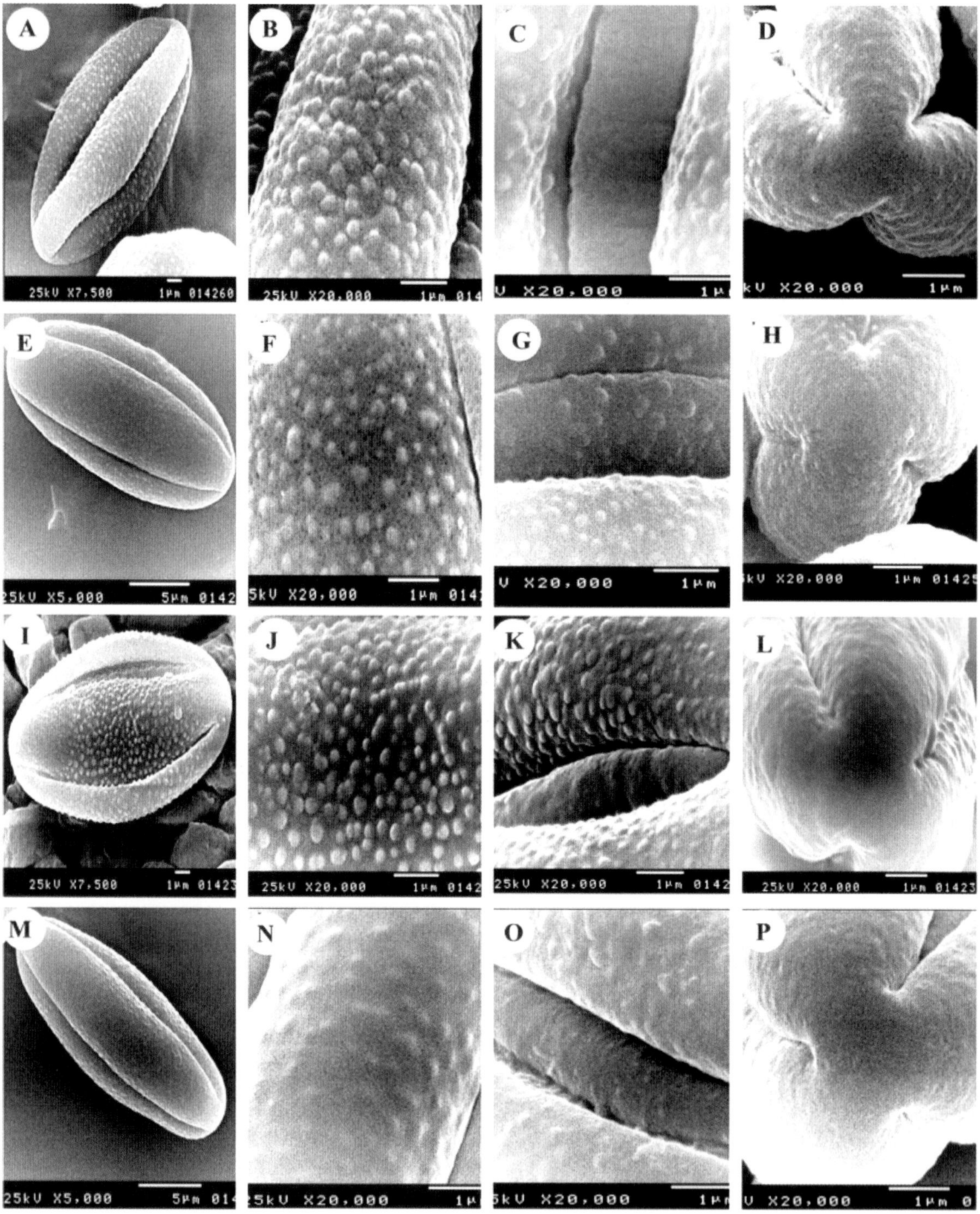

Fig. 1. Pollen grains of *Teucrium*: A–D, *T. barbeyanum*; E–H, *T. apollinis*; I–L, *T. capitatum*; M–P, *T. davaeanum*.

IUCN Red List of Threatened Plants (1998) recorded three endemic species; *T. apollinis, T. barbeyanum* and *T. zanonii* in addition *T. davaeanum* and *T. lini-vaccari* recorded by Marzouk *et al.* (2016). Consequently, the efforts must be intensified giving priority to study both endangered and endemic species through various characters.

In the present study, *Teucrium* pollen grains are monads, isopolar, radio symmetric, and tricolpate with opercula. The pollen shape is subprolate in *T. barbeyanum* (Fig. 1A), *T. campanulatum* (Fig. 3A) and *T. polium* subsp. *flavovirens* (Fig. 2L) and prolate in the rest of species. However, two shapes were recorded in *T. fruticans*; perprolate (form a) and prolate (form b) (Fig. 3I, O). That is in congruent with Ojeda and Díez (1992) for the recognition of the dimorphism phenomenon in this species. The polar axis length ranges from 30.02-31.2 μm in section Chamaedrys, from 30.41-37.01 μm in section Polium, and from 38.74-60.96 μm in section

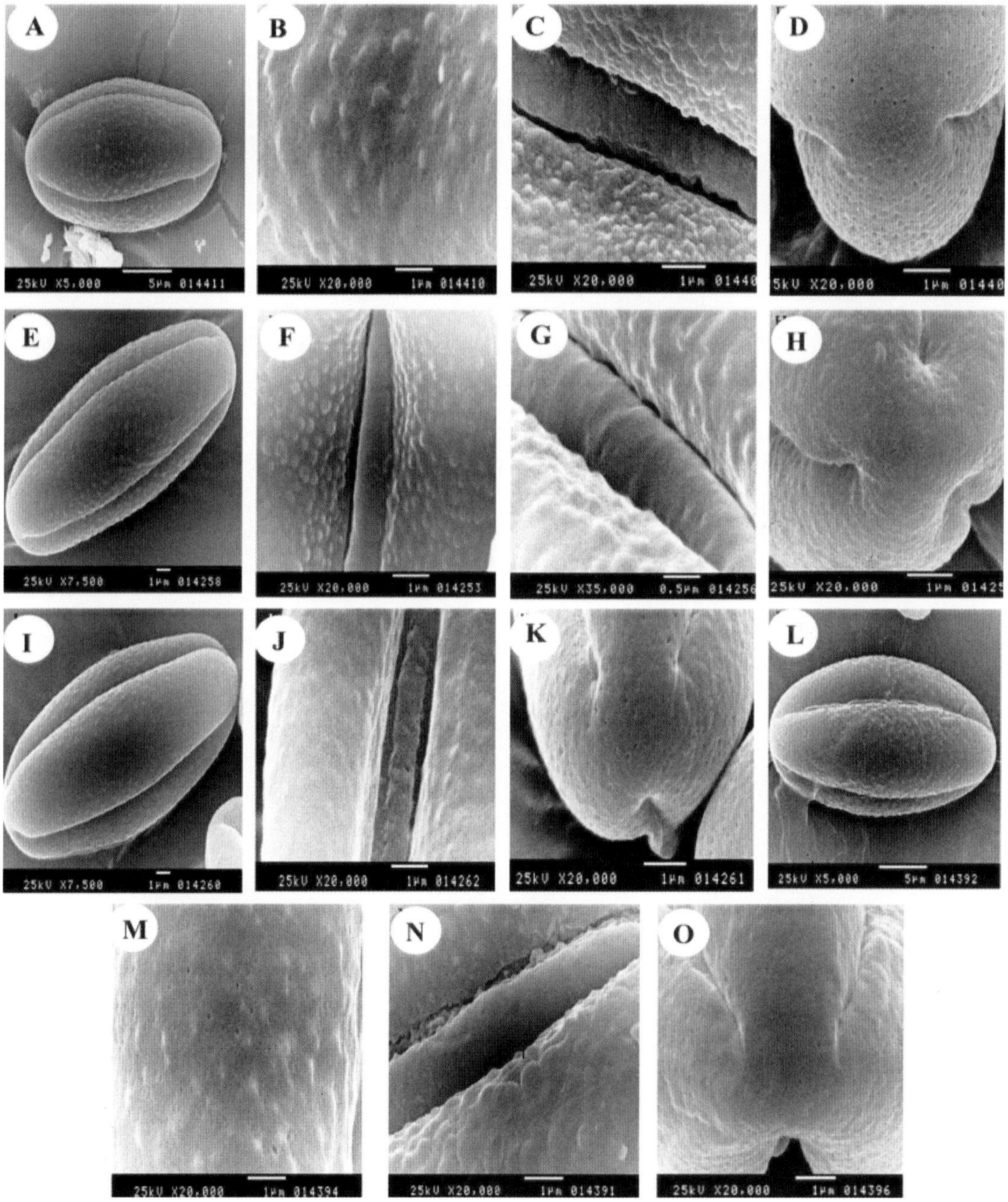

Fig. 2. Pollen grains of *Teucrium*: A–D, *T. lini-vaccarii*; E–H, *T. polium*; I–K, *T. polium* subsp. *flavovirens*; L–O, *T. zanonii*.

Teucrium (Table 2). The equatorial diameter varies from 23.47-28.28 µm in both sections Chamaedrys and Polium, and 19.63-39.23 µm in section Teucrium. Polar axis and equatorial diameter were found useful in separating two closely related taxa where it is large in *T. polium* than *T. polium* subsp. *flavovirens*. The largest colpus dimensions and mesocolpium diameter attain in section Teucrium while the smallest in section Polium (Table 2). The exine thickness assorts from 1.5-1.65 µm in *T. campanulatum* and the taxa of both sections Chamaedrys and Polium, while from 2.33-3.19 µm in the rest of section Teucrium (Table 2). Oybak-Dönmez and Inceoğlu (1988) and Dinç *et al.* (2008) specified *Teucrium* with either verrucate or verrucate-granulate exine sculpture, meanwhile Navarro *et al.* (2004) declared that the basal groups of *Teucrium* with verrucate sculpturing. The current study discriminates among the exine sculpture at the surface,

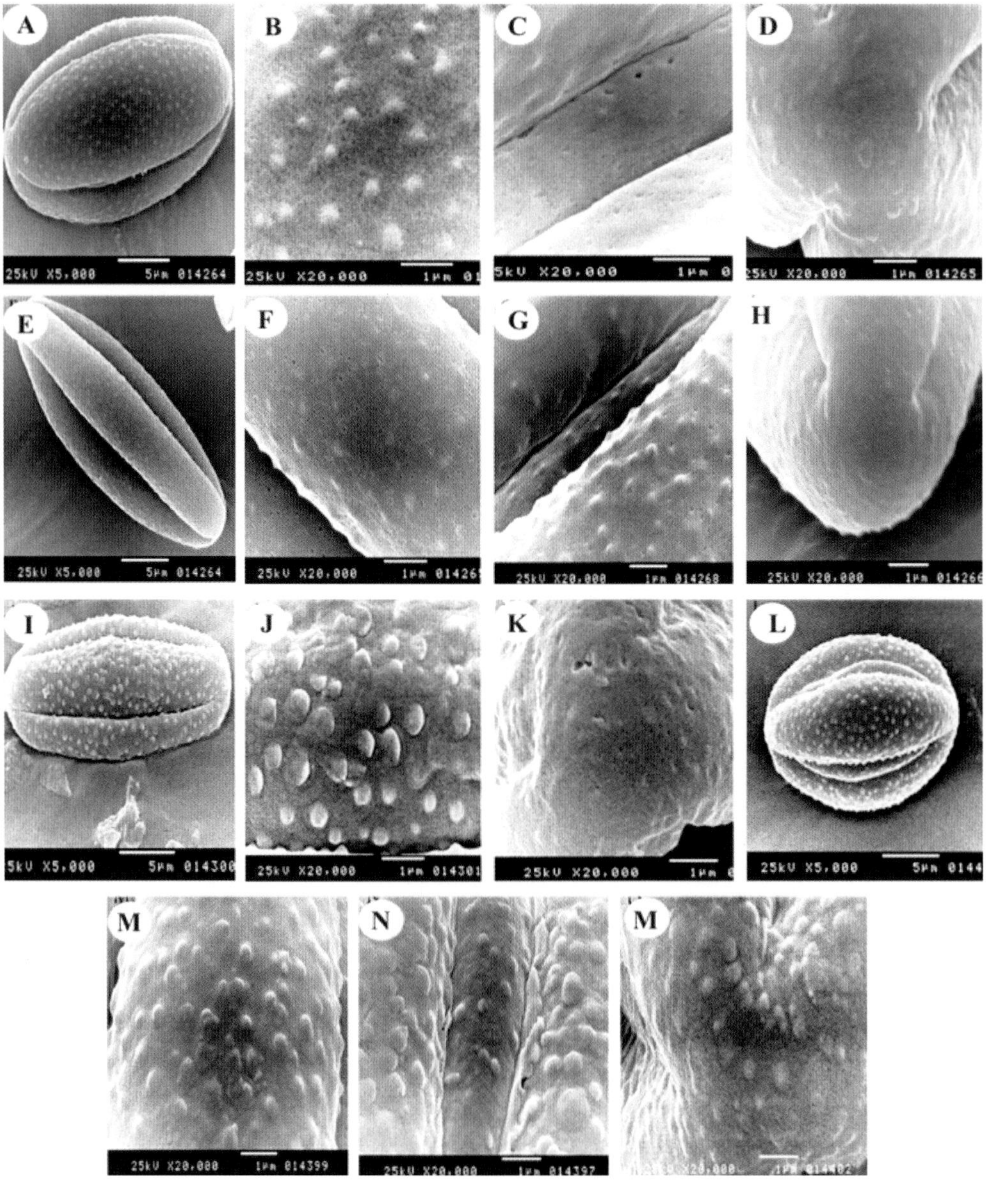

Fig. 3. Pollen grains of *Teucrium*: A–D, *T. brevifolium*; E–H, *T. campanulatum*; I–K, *T. fruticans* form a; L–O, *T. fruticans* form b.

Table 2. Pollen morphological characters for the examined taxa of *Teucrium*.

Section	Species	Polar axis "P" (µ m)	Equatorial diameter "E"(µ m)	P/E ratio	Shape	Colpus length (µ m)	Colpus width (µ m)	Meso-colpium (µ m)	Exine thick-ness (µ m)	Ornamentation		
										Surface	Operculum	Pole
Chamaedrys	*T. barbeyanum*	30.67^{b}	24.75^{a}	1.24^{a}	Subprolate	23.11	3.80	21.85^{a}	1.58	Verrucate-perforate	Perforate	Perforate
Polium	*T. apollinis*	34.87^{a}	22.97^{a}	1.52^{e}	Prolate	23.65	3.16	20.14^{a}	1.57	Verrucate-perforate	Verrucate- perforate	Perforate
	T. capitatum	34.11^{a}	25.13^{a}	1.36^{a}	Prolate	24.45	3.87	21.46^{a}	1.62	Verrucate decreased to scabrate towards the pole	Faint perforate	Faint perforate
	T. davaeanum	34.50^{a}	24.06^{a}	1.44^{d}	Prolate	25.90	3.80	21.14^{a}	1.58	Scabrate-verrucate and faint perforate	Faint perforate	Faint perforate
	T. lini-vaccarii	33.11^{a}	24.35^{a}	1.36^{c}	Prolate	24.42	4.81	19.48^{a}	1.60	Verrucate- perforate	Perforate	Perforate
	T. polium	36.30^{a}	28.43^{a}	1.28^{b}	Subprolate	23.60	3.80	23.26^{a}	1.58	Verrucate- perforate	Perforate	Perforate
	T. polium subsp. *flavovirens*	30.45^{a}	24.83^{b}	1.23^{a}	Subprolate	22.5	3.99	16.84^{a}	1.62	Scabrate-verrucate and faint perforate	Perforate	Perforate
	T. zanonii	36.98^{c}	27.26^{a}	1.36^{a}	Prolate	24.93	3.92	23.17^{a}	1.58	Scabrate-verrucate and perforate	Perforate	Perforate
Teucrium	*T. brevifolium*	43.17^{c}	29.54^{a}	1.49^{a}	Prolate	31.50	3.88	23.75^{a}	2.38	Verrucate- perforate	Perforate	Verrucate
	T. campanulatum	39.40^{d}	32.20^{d}	1.22^{a}	Subprolate	27.54	4.81	24.35^{a}	1.55	Scabrate-verrucate and perforate	Scabrate-verrucate and perforate	Faint perforate
	T. fruticans form a	42.77	19.63	2.18	Perprolate	30.13	4.85	20.11	2.40	Verrucate	Obscure	Scabrate-perforate
	T. fruticans form b	60.96	39.23	1.55	Prolate	41.73	7.14	25.09	3.19	Verrucate and faint perforate	Verrucate	Verrucate
	$LSD_{0.05}$	1.48	1.79	4.37	0.537	-----	-----	1.34	-----			

operculum and pole. In section Chamaedrys, the exine sculpture is verrucate-perforate at the surface and perforate at both operculum and pole (Fig. 1B, D). In section Polium, the verrucate-perforate surface sculpture accomplishes with *T. apollinis* (Fig. 1F), *T. lini-vaccarii* (Fig. 2B) and *T. polium* (Fig. 2F). While in *T. capitatum* (Fig. 1J, K) and *T. polium* subsp. *flavovirens* (Fig. 2J), the sculpture is verrucae to scabrate. Both *T. davaeanum* and *T. zanonii* specify with sculpture of scabrate-verrucate and faint perforate (Fig. 1N) and scabrate-verrucate and perforate (Fig. 2M), respectively. The faint perforate sculpture at both operculum and pole realizes in both *T. capitatum* (Fig. 1K, L) and *T. davaeanum* (Fig. 1O, P), while *T. apollinis* with verrucate-perforate at the operculum (Fig. 1K) and perforate in the rest of species. In section Teucrium, each species attains certain sculpturing, *T. brevifolium* reveals verrucate-perforate, perforate, and verrucate at the surface, operculum and pole, respectively (Fig. 3B, D). In *T. campanulatum*, the sculpture is scabrate-verrucate and perforate at both surfaces and operculum and faint perforate at the pole (Fig. 3F, H). *Teucrium fruticans* accomplishes sculpture dimorphism, the verrucate or verrucate and faint perforate at the surface (Fig. 3J, M), obscure or verrucate at the operculum (Fig. 3I, N), and scabrate-perforate or verrucate at the pole (Fig. 3K, O). The results indicated the validity of pollen characters for taxonomic implications and in the discrimination among *Teucrium* at both section and species levels.

References

Abdel Khalik, K. 2016. A systematic revision of the genus *Plectranthus* L. (Lamiaceae) in Saudi Arabia based on morphological, palynological, and micromorphological characters of trichomes. American Journal of Plant Sciences **7**: 1429–1444.

Abu-Assab, M.S. and Cantino, P.D. 1992. Pollen morphology in subfamily Lamioideae (Lamiaceae) and its phylogenetic implications. *In*: Harley, R.M. and Reynolds, T.(Eds), Advances in Lamiaceae Science. Royal Botanic Gardens, Kew, pp. 97–122.

Al-Idrissi, M., Sbeita, A., Jebriel, A., Zintani, A., Shreidi, A., Ghawawi, H. and Tazi, M. 1996. Libya: Country report to the FAO international technical conference on plant genetic resources. Leipzig, Germany.

Blanca, G., Cueto, M. and Fuentes, J. 2017. *Teucrium teresianum* sp. nov. (Lamiaceae) from southern Spain. Nordic J. Bot. **35**(1): 14–19.

Bukhari, N.A., Al-Otaibi, R.A. and Ibrahim, M.M. 2014. Biodiversity characteristics of *Teucrium polium* species in Saudi Arabia. Saudia J. Biol. Sci. **22**(2): 181–185.

Díez, M.J., Ojeda, F. and Colomer, M. 1993. Contribución a la palinologı a del genero *Teucrium* L. en la Penınsula Iberica e Islas Baleares y suinteretaxonmico. Lagascalia **17**: 119–134.

Dinç, M. and Ozturk, M. 2008. Comparative morphological, anatomical, and palynological studies on the genus *Stachys* L. sect. Ambleia Bentham (Lamiaceae) species in Turkey. Turkish J. Bot. **32**:113–121.

Dinç, M., Duran, A., Pinar, M. and Ozturk, M.2008. Anatomy, palynology and nutlet micromorphology of Turkish endemic *Teucrium sandrasicum* (Lamiaceae). Biologia **63**: 637–641.

El Oualidi, J. and Puech, S. 1993. Quelques marqueurs morphologiques des *Teucrium* section Polium (Lamiaceae) du Maroc: valeurs diagnostiques à différents niveaux d'integration. Acta Botanica Malacitana **18**:163–171.

Erdtman, G. 1952. Pollen Morphology and Plant Taxonomy. Angiosperms. Almqvist and Wiksell, Stockholm, pp. 539.

Harborne, J.B., Tomas-Barberan, F.A., Williams, C.A. and Gil, M.I. 1986. A chemotaxonomic study of flavonoids from European *Teucrium* species. Phytochemistry **25**: 2811–2816.

IUCN Red List of Threatened Plants. 1998. Walter, K.S. and Gillett, H.J. (Eds),1997. IUCN Compiled by the World Conservation Monitoring Centre. IUCN-The World Conservation Union, Gland, Switzerland and Cambridge, UK, pp. 862.

Juani, R., Pastor, J., Milla, F., Alaiz, M. and Vioque, J. 2004. Amino acids composition of *Teucrium* nutlet proteins and their systematic significance. Ann. Bot. **94**: 615–621.

Manzanares, P., Gómez-Campo, G. and Tortosa, M.E. 1983. Estudiossobre el indumento de las especiesibéricasy baleáricas del género *Teucrium* L. (Lamiaceae). Anales del Jardín Botánico de Madrid **40**: 94–106.

Marzouk, R.I., El-Darier, S.M. and Askar A.M. 2016. Nutlet micromorphological characters of *Teucrium* taxa (Lamiaceae) in Libya. Phytotaxa **263**(3): 245–254.

Navarro, T. and El Oualidi, J. 2000a. Sinopsis of *Teucrium* L. (Labiatae) in the Mediterranean region and surrounding areas. Flora Mediterranea **10**: 349–363.

Navarro, T. and El Oualidi, J. 2000b. Trichome morphology in *Teucrium* L. (Labiatae), a taxonomic review. Anales del Jardín Botánico de Madrid **57**: 277–297.

Navarro, T., El Oualidi, J. and Trigo D.M. 2004. Pollen morphology of *Teucrium* (Labiatae) and its taxonomic value. Belgian J. Bot. **137**(1): 70–84.

Ojeda, F. and Díez, M.J. 1992. Pollen dimorphism in three *Teucrium* species (Lamiaceae). Plant Syst. Evol. **183**: 43–49.

Oybak-Dönmez, E. and Inceoğlu, Ő. 1988. Pollen morphology of some *Teucrium* L. (Labiatae) species. Communications Faculty of Sciences University of Ankara Series C: Biology **6**: 133–146.

Oybak-Dönmez, E., Inceoğlu, Ő. and Pinar, N.M. 1999. Scanning electron microscopy study of pollen in some Turkish *Teucrium* L. (Labiatae). Turkish J. Bot. **23**: 379–382.

Punt, W., Hoen, P.P., Blackmore, S., Nilsson, S. and Le Thomas, A. 2007. Glossary of pollen and spore terminology. Review of Palaeobotany & Palynology **143**: 1–81.

Radulović, N., Dekić, M., Joksović, M. and Vukićević, R. 2012. Chemotaxonomy of Serbian *Teucrium* species inferred from essential oil chemical composition: the case of *Teucrium scordium* L. ssp. *scordioides*. Chemistry and Biodiversity **9**:106–122.

Siddiqi, M.A. 1985. 3: *Teucrium*. *In*: Jafri, S.M.H. and El-Gadi, A. (Eds), Flora of Libya. Al Faateh University, Faculty of Science, Department of Botany, Tripoli, pp. 13–30.

Valdés-Bermejo, E. and Sánchez-Crespo, A. 1978. Datos cariologicos sobre el genero *Teucrium* L. (Labiatae) en la Peninsula Ibérica. Acta Botánica Malacitana **4**: 27–54.

Velasco-Negueruela, A. and Pérez-Alonso, M.J. 1990. The volatiles of six *Teucrium* species from the Iberian Peninsula and the Balearic Islands. Phytochemistry **29**: 1165–1169.

Yasaman, S., Stefan, K., Günther, H. and Christian, B. 2016. Phylogeny of non-monophyletic *Teucrium* (Lamiaceae: Ajugoideae): implications for character evolution and taxonomy. Taxon. **65**(4): 805–822.

18

MORPHOLOGICAL, ANATOMICAL AND CYTOLOGICAL INVESTIGATIONS ON THREE TAXA OF *CENTAUREA* L. (ASTERACEAE) FROM TURKEY

NESLİHAN TAŞAR, GÜLDEN DOĞAN[1], YAŞAR KIRAN[1], M. OLIUR RAHMAN[2] AND UĞUR ÇAKILCIOĞLU[3]

Munzur University, Tunceli Vocational School, Department of Organic Agriculture, Tunceli, Turkey

Keywords: Centaurea; Anatomy; Cytology; Taxonomy; Asteraceae; Turkey.

Abstract

Morphological, anatomical and cytological features of three Turkish taxa of *Centaurea* L., viz. *C. polypodiifolia* Boiss. var. *polypodiifolia*, *C. urvillei* DC. subsp. *urvillei* and *C. urvillei* subsp. *armata* Wagenitz were investigated. Stem anatomy revealed the presence of a thick cuticle layer outside the stem, and epidermis with dense hairs was observed in the lower part. Investigation on leaf anatomy showed that different types of hairs on the outside of the leaf were very intense. Palisade parenchyma was observed below the upper and lower epidermis. In *Centaurea polypodiifolia* var. *polypodiifolia*, chromosome number was found to be 2n=16, while in *C. urvillei* subsp. *urvillei* and *C. urvillei* subsp. *armata* 2n=20. Total karyotype length of *C. polypodiifolia* var. *polypodiifolia, C. urvillei* subsp. *urvillei* and *C. urvillei* subsp. *armata* was 22.9 µm, 37.84 µm and 40.01 µm, respectively. Among the investigated taxa the karyotype asymmetry index was found lowest in *C. urvillei* subsp. *armata*. Satellite was detected in *C. urvillei* subsp. *urvillei* and subsp. *armata*, while it was absent in *C. polypodiifolia* Boiss. var. *polypodiifolia*.

Introduction

Centaurea L. (Asteraceae) consists of about 700 species and distributed in the Mediterranean region and the Near East with a few species reaching northern Eurasia, north and east Africa, North America, and Australia (Bancheva *et al*., 2014; Behçet *et al*., 2017). Turkey is one of the main centers of this genus, particularly the Southwest and East of the country (Wagenitz, 1986), and regarded as the third largest genus in Turkey (Davis, 1975). In Turkey, *Centaurea* is represented by 194 species, of which 118 are endemic (Guner *et al*., 2012). The systematics of the genus *Centaurea* is problematic, and the sectional classification of *Centaurea* relies heavily on the morphology of the appendage of phillary and achenes (Garcia-Jacas *et al*., 2001). The taxonomic coplexity of *Centaurea*, especially in the Near East, has stirred much research (Duran and Duman, 2002; Türkoglu *et al*., 2003; Yüzbaşıoğlu *et al*., 2015). The status of species and infraspecific taxa included in *Centaurea* has been revised in several taxonomic treatments (Rahiminejad *et al*., 2010; Ranjbar and Negaresh, 2013; Negaresh and Rahiminejad, 2014). In the recent past, several authors concentrated on morphological studies of *Centaurea* alongside describing new species from Turkey (Kose *et al*., 2010; Hayta *et al*., 2016; Behçet *et al*., 2017).

Anatomical and cytological characters play pivotal role in plant systematics, and quite often offer tools for species delimitation. Stem and leaf anatomy provide many characters which have

[1]Fırat University, Science Faculty, Biology Department, Elazığ, Turkey.
[3]Department of Botany, University of Dhaka, Dhaka 1000, Bangladesh.
[3]Munzur University, Pertek Sakine Genç Vocational School, Tunceli, Turkey. Correspending author. Email: ucakilcioglu@yahoo.com

already been proven to be of importance in classification (Lu *et al.*, 2008), and have been widely used in taxonomic treatments and systematic studies (Noman *et al.*, 2014). Though some taxa of *Centaurea* were investigated based on anatomical (Ozcan, 2013; Uysal *et al.*, 2016) and cytological (Martin *et al.*, 2006; Uysal *et al.*, 2009) characters, many taxa of this genus occurring in Turkey remain unexplored using these important tools. Since *Centaurea* is taxonomically a complicated genus as it contains many taxa that show a great degree of morphological variations, anatomical and cytological data should be involved to resolve taxonomic limits. Therefore, the present study aimed at exploring the anatomical and cytological features of three *Centaurea* taxa from Turkey and to enrich the knowledge of these features for better understanding of systemaics of *Centaurea*.

Materials and Methods

Plant materials

Plant materials were collected from natural habitats from 2011 to 2012. The voucher specimens have been deposited at the Firat University Herbarium (FUH). The taxonomic identifications of the taxa were confirmed following Davis (1975). The list of the taxa investigated in this study along with their localities and vouchers are provided in Table 1.

Table 1. List of the investigated taxa with their localities and voucher specimens.

Taxa	Locality	Vouchers
Centaurea polypodiifolia Boiss. var. *polypodiifolia*	B7/Elazığ; Çemişgezek, Danbüken, Avşan Köyü, 1090 m. 16.07.2012	Tasar, 1005
C. urvillei DC. subsp. *urvillei*	B7/Elazığ; Harput, Anguza Baba, Kayalık alan, 1400 m. 13.06.2011	Tasar, 1008
C. urvillei subsp. *armata* Wagenitz	B7/ Elazığ, Baskil, Yukarı Kuluşağı Köyü, Kayalık alan, 1400 m. 13.06.2011	Tasar, 1009

Anatomical investigation

Cross sections of stem, root, leaf and surface sections of leaf of three taxa of *Centaurea* were made after fixing in 70% ethanol (Yakar-Tan, 1982). The sections were observed under light microscope Olympus BX51 and photographed with Olympus Camedia C-4000 digital camera.

Cytological study

The cytological studies were conducted on root tips meristematic cells. The seeds were germinated on moist filter paper in Petri dishes at 25°C. The actively growing root tips were pretreated with aqueous colchicine (0.05%) for 3–3.5 h at room temperature. Afterwards, the root tips were fixed with Carnoy (1:3 glacial acetic acid–absolute ethanol) for at least 24 h at 4°C, hydrolysed in 1 N HCl at 60°C for 15 min, then rinsed in tap water for 3–5 min. Finally, they were stained in Feulgen for 1 h and mounted in 45% acetic acid. Digital microphotographs from at least 5 well-spread metaphase plates were taken using an Olympus BX51 microscope, and were recorded with an Olympus Camedia C-4000 digital camera. Diploid chromosome number (2n), ploidy level, karyotype formula, chromosome length range and total karyotype length (TKL) were determined. Chromosomes were classified according to nomenclature given by Levan *et al.* (1964). The intrachromosomal asymmetry index (A1) and the interchromosomal asymmetry index (A2) followed Romero-Zarco (1986), while the karyotype symmetry nomenclature followed Stebbins (1971). For calculating coefficient of variation of chromosome length (CV_{CL}), coefficient of variation of the centromeric index (CV_{CI}) and asymmetry index (AI) Paszko (2006) was tailed.

Results and Discussion

Morphological characteristics

Centaurea polypodiifolia Boiss., Diagn. Pl. Orient. Ser. 1, 6: 126 (1846), var. **polypodiifolia.** *C. euphratica* Boiss., Diagn. Pl. Orient. Ser. 1, 6: 125 (1845). *Microlophus polypodiifolius* (Boiss.) Agadshanov, Fl. Azerb. 8: 443 (1961). **(Fig. 1A-B)**.

Biennial or perennial, 33-44 cm long, with numerous branches and thickened taproot. Root 16-23 cm long. Stem, branches and lower leaves with crisp articulate hairs, other leaves almost glabrescent. Lower leaves 120-180×50-80 mm, median leaves 50-90×10-18 mm. Capitula corymbosely arranged, 2.5-3.5 cm long. Flowers yellow. Appendage very small (0.5-2.5 mm) and easily deciduous, a simple spinule or with 1-2 pairs of minute teeth. Involucre 16-20×10-13 mm, rarely broader, upper leaves narrowly lanceolate, not concealing involucre. Outer phyllaries 8-12 x 5-7 mm, median phyllaries 10-14×4-7 mm, inner phyllaries 15-18×2-4 mm. Achenes 4-5 mm, pappus 5-8 mm long.

Flowering: June to August.

Habitat: Rocky slopes, steppe, fallow fields, at 800-2500 m above the sea level.

Centaurea urvillei DC., Prodr. 6: 592 (1838), subsp. **urvillei**. *Aegialophila longispina* Cand. in Bull. Soc. Bot. Fr. 44: 146 (1897). *C. urvillei* DC. var. *leptacantha* Bornm. in Beih. Bot. Centr. 38(2): 464 (1921). *C. chiosicola* Beauv. & Topali in Bull. Soc. Bot. Geneve Ser. 2, 26: 156 (1936). **(Fig. 1C-D)**.

Short-lived perennial, 12-28 cm tall. Root 3-5 cm long. Stem simple or branched from near base, very short or elongate, mostly rather slender. Lower leaves 90-12 x30-50 mm, median leaves 150-210 x 23-42 mm, slightly arachnoid to distinctly tomentose, lyrate with triangular, rhombic or oblong-ovate terminal segments, few to numerous lanceolate, lateral segments of leaves entire or dentate. Capitula 1-4, c. 5-7 cm long. Flowers rose-purple or whitish. Appandages (excluding cilia) 2-4(-5) mm broad at the base, rarely more than 20 mm long. Involucre 30-40×17-38 mm, ovoid to globose. Outer phyllaries 7-12×5-9 mm, median phyllaries 13-15×5-7 mm, inner phyllaries 15-22 × 5-7 mm. Achenes 4-6 mm long, pappus 7-12 mm long.

Flowering: June to July.

Habitat: Rocky slopes, macchie, open *Pinus* forests, at 2000 m above sea level.

Centaurea urvillei DC., Prodr. 6: 592 (1838), subsp. **armata** Wagenitz., Willdenowia 6(3): 491 (1972). *C. urvillei* DC. var. *platyacantha* Bornm. in Beih. Centr. 38(2): 464 (1921) **(Fig. 1E-F)**.

Short-lived perennial or biennial, 11-20 cm long. Root 4-10 cm long. Stem simple or branched from near the base, short or elongated, stout. Lower leaves 90-120×25-30 mm, median leaves 110-160×30-50 mm, slightly arachnoid to distinctly tomentose, lyrate with triangular, rhombic or oblong-ovate terminal segments, few to numerous lanceolate, oblong or lyrate, interspersed with small lobes. Capitulum 5.0-5.5 cm long. Flowers rose-purple or whitish. Appandages 4-7 mm broad at the base, with 8-12 cilia on each side. Involucre 30-35×30-40 mm, ovoid to globose. Outer phyllaries 15-24×8-10 mm, median phyllaries 15-18×5-10 mm, inner phyllaries 20-25×6-10 mm. Achenes 4-6 mm long, pappus 6-13 mm long.

Flowering: June to August.

Habitat: Dry stony slopes, scree, rocks, at 50-2800 m above sea level.

Notes: The present morphological investigation deals with in-depth study including the several quantitative characters of three taxa of *Centaurea*, viz, size of root, lower and median leaves, outher, median and inner phyllaries and capitulum, where these characters are missing in the Flora of Turkey (Davis, 1975).

Fig. 1. A-B: *Centaurea polypodiifolia* var. *polypodiifolia*; C-D: *C. urvillei* subsp. *urvillei*, E-F: *C. urvillei* subsp. *armata*; A,C,E: Habitat, B,D,F: Herbarium specimens.

Anatomical characters

Root anatomy

The outer layer contains a periderm in the form of an epidermis (protective tissue). The crushed epidermis is found below the peridermis. Under the epidermis is a cortex layer consisting of small, ovoid, thick-walled parenchyma cells. Sclerenchymatous bundles are partially embedded in the cortex layer. Under the cortex, there is cambium trapped in a narrow area. Next to the cortex there are well developed vascular bundles. Although the number of phloem cells is less, xylem cells occupy a broader space. The xylem expanded to cover its own pith region (Fig. 2).

Stem anatomy

Stem anatomy revealed the presence of a thick cuticle layer outside the stem, and in the lower part of it, epidermis with dense hairs was observed. Collenchyma and chlorenchyma occurred in alternating segments below epidermis. There are sclerenchymatous bundles between chlorenchyma and collenchyma. Scleranchyma cells are in the order of 4-5, and there are intercellular spaces between these cells. Under this layer, phloem and xylem are present, respectively. It has been observed that the vascular bundles under the collenchymatic tissue are larger. The vascular bundles in the stem of studied taxa are arranged in two rings and the type of vascular bundles is bicollateral. The cambium is seen as crushed. The phloem in the inner side of xylem is less visible. The collenchyma tissue was generally located very close to the epidermis with 3-4 rows in *C. urvillei* subsp. *urvillei* and *C. urvillei* subsp. *armata* and 4–6 rows in *C. poypodiifolia* var. *polypodiifolia.* In addition, a chlorenchymatous tissue below the epidermis was observed in the stem cortex with 3-4 rows in *C. urvillei* subsp. *urvillei* and *C. urvillei* subsp. *armata* and 4-5 rows in *C. poypodiifolia* var. *polypodiifolia* (Fig. 2).

Leaf anatomy

Investigation on leaf anatomy showed different types of hairs on the outside of the leaf that were very intense. On the outer side, there is a cuticular layer framing the epidermis. Under the upper and lower epidermis, there is palisade parenchyma. Because the palisade parenchyma is present on both sides, it can be concluded that the leaf type is isolateral (equifacial). Between the two palisade parenchyma, there is an irregularly arranged spongy parenchyma which occupies less area. All three taxa have equifacial leaves. In *C. polypodiifolia* var. *polypodiifolia*, mesophyll tissue consists of 2-3 layers of palisade and 2-3 layers of spongy parenchymatous cells, whereas in *C. urvillei* subsp. *urvillei* and *C. urvillei* subsp. *armata* it is composed of 3 layers of palisade parenchymatous cells and 2 layers of spongy parenchymatous cells (Fig. 3). In surface section, stoma cells were found as amaryllis type. Stomata are surrounded by usually 3, and rarely 4 neighboring cells. The type of stomata is anisocytic (Fig. 3).

Cytological characters

Chromosome numbers and detailed chromosome morphology of three *Centaurea* taxa are reported in this study. In *Centaurea polypodiifolia* Boiss. var. *polypodiifolia*, the chromosome number was found to be 2n=16 and the basic chromosome number of x=8. The shortest chromosome length is 1.92 µm, while the longest is 4.66 µm, and haploid chromosome length is 22.9 µm. The karyotype formula of this taxon is 1M+5m+2sm (Table 2).

In *C. urvillei* subsp. *urvillei*, chromosome number 2n=20 and the basic chromosome number of x=10 is reported. The shortest chromosome length is 2.80 µm, the longest is 5.28 µm and haploid chromosome length is 37.84 µm. The karyotype formula of this taxon is 2M+5m+3sm (Table 2). Satellite was detected on the short arm of chromosome 5 in this taxon. The chromosome number of *C. urvillei* subsp. *armata* is 2n=20, where the basic chromosome number of x=10. The shortest chromosome length is 3.05 μm, the longest 4.74 μm and haploid chromosome length is 41.01 μm. The karyotype formula of this taxon is 2M+5m+3sm (Table 2). Satellite was detected on the short arm of chromosome 3. Other karyotype parameters and asymmetries are given in Table 3. Somatic metaphase chromosomes and idiograms of all three taxa of *Centaurea* are presented in Figures 4 and 5.

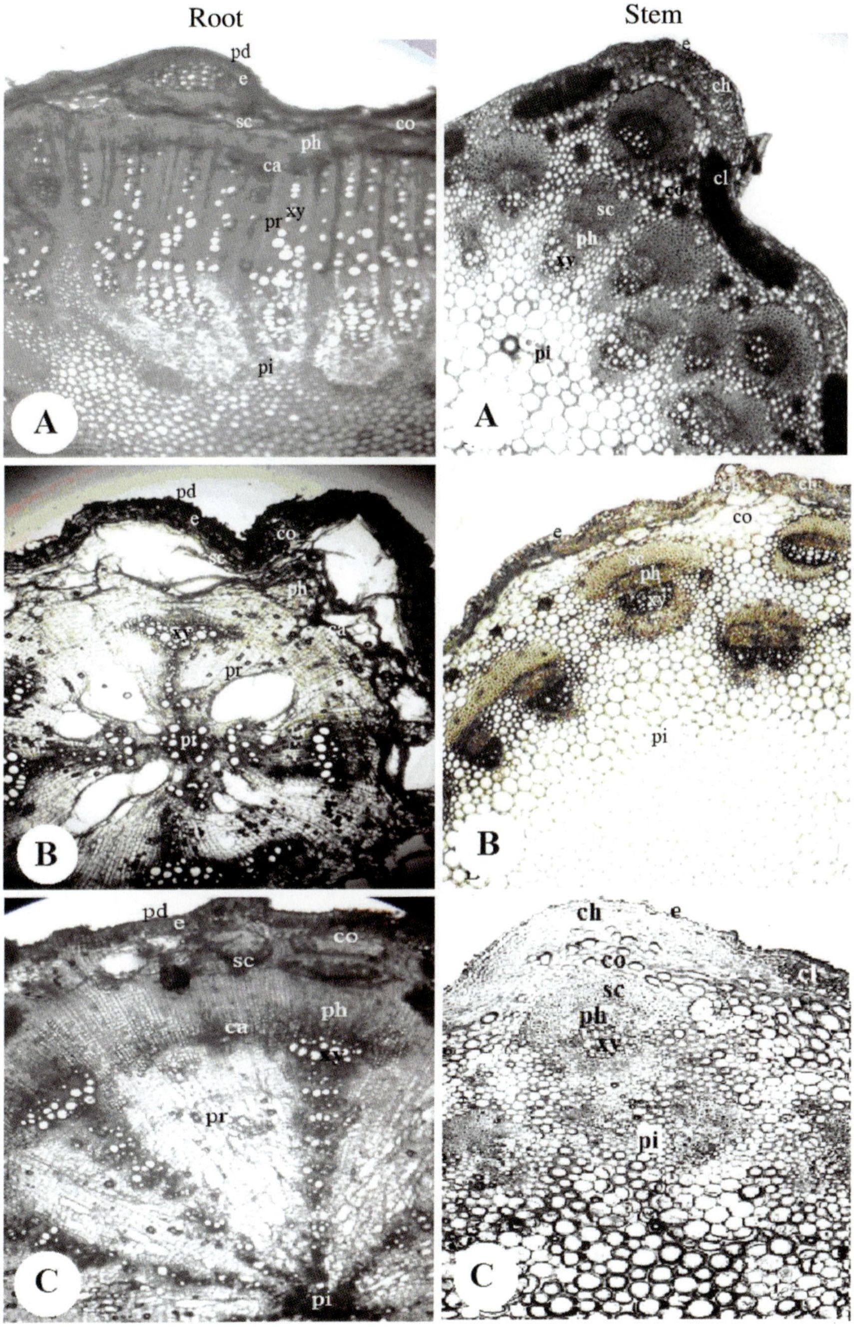

Fig. 2. Cross-sections of stems and roots. A. *Centaurea polypodiifolia* var. *polypodiifolia*; B. *C. urvillei* subsp. *urvillei;* C. *C. urvillei* subsp *armata.* (pd=periderm, e=epidermis, co=cortex, sc=sclerenchyma, ph=phloem, ca=cambium, xy=xylem, pr=pith ray, pi=pith, ch=collenchyma, cl=chlorenchyma).

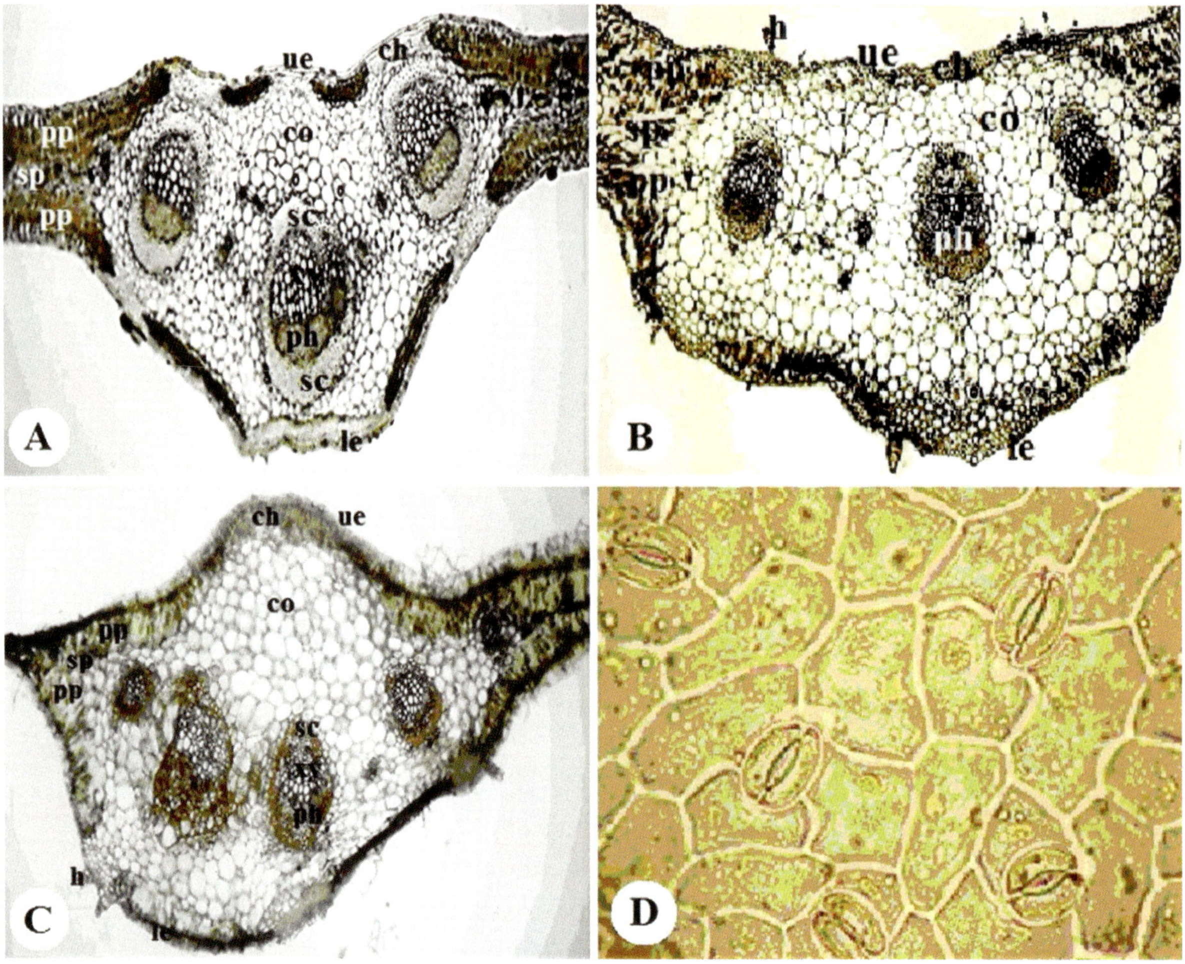

Fig. 3. Cross-sections of leaves of three *Centaurea* taxa. A. *C. polypodiifolia* var. *polypodiifolia*; B. *C. urvillei* subsp. *urvillei;* C. *C. urvillei* subsp. *armata;* D. Surface section of leaf of *C. polypodiifolia* var. *polypodiifolia* (other two taxa similar) (h=hair, ue=upper epidermis, ch=collenchyma, co=cortex, sc=sclerenchyma, ph=phloem, xy=xylem, pp=palisade parenchyma, sp=spongy parenchyma, le=lower epidermis).

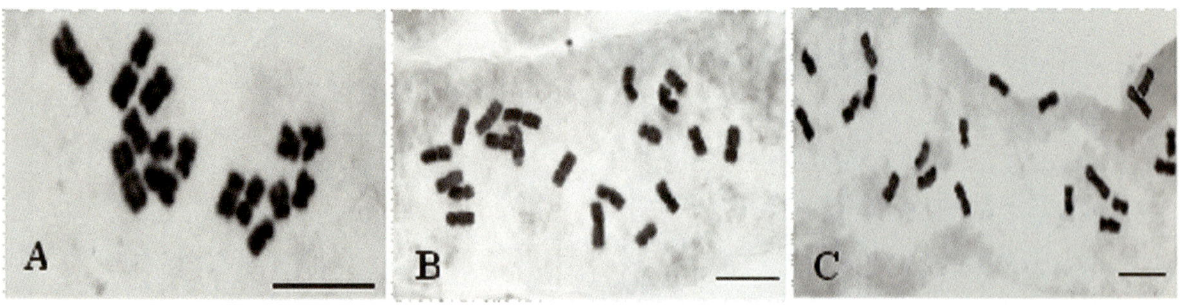

Fig. 4. Somatic metaphase chromosomes of three *Centaurea* taxa. A. *C. polypodiifolia* var. *polypodiifolia* (2n=16); B. *C. urvillei* subsp. *urvillei* (2n=20); C. *C. urvillei* subsp. *armata* (2n=20) (Scale bars: 10 μm).

Table 2. Karyological features of three taxa of *Centaurea* studied.

Cytological characters	*C. polypodiifolia* var. *polypodiifolia*	*C. urvillei* subsp. *urvillei*	*C. urvillei* subsp. *armata*
Chromosome number (2n)	16	20	20
Ploidy level	2x	2x	2x
Karyotype formule	1M+5m+2sm	2M+5m+3sm	2M+5m+3sm
Chromosome length (µm)	1.92-4.66	2.80-5.28	4.74-3.05
Total karyotype length (TKL) (µm)	22.9	37.84	41.01
Intrachromosomal asymmetry index (A_1)	0.29	0,31	0.27
Interchromosomal asymmetry index (A_2)	0.32	0.21	0.14
Karyotype symmetry nomenclature followed Stebbins (SC)	3B	3A	3A
Coefficient of variation of chromosome length (CV_{CL})	32.75	21.70	14.36
Coefficient of variation of centromeric index (CV_{CI})	15.02	22.57	19.80
Karyotype asymmetry index (AI)	4.92	4.89	2.84

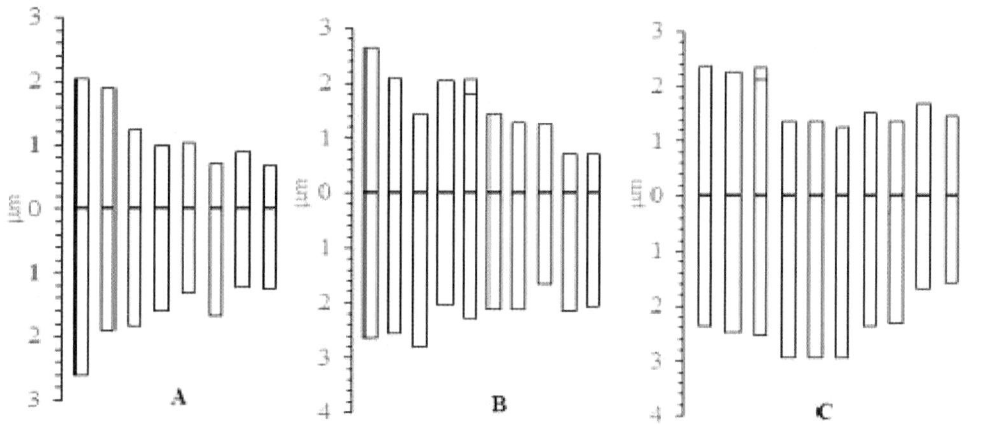

Fig. 5. Idiogram of three *Centaurea* taxa. A. *C. polypodiifolia* var. *polypodiifolia*; B. *C. urvillei* subsp. *urvillei,* C. *C. urvillei* subsp. *armata.*

The present study was carried out in order to provide useful and additional morphological, anatomical and cytological data for three *Centaurea* taxa of Turkey, namely *C. polypodiifolia* var. *polypodiifolia, C. urvillei* subsp. *urvillei* and *C. urvillei* subsp. *armata.* Several attempts have been made to subdivide the genus *Centaurea* (*s.l.*) comprising about 300 problematic species (Garcia-Jacas *et al.*, 2006). Wagenitz (1975) placed these taxa in the two distinct sections of the genus *Centaurea*, viz. section *Acrocentron* (*C. urvillei* subsp. *urvillei*, *C. urvillei* subsp. *armata*) and section *Microlopus* (*C. polypodiifolia* var. *polypodiifolia*) in the Flora of Turkey. *Centaurea polypodiifolia* Boiss. is one of the six species included in the section *Microlophus* (Cass.) DC., and is divided into three varieties based on length of appendages and width of involucre. The appandages of *C. polypodiifolia* var. *polypodiifolia* are very small and easily decidous. The type has narrow upper leaves but larger involucres than most of the other material of this variety and approaches var. *szovitsiana* in this respect. The appandages of *C. polypodiifolia* var. *pseudobehen* are very variable and the upper leaves are narrowly lanceolate. The following gathering is almost intermediate between var. *polypodiifolia* and var. *pseudobehen* (Davis, 1975).

Centaurea urvillei DC. is one of the seventeen species included in the section *Acrocentron* (Cass.) DC. and is a very polymorphic species with five distinct subspecies, connected by transitional forms although the extrems (subsp. *urvillei* and subsp. *hayekiana*) look very different and could easily be placed at species rank. The geographical differentiation is only partial, especially where the areas of subsp. *urvillei* and subsp. *armata* overlap. *C. urvillei* is endemic to Turkey and closely related to *C. lydia* and *C. raphanina sensu lato* (Davis, 1975). *C. urvillei* subsp. *urvillei* shows considerable variation in the length of the appendages (spines) and appendages have been found shortest in some samples collected from Sout-western Anatolia of Turkey.

Wagenitz (1975) explained morphological characters in the genus *Centaurea,* and Metcalfe and Chalk (1979) provided information on anatomical characteristics of the family Asteraceae. In addition to some morphological and anatomical investigations, studies on palynology, phytogeographic distribution and ecology of some *Centaurea s.l.* species in Turkey were presented (Ozler *et al.*, 2009). However, no detailed morphological and anatomical studies of the investigated taxa are available so far.

In the present study we investigated root, stem and leaf anatomy of *C. polypodiifolia* var. *polypodiifolia, C. urvillei* subsp. *urvillei* and *C. urvillei* subsp. *armata*. The general root, stem and leaf structure are almost similar in all examined taxa, but a few differences were determined as shown in Figures 2 and 3. The anatomical characters of all studied taxa have been presented here for the first time. In root of all examined taxa, the crushed epidermis is found below the peridermis. Although the number of phloem cells is less, xylem cells occupy a broader space. The xylem expanded to cover its own pith region.

The collenchyma tissue was generally located very close to the epidermis with 3-4 rows in *C. urvillei* subsp. *urvillei* and subsp. *armata* and 4–6 rows in *C. poypodiifolia* var. *polypodiifolia.* It is well established that the position and the average number of row in collenchyma tissue is important for comparative anatomical studies in plants (Lersten and Curtis, 1997; Makbul *et al.*, 2008). Additionally, a chlorenchymatous tissue below the epidermis was observed in the stem cortex with 3-4 rows in *C. urvillei* subsp. *urvillei* and subsp. *armata* and 4-5 rows in *C. poypodiifolia* var. *polypodiifolia.* This tissue was reported in the genus *Centaurea* in some earlier studies (Uysal *et al.*, 2005; Celik *et al.*, 2005, 2008; Kaya *et al.*, 2010). The vascular bundles in stem of *Centaurea* are generally arranged in 2 rings. Our investigated taxa revealed 2 rings, however, they are sometimes arranged in 1 ring in some taxa of *Centaurea.* Celik *et al.* (2005, 2008) and Kaya *et al.* (2010) reported that vascular bundles are scattered in a circular manner in a single ring in the stem of some *Centaurea* species.

Leaf anatomical properties are used as significant distinctive characters in plant taxonomy (Uysal *et al.*, 2005; Kaya *et al.*, 2010). In cross-sections of leaves we found that vascular bundle was capped by sclerenchymatic fibres at both sides in all the examined taxa (Fig. 3). However, the sclerenchymatic fibres were more prominent in *C. polypodifolia* var. *polypodifolia.* All studied taxa have equifacial leaves, with a mesophyll tissue that consists of 2-3 layers of palisade and 2-3 layers of spongy parenchymatic cells of *C. polypodiifolia* var. *polypodiifolia*, 3 layers of palisade parenchymatic cells and 2 layers of spongy parenchymatic cells in *C. urvillei* subsp. *urvillei* and *C. urvillei* subsp. *armata*, respectively.

The importance of karyomorphology in *Centaurea* has long been realized (Garcia-Jacas and Susanna, 1992). In the genus *Centaurea* in Turkey the somatic chromosome number varies from 2n=16 to 66 (Wagenitz, 1975; Martin *et al.*, 2006; Inceer *et al.*, 2007). The basic chromosome number of *Centaurea* species varies from x=7 to 16 and four ploidy levels (2x, 3x, 4x, 6x) have been determined (Gomurgen *et al.*, 2010; Meric *et al.*, 2010; Kocyigit and Bona, 2013). The cromosome number of *C. polypodiifolia* var. *polypodiifolia* in the present study was found to be 2n=2x=16, and this finding was found consistent with Ghaffari and Shahraki (2001). In *C. urvillei* subsp. *urvillei* and subsp *armmata* 2n=2x=20 were found and our results were supported by previous studies (Gardou and Tchehrehgosha, 1975; Garcia-Jacas *et al.*, 1997), although different chromosomal counts for *C. urvillei* subsp. *urvillei* (2n=40) have also been made (Martin *et al.*, 2009). In this study we found anatomical and cytological characters as useful in the taxonomy of three *Centaurea* taxa in Turkey. Investigation of anatomical features were first of its nature among the studied taxa. The findings of the present study would be useful for contributing to systematics of this genus *Centaurea* and to infer interspecific relationships based on anatomical and karyological characters.

Acknowledgements

The financial support from the Scientific Investigation Project Coordinator of Firat University (Project number: FF.12.02) is gratefully acknowledged.

References

Bancheva, S., Kaya, Z. and Binzet, R. 2014. Morphological, cytological and palynological features of three closely related *Centaurea* species (Asteraceae) from Turkey. Mod. Phytomorphol **5**: 79–84.

Behçet, L., Ilçim. A. and Yapar, Y. 2017. *Centaurea bingoelensis* (Asteraceae), a new species from Turkey. Turk. J. Bot. **41**: 180–188.

Celik, S., Uysal, T., Menemen, Y. and Karabacak, E. 2005. Morphology, anatomy, ecology, pollen and achene structure of *Centaurea consanguinea* DC. (Sect. *Acrolophus*) in Turkey. Int. J. Bot. **1**: 85–89.

Celik, S., Uysal, T. and Menemen, Y. 2008. Morphology, anatomy, ecology and palynology of two *Centaurea* species from Turkey. Bangladesh J. Bot. **37**: 67–74.

Davis, P.H. (Ed.). 1975. Flora of Turkey and the East Aegean Islands. Vol. **5**. Edinburgh University Press, UK.

Duran, A. and Duman, H. 2002. Two new species of *Centaurea* (Asteraceae) from Turkey. Ann. Bot. Fenn. **39**: 43–48.

Garcia-Jacas, N. and Susanna, A. 1992. Karyological notes on *Centaurea* sect. *Acrocentron*. Plant Syst. Evol. **179**(1): 1–18.

Garcia-Jacas, N., Susanna, A., Ilarslan, R. and Ilarslan, H. 1997. New chromosome counts in the subtribe Centaureinae (Asteraceae, Cardueae) from West Asia. Bot. J. Linn. Soc. **125**: 343–349.

Garcia-Jacas, N., Susanna, A., Garnatje, T. and Vilatersana, R. 2001. Generic delimitation and phylogeny of the subtribe Centaureinae (Asteraceae): a combined nuclear and chloroplast DNA analysis. Ann. Bot. **87**: 503–515.

Garcia-Jacas, N., Uysal, T., Romashchenko, K., Suarez-Santiago, V.N., Ertugrul, K. and Susanna, A. 2006. *Centaurea* Revisited: A molecular survey of the Jacea Group. Ann. Bot. **98**: 741–753.

Gardou, C. and Tchehrehgocha, I. 1975. IOPB chromosome number reports XLIX. *In*: Love, A. (Ed). Taxon **24**: 501–516.

Ghaffari, S.M. and Shahraki, M.A. 2001. Some chromosome counts and meiotic behavior in *Centaurea* species from Iran. Iranian J. Bot. **9**: 11–18.

Gomurgen, A.N., Potoglu-Erkara, I. and Altınozlu, H. 2010. Chromosome and pollen morphology of the rare endemic *Centaurea lycopifolia* Boiss. & Kotschy. Bangladesh J. Bot. **39**: 223–228.

Guner, A., Aslan, S., Ekim, T., Vural, M. and Babac, M.T. 2012. Türkiye bitkileri listesi (Damarlı bitkiler). Nezahat Gökyiğit Botanik Bahçesi ve Flora Araştırmaları Derneği Yayını: İstanbul.

Hayta, S., Tasar, N., Kiran, Y., Cakılcıoglu, U. and Bagci, E. 2016. Morphological, karyological and palynological investigation of endemic *Centaurea kurdica* Reichardt from Turkey. Plant Biosyst. **150**(3): 484–489.

Inceer, H., Hayırlıoglu-Ayaz, S. and Ozcan, M. 2007. Chromosome numbers of the twenty-two Turkish plant species. Caryologia **60**: 349–357.

Kaya, Z., Orcan, N. and Binzet, R. 2010. Morphological, anatomical and palynological study of *Centaurea calcitrapa* L. ssp. *cilicica* (Boiss. & Bal.) Wagenitz and *Centaurea solstitialis* L. ssp. *carneola* (Boiss.) Wagenitz endemic for Turkey. Pakistan J. Bot. **42**: 59–69.

Kocyigit, M. and Bona, M. 2013. Chromosome numbers of five Turkish *Centaurea* L. (Asteraceae) species. Plant Biosyst. **147**: 970–978.

Kose, Y.B., Alan, S. and Yucel, E. 2010. Comparative investigation of the morphological characteristics of species belonging to the *Centaurea* L. section *Phalolepis* (Cass.) DC. Biol. Div. Conser. **3**: 10–22.

Lersten, N.R. and Curtis, J.D. 1997. Anatomy and distribution of foliar idioblasts in *Scrophularia* and *Verbascum* (Scrophulariaceae). Am. J. Bot. **84**: 1638–1645.

Levan, A., Fredga, K. and Sandberg, A.A. 1964. Nomenclature for centromeric position on chromosomes. Hereditas **52**: 201–220.

Lu, H.F., Jiang, B., Shen, Z.G., Shen, J.B., Peng, Q.F. and Cheng, C.G. 2008. Comparative leaf anatomy, FTIR discrimination and biogeographical analysis of *Camellia* section *Tuberculata* (Theaceae) with a discussion of its taxonomic treatments. Plant Syst. Evol. **274**: 223–235.

Makbul, S., Turkmen, Z., Coskuncelebi, K. and Beyazoglu, O. 2008. Anatomical and pollen characters in the genus *Epilobium* L. (Onagraceae) from Northeast Anatolia. Acta Bio. Crac. Ser. Bot. **50**: 57–67.

Martin, E., Dinc, M., Duran, A. and Ozturk, M. 2006. Karyological studies on *Lotus strictus* Fisher and C.A. Mey. (Leguminosae), *Centaurea amanicola* Hub.-Mor. (Compositae) and *Teucrium lamiifolium* Durv. subsp. *lamiifolium* (Labiatae). Am-Eur. J. Sci. Res. **1**: 12–17.

Martin, E., Dinc, M. and Duran, A. 2009. Karyomorphological study of eight *Centaurea* L. taxa (Asteraceae) from Turkey. Turk. J. Bot. **33**: 97–104.

Meric, C., Arda, H., Guler, N. and Dayan, S. 2010. Chromosome number and nuclear DNA content of *Centaurea kilaea* (Asteraceae), an endemic species from Turkey. Phyt. Balcanica **16**: 79–84.

Metcalfe, C.R. and Chalk, L. 1979. Anatomy of Dicotyledones I. Oxford University Press, Oxford, pp. 783–803.

Negaresh, K. and Rahiminejad, M.R. 2014. A contribution to the taxonomy of *Centaurea* sect. Cynaroides (Asteraceae, Cardueae–Centaureinae) in Iran. Phytotaxa **158**: 229–244.

Noman, A., Ali, Q., Hameed, M., Mehmood, T. and Iftikhar, T. 2014. Comparison of leaf anatomical characteristics of *Hibiscus rosa-sinensis* grown in Faisalabad Region. Pakistan J. Bot. **46**: 199–206.

Ozcan, M. 2013. Türkiye'de yetişen *Psephellus pulcherrimus* (syn: *Centaurea pulcherrima* var. *freynii*) (Cardueae, Asteraceae)'un morfolojik ve anatomik özellikleri. Artvin Çoruh Univ. Orman Fak. Der. **14**: 104–112.

Ozler, H., Kaya, Z. and Pehlivan, S. 2009. Pollen morphology of some *Centaurea* L., *Psephellus* Cass. and *Cyanus* Miller taxa. Acta Bio. Crac. Ser. Bot. **51**: 53–66.

Paszko, B. 2006. A critical review and a new proposal of karyotype asymmetry indices. Plant Syst. Evol. **258**: 39–48.

Rahiminejad, M.R., Mozaffarian, V. and Montazerolghaem, S. 2010. A taxonomic revision of *Centaurea* section *Acrocentron* (Asteraceae) in Iran. Bot. J. Linn. Soc. **163**: 99–106.

Ranjbar, M. and Negaresh, K. 2013. A revision of *Centaurea* sect. *Phaeopappus* (Asteraceae, Cardueae). Phytotaxa **123**(1): 1–40.

Romero-Zarco, C. 1986. A new method for estimating karyotype asymmetry. Taxon **35**: 526–530.

Stebbins, G.L. 1971. Chromosomal Evolution in Higher Plants. Edward Arnold. London, UK.

Türkoglu, I., Akan, H. and Civelek, E. 2003. A new species of *Centaurea* (Asteraceae: sect. *Psephelloideae*) from Turkey. Bot. J. Linn. Soc. **143**: 207–212.

Uysal, I., Celik, S. and Menemen, Y. 2005. Morphology, anatomy, ecology, pollen and achene features of *Centaurea polyclada* DC. (Sect. *Acrolophus*) in Turkey. J. Bio. Sci. **5**: 176–180.

Uysal, T., Ertugrul, K., Susanna, A. and Garcia-Jacas, N. 2009. New chromosome counts in the genus *Centaurea* (Asteraceae) from Turkey. Bot. J. Linn. Soc. **159**: 280–286.

Uysal, T., Hamzaoglu, E., Ertugrul, K. and Bozkurt, M. 2016. A new species of *Centaurea* (Asteraceae) from Turkey. Phytotaxa **275**(2): 149–158.

Wagenitz, G. 1975. *Centaurea* L. *In*: Davis, P.H. (Ed.), Flora of Turkey and the East Aegean Islands. Vol. **5**. Edinburgh University Press, UK.

Wagenitz, G. 1986. *Centaurea* in South-West Asia: Patterns of distribution and diversity. Edinburgh Proceedings of the Royal Society **89**: 11–21.

Yakar-Tan, N. 1982. Bitki Mikroskopisi Klavuz Kitabı. İstanbul University, Faculty of Science Publishing, İstanbul, Turkey, 166 pp.

Yüzbaşıoğlu, İ.S., Bona, M. and Genç, İ. 2015. A new species of *Centaurea* sect. *Pseudoseridia* (Asteraceae) from north-eastern Turkey. PhytoKeys **53**: 27–38.

MORPHOLOGICAL AND ANATOMICAL STUDIES OF THE NEWLY RECORDED *RHUS CHINENSIS* MILL. (ANACARDIACEAE) FROM TURKEY

Özgür Eminagaoglu[1] and Melahat Ozcan

Department of Forest Engineering, Faculty of Forestry, Artvin Coruh University, 08000 Artvin, Turkey

Keywords: *Rhus chinensis*; Anatomy; Chromosome number; SEM; New records; Turkey.

Abstract

Rhus chinensis Mill. (Anacardiaceae) has been reported as a new record for the flora of Turkey. Detailed morphological description and leaf anatomical properties are provided. Capitate glandular and nonglandular trichomes, and also epicuticle hairs have been observed in the leaf surfaces of *R. chinensis*. Fruit micromorphology and chromosome number of this species have also been evaluated.

Introduction

Anacardiaceae Lindl., the Cashew family, includes more than 700 species in the world belonging to 82 genera that are primarily distributed pantropically. The genus *Rhus* L., the sumac genus (Spondiadoideae, Rhoeae), is one of the most widespread and recognizable genera in the north temperate zone, includes approximately 250 species from subtropical and warm temperate regions of the world. The genus is divided into two subgenera: *Rhus* L. and *Lobadium* (Raf.) Torr. & A.Gray (Barkley, 1937; Yi *et al.*, 2007).

During field survey in Artvin Province (Turkey), some interesting specimens belonging to Anacardiaceae were collected. After critical study and consultation with relevant literature (Davis *et al.*, 1967; Linchevskii, 1974; Davis, 1988; Güner *et al.*, 2000; Hsu and Su, 2013; Eminağaoğlu, 2015), these specimens were identified as *Rhus chinensis* Mill. The species was not reported earlier from Turkey (Eminağaoğlu and Anşin, 2003; Eminağaoğlu and Anşin, 2004; Eminağaoğlu *et al.*, 2008; Özhatay *et al.*, 2011; Eminağaoğlu and Ozcan, 2013, 2014; Yuksel and Eminağaoğlu, 2017). The number of species of *Rhus* in the flora of Turkey is increased to 2 with the addition of this species. In this study, we describe detailed morphological characters of *Rhus chinensis*, and investigate the anatomical and cytological properties of the species.

Material and Methods

Morphological analysis

Plant materials were collected from different parts of Artvin, Turkey, at different altitudes between 2013 and 2017. The collected materials were critically studied. The voucher specimens have been deposited at the Herbarium of Artvin Coruh University (ARTH), Artvin, Turkey.

Anatomical preparation

For anatomical investigation leaf samples were stored in 70% alcohol. Transverse sections of leaf, and paradermal sections of upper and lower epidermis of leaves were prepared manually using commercial razor blades and stained in Haematoxylin for about 15 min. To remove the excess stain, sections were washed in water several times (Algan, 1981). Semi-permanent slides were mounted in glycerin or permanent slides were covered with glycerin-gelatin (Vardar, 1987).

---

[1]Artvin Çoruh Üniversitesi, Orman Fakültesi, Orman Mühendisliği, Orman Botaniği Anabilim Dalı, Seyitler Yerleşkesi, 08000 Artvin, Turkey. Corresponding author. Email: oeminagaoglu@artvin.edu.tr

Well stained sections were examined under a light microscope and photographed using an Olympus BX-53 microscope with digital camera attachment DP-73.

Micromorphological analysis

Micromorphological features of the fruits were studied using a stereomicroscope (Leica M60 with a digital camera attachment DFC 295) and a scanning electron microscope (Zeiss Evo LS 10, ACU-Biltekmer). For scanning electron microscopy, fruits and seed covered with endocarp were separately placed on stubs using double-sided adhesive tape, and coated with gold in a Cressington sputter coater 108 auto coating apparatus for 2 min. Fruits and endocarp were examined and photographed from the same region (from the middle part of the lateral region).

Chromosome count

For mitotic chromosome observation, root tips were cut off and pretreated with 1-Bromonaphtalen at 4 °C for 16 h (Ozcan *et al.*, 2011), then fixed in fresh Carnoy absolute alcohol-glacial acetic acid (3:1) for 24 h at 4ºC. For chromosome counts, root tips were hydrolyzed in 5N HCl for 3-5 min at room temperature and then rinsed with distilled water for 2-3 min. Staining was carried out in lacto-propionic orcein at least for 3 h at room temperature. Permanent slides were prepared from at least ten well-spread cells. The best metaphase plates were photographed with Olympus BX-53 microscope with digital camera attachment DP-73.

Results and Discussion

Rhus chinensis Mill. Gard. Dict. ed. 8, n. 7 (1768). *R. osbeckii* Carrière, Rev. Hort. [Paris]: 111 (1887); *R. osbeckii* Steud. Nomencl. Bot. [Steudel], ed. 2, 2: 452 (1841); *R. semialata* Murray, Commentat. Soc. Regiae Sci. Gott. 6: 27, t 3 (1784); *Schinus indicus* Burm. f., Fl. Ind.: 215 (err. typ. 315) (1768). **(Fig. 1)**.

Shrub to small deciduous tree, 2–10 m tall; branchlets ferruginous pubescent, lenticellate. Leaves sessile, imparipinnately compound; rachis broadly winged, ferruginous pubescent; leaflets 9-13, leaflet blade ovate to oblong, 7–13×3–7 cm, increasing in size towards apex, adaxially dark-green, sparsely pubescent or glabrescent, abaxially pale-green, glaucous, and ferruginous pubescent, base rounded to cuneate in terminal leaflet, margin dentate, often crenate, apex acute, lateral veins and reticulate venation impressed adaxially and prominent abaxially; petiole 4–9 cm long, densely pubescent. Flowers dioecious; inflorescence panicle, densely ferruginous pubescent. Staminate flowers 35–40 mm long, pedicel short, c. 1 mm, minutely pubescent; calyx pubescent, lobes ovate, c. 1 mm long, margin ciliate; corolla obovate-oblong, white, c. 2 mm long; filaments c. 2 mm long; anthers ovoid, c. 0.7 mm; disk annular, yellow; ovary very reduced or absent. Pistilate flowers: corolla elliptic-ovate, white, c. 1.6–2.0 mm long; calyx lobes c. 0.6 mm long; staminodes 4 or 5, disk annular, yellow; ovary ovoid, c. 1 mm long, densely white pubescent; styles 3; stigma capitate. Drupe globose, slightly compressed, 4–5 mm in diam., mixed pilose and glandular-pubescent, red at maturity.

Flowering period: June to August; *fruiting period*: September to November.

Specimens examined: A8 Artvin, Hopa, Kemalpaşa, damp roadside, grassland, 15m, 27 July 2013, 41°30′33″N, 41°32′15″E, *Ö. Emin.* 18677 (ARTH 5250); Artvin, Hopa, Kemalpaşa, 36m, 29 September 2013, 41°27′52″N, 41°29′9″E, *Ö. Emin.* 19101 (ARTH 5251); Artvin, Hopa, Kemalpaşa, damp roadside, grassland, 10m, 8 May 2014, 41°28′22″N, 41°30′17″E, *Ö. Emin.* 19364 (ARTH 5269); Artvin, Arhavi, roadside, 40m, 5 June 2017, 41°20′47″N, 41°16′44″E, *Ö. Emin.* 22360 (ARTH 11342).

Fig. 1. *Rhus chinensis* Mill.: A. Winged rachis with pinnate leaves; B. Flowering branch; C. Pistillate flowers; D. Staminate flowers; E. Fruits. Scale bars: A=3 cm, B=5 mm, C=2 mm, D=8 mm, E=2 mm.

Distribution: China, Manchuria, India, Thailand, Korea, Japan, Tibet, Taiwan, North Korea, South Korea, Malaysia, India, Bhutan, Myanmar, Pakistan, Jammu and Kashmir, Laos, Thailand and Cambodia (Hassler, 2017). New record for Turkey.

Ecology: *R. chinensis* grows on damp, humid and gravelly soil including roadsides and forest sides at 20-100 m with *Carex pendula* Huds., *Corylus avellana* L., *Equisetum fluviatile* L., *Hypericum xylosteifolium* (Spach.) N. Robson, *Juncus effusus* L., *Plantago major* L., *Polygonum aviculare* L, *Oxalis corniculata* L., *Sambucus nigra* L. and *Sisyrinchium angustifolium* Mill.

Uses: Different parts of *R. chinensis* including root, bark, stem, leaf, flowers, fruit, seed and gall are used in the treatment of hemoptysis, inflammations, cough, dysentery, fever, jaundice, malaria, rheumatism, laryngitis, snakebite, stomach-ache and traumatic fractures. Galls on the plant are also used for treatment of diarrhoea, hemorrhage, ulcer of mouth, diabetes, and rectal and intestinal cancer (Djakpo and Yao, 2010).

Leaf anatomical characteristics

Midrib: It is semi-circle and covers a large area. Under the upper and lower epidermis, several layers of collenchyma cells are observed. 12-15 collateral vascular bundles arranged in a circle. The larger bundles present near to upper epidermis are thin walled parenchymatous cells filled in the pith. At least one large resin duct is present in each vascular bundle. Sclerenchymatous sheath can be visible in phloem of vascular bundle as a cap (Fig. 2).

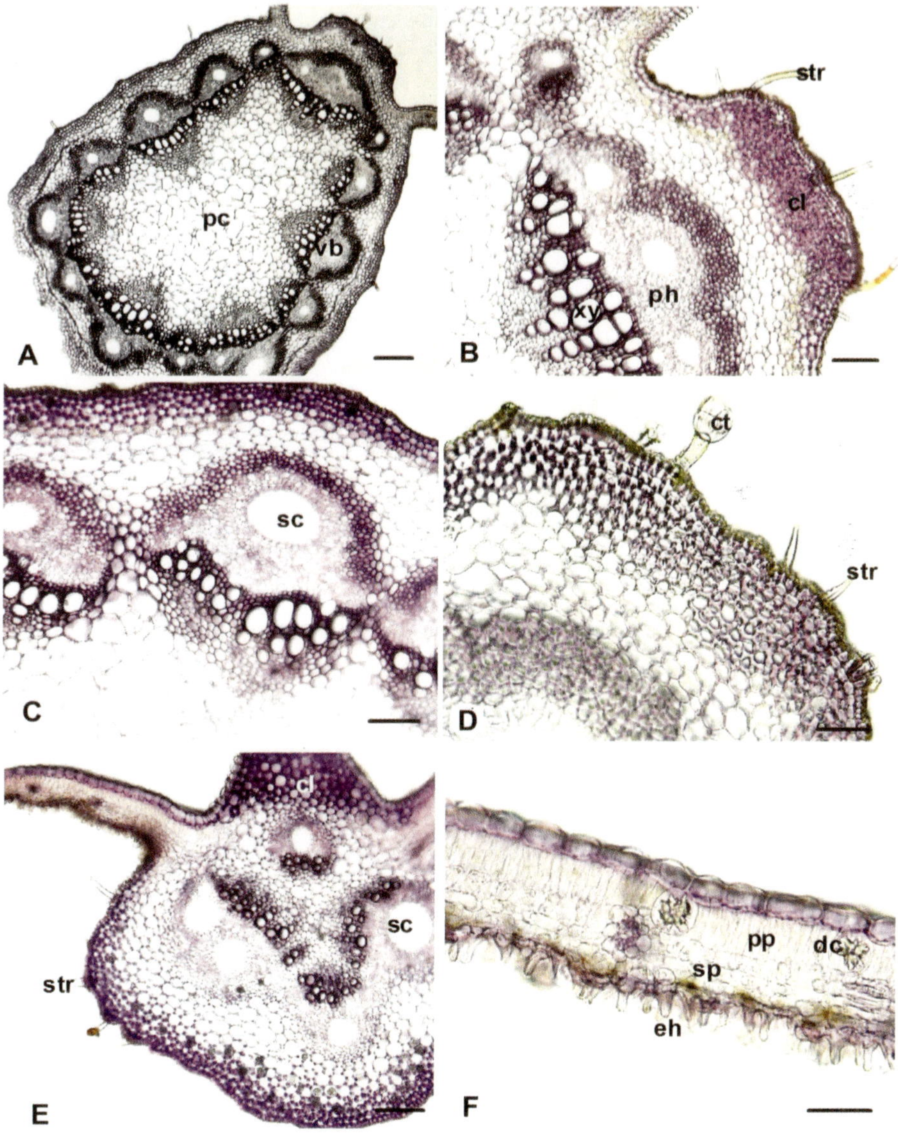

Fig. 2. Leaf anatomy of *Rhus chinensis*: A-B. Midrib; C. Vascular bundles; D. Trichomes; E. Veinlet; F. Lamina. cl: collenchyma, ct: capitate trichome, eh: epicuticular hair, pc: pith cell, ph: phloem, pp: palisade parenchyma, sc: secretory canal (duct), dc: druse crystals, sp: spongy parenchyma, str: simple trichome, xy: xylem, Scale bars: A = 200 μm; B,C,E = 100 μm; D,F = 50 μm.

Lamina: The leaves show dorsiventral mesophyll. It is composed of single palisade layer and 3-4 spongy layers. The palisade tissue covers in equal areas to spongy parenchyma. Epidermis is covered with a thick cuticle. Upper epidermal cells with straight walls are distinctly larger than the lower ones (Fig. 2F). Non-glandular and capitate glandular trichomes are sparsely observed in the adaxial surface, but abaxial one included densely non-glandular multicellular and capitate trichomes with 4-8 head cells and epicuticle hairs. It bears hypostomatic type stomata which are found only in abaxial side of leaf (Fig. 3). They are sunken in lower epidermis (Fig. 2C, D).

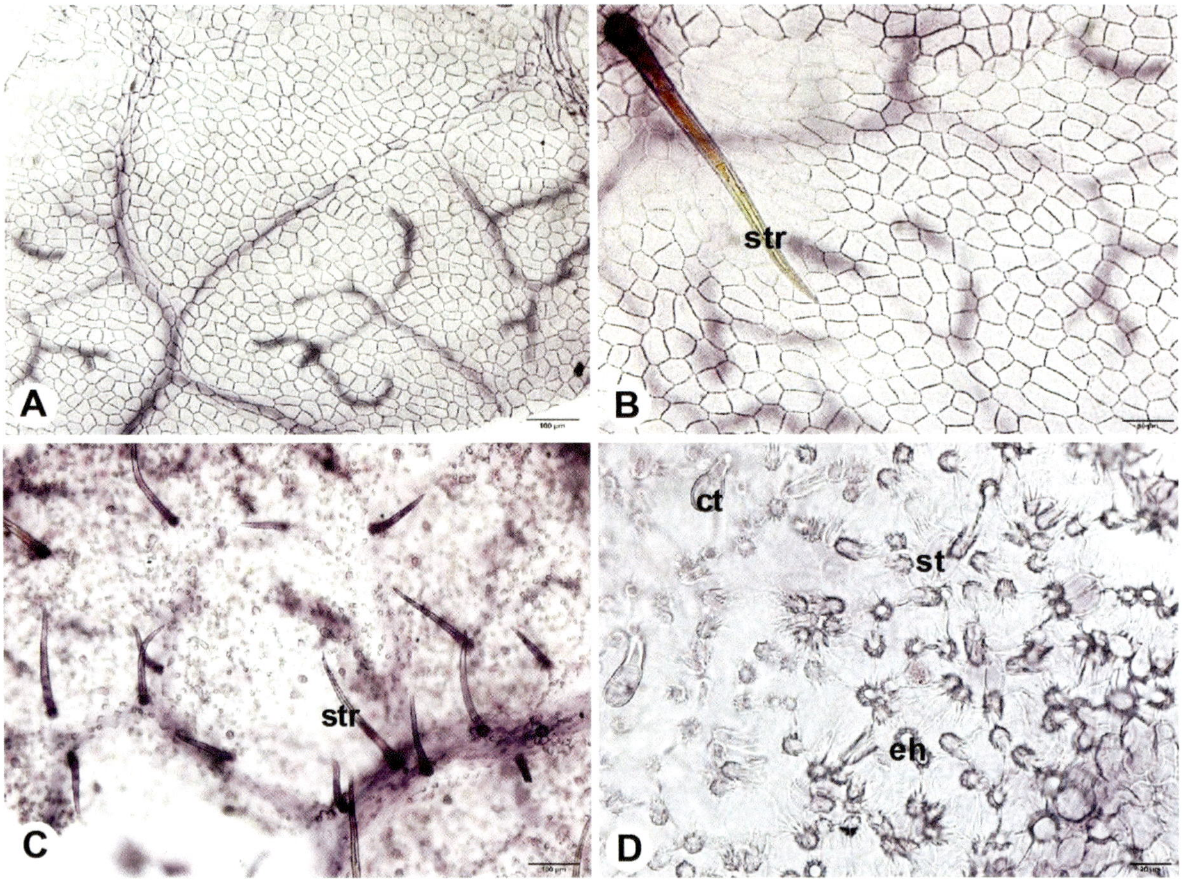

Fig. 3. Paradermal section of leaf. A,B. Adaxial epidermis; C,D. Abaxial epidemis. eh: epicuticular hair; st: stomata; str: simple trichome. Scale bars: A,C. = 100 μm; B,D. = 50 μm.

Crystals: Druse crystal compounds are present in some collenchymatous cells of midrib and inside of the palisade cells of lamina. They are much bigger (Fig. 2F)

R. chinensis has dorsiventral mesophyll with single layer of palisade cells and 3-4 layers of spongy cells. Yanping *et al.* (2001) investigated formation of gall in this species and showed the differences in the ratio of palisade tissue and spongy tissue. In the present study, we observed three types of trichomes in the leaf of *R. chinensis* (Fig. 3). Mobius (1899), and Rost and Gilg (1912) reported two different types of trichomes (thick-walled bristle hairs and club-shaped trichomes) in *R. vernicifera* L. and *R. toxicodendron* L., respectively. Liu *et al.* (2008) found epicuticular hairs in *R. chinensis*. In the present study, crystal compounds were observed in *R. chinensis* (Fig. 2). McNair (1921) reported these types of crystals in *R. diversiloba* Torr. & A. Gray [*Toxicodendron diversilobum* (Torr. & A.Gray) Greene]. McNair (*l.c.*) also reported resin ducts in *R. diversiloba*. Harada (1932) reported resin ducts in the petiole and veinlets of *R. succedanea* L. In the present study similar results were found for *R. chinensis* as like in *R. diversiloba*. Therefore our results are in accordance with previous studies.

Fruit micromorphology

Fleshy fruit has reddish colour with round shape, and endocarp is laterally compressed. Fruit length ranged from 4.5 to 4.7 mm. Results of fruit micromorphology are shown in Fig. 4. Exocarp cells are undulate and more or less inflated with striate surface. Epidermal surface was covered with glandular trichomes. Endocarp has regular sclerified cells.

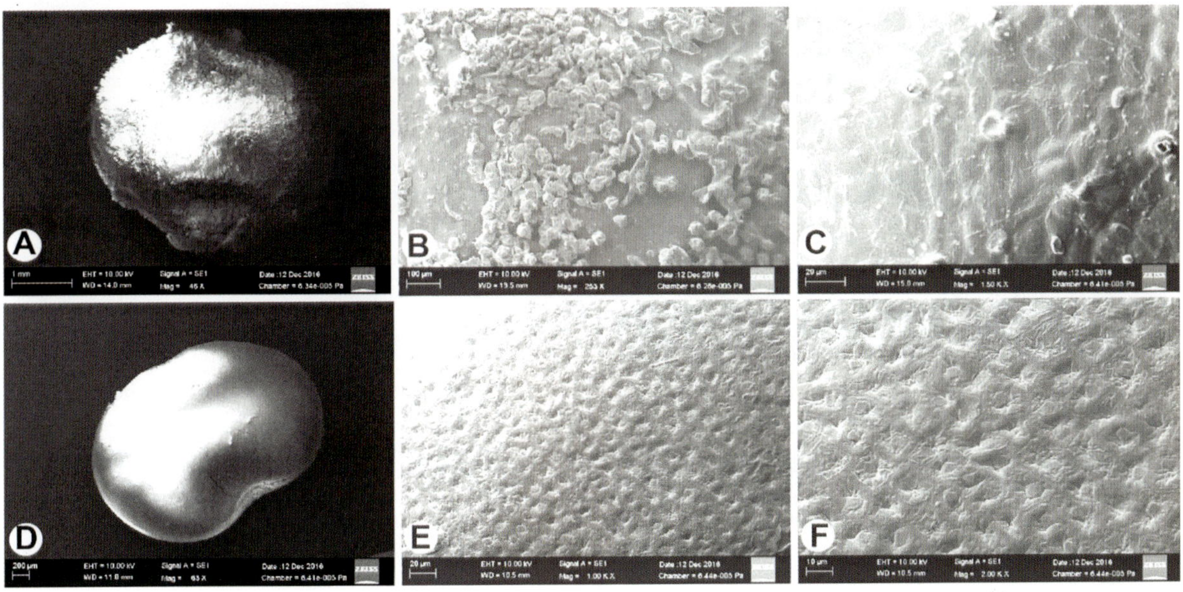

Fig. 4. SEM micrography of fruit of *Rhus chinensis*. A-C. Fruit; D-F. Endocarp.

Chromosome number

Somatic chromosome number of *R. chinensis* has been determined as $2n=2x=30+0\text{-}2B$ (Fig.5).

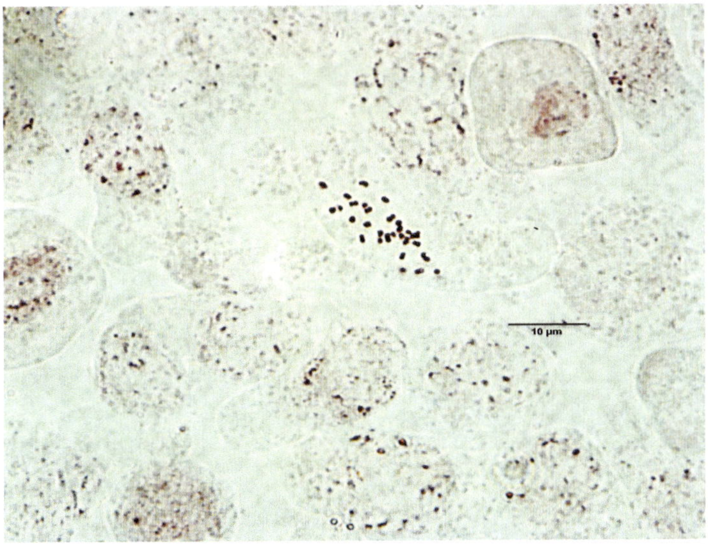

Fig. 5. Somatic metaphase of *Rhus chinensis*.

Two B chromosome have also been found. In a previous report, Shang *et al.* (1990) documented chromosomes of four *Rhus* species, and reported chromosome number of *R. chinensis* as $2n = 30 + 0\text{-}5B$ and chromosome numbers of other three taxa as $2n = 30$. However, triploid cultivar was also detected in their study for *R. verniciflua* as $2n = 3x = 45$. Parfitt *et al.* (1990) found gametic chromosome number of *R. aromatica* Aiton var. *pilosissima* (Engl.) Shinners as $n = 15$. Our results are in accordance with those previous reports and only diploid number was observed for *R. chinensis* in this study.

Acknowledgements

The authors are thankful to the Ziraat Bank in Turkey for their financial supports.

References

Algan, G. 1981. Bitkisel Dokular İçin Mikroteknik, İstanbul. Fırat University Science Faculty Press (in Turkish), 94 pp.

Barkley, F.A. 1937. A monographic study of *Rhus* and its immediate allies in North and Central America, including the West Indies. Ann. Miss. Bot. Gard. **24**: 265–498.

Davis, P.H. (Ed.). 1988. Flora of Turkey and the East Aegean Islands. Vol. **10**. Edinburgh University Press, Edinburgh, 590 pp.

Davis, P.H., Coode, M.J.E. and Cullen, J. 1967. *Rhus* L. *In*: Davis, P.H. (Ed.), Flora of Turkey and the East Aegean Islands, Vol. **2**. Edinburgh University Press, Edinburgh, pp. 543–544.

Djakpo, O. and Yao, W. 2010. *Rhus chinensis* and *Galla chinensis* folklore to modern evidence: review. Phytother. Res. **24**: 1739–1747.

Eminağaoğlu, Ö. (Ed.) 2015. Artvin'in Doğal Bitkileri (Native Plants of Artvin). Promat Press, Istanbul, 456 pp.

Eminağaoğlu, Ö, Kutbay, H.G., Özkan, Z.C. and Ergül, A. 2008. Flora of the Camili Biosphere Reserve Area (Borçka, Artvin, Turkey). Turk. J. Bot. **32**: 43–90.

Eminağaoğlu, Ö. and Anşin, R. 2003. The flora of Hatila Valley National Park and its close environs (Artvin). Turk. J. Bot. **27**: 1–27.

Eminağaoğlu, Ö. and Anşin, R. 2004. Flora of the Karagol-Sahara National Park (Artvin) and its environs. Turk. J. Bot. **28**: 557–590.

Eminağaoğlu, Ö. and Ozcan, M. 2013. *Euonymus leiophloeus* (Celastraceae) - A new record for the flora of Turkey. Bangladesh J. Plant Taxon. **20**(2): 263–266.

Eminağaoğlu, Ö. and Ozcan, M. 2014. Systematics of *Sisyrinchium angustifolium* Mill. (Iridaceae), a newly recorded species from Turkey. Bangladesh J. Plant Taxon. **21**(2): 175–180.

Güner, A., Özhatay, N., Ekim, T. and Başer, K.H.C. (Eds). 2000. Flora of Turkey and the East Aegean Islands. Vol. **11**. Edinburgh University Press, Edinburgh, 680 pp.

Harada, M. 1932. Anatomical study on the leaves of *Rhus succedanea*, especially on the distribution of the resin canal. J. Soc. Forestry **14**(4): 238–250.

Hassler, M. 2017. World Plants: Synonymic Checklists of the Vascular Plants of the World. *In*: Roskov, Y., Abucay, L., Orrell, T., Nicolson, D., Bailly, N., Kirk, P.M., Bourgoin, T., DeWalt, R.E., Decock, W., De Wever, A., Nieukerken, E. van Zarucchi, J. and Penev, L. (Eds), Species 2000 and ITIS Catalogue of Life. Naturalis, Leiden.www.catalogueoflife.org [Accessed on 7 June 2017].

Hsu, T.W. and Su, M.H. 2013. A taxonomic revision of *Rhus chinensis* Mill. (Anacardiaceae) in Taiwan. Taiwan J. For. Sci. **28**(3): 145–151.

IUCN 2017. The IUCN Red List of Threatened Species. Version 2017-1. http://www.iucnredlist.org. [Accessed on 07 June 2017].

Linchevskii, I.A. 1974. *Rhus* L. *In*: Komarov, V.L. (Ed.), Flora of the U.S.S.R, Vol. **14**, Jerusalem, Israel Program for Scientific Translation, pp. 530–531.

Liu, Y., Zhao, M., Li, Z. and Jiang, J. 2008. Adaptation of *Rhus chinensis* to drought stress in Karst area: Insights from anatomical structure of leaf. Carsologica Sinica **27**(3): 235–239.

McNair, J.B. 1921. A study of *Rhus diversiloba* with special reference to its toxicity. Am. J. Bot. **8**: 127–146.

Mobius, M.A. 1899. Der Japanische Lackbaum. Abhandl. Senckenberg. Naturforsch. Ges. **20**: 201–247.

Ozcan, M., Hayırlıoğlu-Ayaz, S. and Inceer, H. 2011. Chromosome reports in some *Cirsium* (Asteraceae, Cardueae) taxa from north-east Anatolia. Caryologia **64**(1): 55–66.

Özhatay, F.N., Kültür, Ş. and Gürdal, M.B. 2011. Checklist of additional taxa to the supplement Flora of Turkey V. Turk. J. Bot. **35**: 589–624.

Parfitt, B.D., Pinkava, D.J., Rickel, D., Fillipi, D., Eggers, B. and Keil. D.J. 1990. Documented chromosome numbers 1990: 1. Miscellaneous North American vascular plants. Sida **14**: 305–308.

Rost, E. and Gilg, E. 1912. Der Giftsumach, *Rhus toxicodendron* L., und seine Giftwirkungen. Ber Deutsch. Pharm. Ges. **22**: 296–358.

Shang, Z.Y., Zhang, J.Z., Li, R.J. and Liu, Q.H. 1990. The chromosome observation on four species in the genus *Rhus*. J. Wuhan Bot. Res. **8**: 13–17.

Vardar, Y. 1987. Botanikte Preparasyon Teknigi. Izmir, Ege University Science Faculty Press (in Turkish), pp. 1–66.

Yanping, Z., Xingfeng, Z., Lizhen, Y. and Juan, L. 2001. A study on anatomical structure of horned gall and the leaf of *Rhus chinensis*. J. Nanjing Forestry Univ. (Natural Science Edition) **25**(6): 6–10.

Yi, T., Miller, A.J. and Wen, J. 2007. Phylogeny of *Rhus* (Anacardiaceae) based on sequences of nuclear Nia-i3 intron and chloroplast *trn*C-*trn*D. Syst. Bot. **32**(2): 379–391.

Yuksel, E. and Eminağaoğlu, Ö. 2017. Flora of the Kamilet Valley (Arhavi, Artvin, Turkey). Int. J. Ecosyst. Ecol. Sci. **7**(4): 905–914.

20

MUSA × PARAHAEKKINENII (MUSACEAE): A NEW ARTIFICIAL INTERSPECIFIC HYBRID FROM KERALA, INDIA

KOMBAN PARAMESWARAN SMISHA AND MAMIYIL SABU[1]

Angiosperm Taxonomy and Floristics Division, Department of Botany, University of Calicut, Kerala 673 635, India

Keywords: *Musa coccinea*; *Musa haekkinenii*; Wild parents; Manual crosses; New artificial hybrid.

Abstract

Musa × *parahaekkinenii* (Musaceae), a new manually crossed interspecific hybrid of two wild parent plants *Musa coccinea* Andrews (female) and *Musa haekkinenii* N.S. Lý & Haev. (male), is described and illustrated. A comparison of characters with its parents and a key to the new hybrid *M.* × *parahaekkinenii* are provided.

Introduction

The Musaceae (Commelinids: Zingiberales) is a tropical family comprising three genera, namely *Ensete* Horaninow, *Musa* L. and *Musella* (Franchet) Wu. *Musa* L. is the largest genus of the family (ca. 65 species) distributed in tropical Asia from southern India to eastern Himalayas to northern Australia, Sri Lanka and Africa. Globally, bananas (*Musa* spp.) form the fourth-most important food crop (Novak *et al*., 2014). Moreover, the bananas have got much attention for their medicinal, ornamental and socio-economic value (Cordeiro *et al*., 2004; Aziz *et al*., 2011; Joe and Sabu, 2016). The two new artificial hybrids of Musaceae, *Musa* × *georgiana* Rich. H. Wallace (Wallace and Hakkinen, 2009) and *Musa* × *formobisiana* H.-L.Chiu, C.-T. Shii & T.-Y.A. Yang. (Chiu *et al*., 2017) were developed earlier to explore the breeding relevance and ornamental potential value for the banana breeding programmes and landscaping applications.

As part of our studies on reproductive biology of Musaceae with the purpose of breeding evaluation and finding potential ornamental value, intra and intersectional hybridization have been done. The paper focuses on the new artificial hybrid, developed from artificial breeding of the Scarlet Banana *M. coccinea* Andrews (female parent) and *M. haekkinenii* N.S. Lý & Haev. (male parent) which shared the vegetative and floral characters of both the parents and expressed some of its own.

Materials and Methods

The present study was conducted at Calicut University Botanical Garden (11°25'45''N, 75°45'50''E) during 2013–2016. The stigma receptivity of *Musa coccinea* (female parent) and pollen viability of *M. haekkinenii* (male parent) were assessed at different time intervals from anthesis to flower closing using MTT (Dafni *et al*., 2005) and TTC (Shivanna and Rangaswamy, 1992) tests, respectively. Manual cross pollination was done in a period of maximum stigma receptivity and pollen viability which coincides. The self-compatibility was assessed by a method suggested by Dafni *et al.* (2005). The seed set and seed germination were observed for 3–4 months. The phenological events of new interspecific hybrid were observed with 10 plants of F1 hybrid. Morphometric analysis of vegetative and floral characters was done with a scale and

[1]Corresponding author. Email: msabu9@gmail.com

LEICA M80 stereomicroscope. Colour comparison of new hybrid with that of parents was referred by colour code (Kornerup and Wanscher, 1978). Descriptions of a new hybrid and parent plants were given using INIBAP *Musa* descriptor list (IPGRI-INIBAP/CIRAD, 1996). Photographs were taken with Sony DSC-HX400V digital camera. Voucher specimens were deposited at CALI and MH.

Results and Discussion

Musa × parahaekkinenii K.P. Smisha & M. Sabu, **hybrid nov.** **(Figs 1&2)**.

Diagnosis: *Musa × parahaekkinenii* differs from the female parent *M. coccinea* in having orange red bracts (vs. scarlet red) and presence of yellow fruits (vs. creamy fruits). The *M. × parahaekkinenii* shows distinct characters from the male parent *M. haekkinenii* by the presence of bracts obliquely erect to axis (vs. bracts curving downward). Moreover, the hybrid *M. × parahaekkinenii* exhibits the presence of horn-like appendages on lateral lobes of compound tepal. These appendages are present on the five lobes of compound tepal of *M. coccinea* while totally absent in all lobes of compound tepal of *M. haekkinenii*.

Type: India, Kerala, Malappuram district, Thenhipalam, Calicut University Botanical Garden (11°25'45''N, 75°45'50''E), 09 Dec 2016, *K.P. Smisha* 147908 (*Holotype*: CALI; *Isotype*: MH).

Clump forming; plants slender, herbaceous, suckering freely with 8–10 suckers of 10–15 cm long, oriented vertically. Mature pseudo-stem slender, 90–100 cm high, 13–15 cm in diam. at the base; sap milky. Leaf green, dorsiventral, 100–130 cm length; petiole 28–30 cm long, green with sparse brown blotches at the base, petiole canal margins incurved, narrow, 0.6–0.7 cm wide, scarious, clasping pseudo-stem at the base; lamina oblong-lanceolate, 70–100×23.5–24 cm, apex obtuse, margin corrugated, midrib 70–100×0.6–0.7cm, greyish green adaxially, pale greyish green abaxially, one side rounded and other pointed, adaxial surface dark green and dull, abaxial surface deep green and shiny, insertion point of leaf bases asymmetric on both sides. Inflorescence erect; peduncle 5–8 cm long, 5.0–5.5 cm in diam., glabrous, yellowish cream. Flag leaf with colourful bract like base and leafy apex persistent, 45–55 cm long. Sterile bracts lanceolate, 20–21×4.0–4.5 cm, adaxial surface dull, orange red with yellow tinge at base, abaxial surface shiny, orange red with yellow tinge at base, apex acute, greenish, base greyish orange, small shouldered, persistent. Male bracts lanceolate, 11–12×4.3–4.5 cm, bract lifting one at a time, persistent, adaxial surface dull, orange red, abaxial surface shiny, orange red, apex intermediate, green tinted with yellow, base greyish orange, small shouldered, margin not revolute. Female bracts lanceolate, 12.5–13.5×4.0–4.5 cm, adaxial surface dull, orange red, abaxial surface shiny, orange red, apex acute to obtuse, green tinted with yellow, base greyish orange, small shouldered, margin not revolute, lifting one bract at a time, imbricate. Basal flowers female, yellow, 1 flower per bract, 4–7 cm long. Compound tepal 3–4×1.2–2.3 cm, lower half deep yellow and upper half olive green, ribbed on either side, apex 5-lobed, rounded, with one horn-like appendages on lateral lobes, two lobes larger and exserted, 0.1×0.2 cm, middle and lateral lobes curved backward, 0.1×0.2 cm. Free tepal 3.0–3.8×1.0–1.2 cm, ovate, as long as the style, closely appressed to the stigma, translucent, opaque yellow, margin entire, apex corrugated with short acumen, adaxial surface smooth, abaxial surface ribbed. Staminodes 5, lanceolate, creamy yellow, 0.8–1.7×0.1–0.2 cm. Ovary straight, 2.2–4.0 cm long, lemon yellow, waxy, 3-locular; style straight, 2–3×0.1–0.2 cm, pale yellow with olive green tinge at apex. Stigma yellow, terete, 0.8×0.5 cm. Male flowers yellow, 2 flowers per bract, 4.7–5.0 cm long. Compound tepal 4.5–5 ×1.8–2.3 cm, lower half deep yellow and upper half olive green, ribbed on either side, apex 5-lobed, rounded, with horn-like appendages on lateral lobes, three middle lobes larger and exserted, 0.2×0.1 cm, 3 central lobes curved backward, 0.1×0.1 cm. Free tepal 4.5–4.7×1.0–1.3 cm, ovate, 3/4th of compound tepal, translucent, opaque yellow, margin entire, apex corrugated with short acumen, adaxial surface smooth, abaxial surface ribbed. Stamens 5, 4.2–4.3 cm long; anther greyish yellow, fertile, 1.8–2×0.05–0.1cm; filament

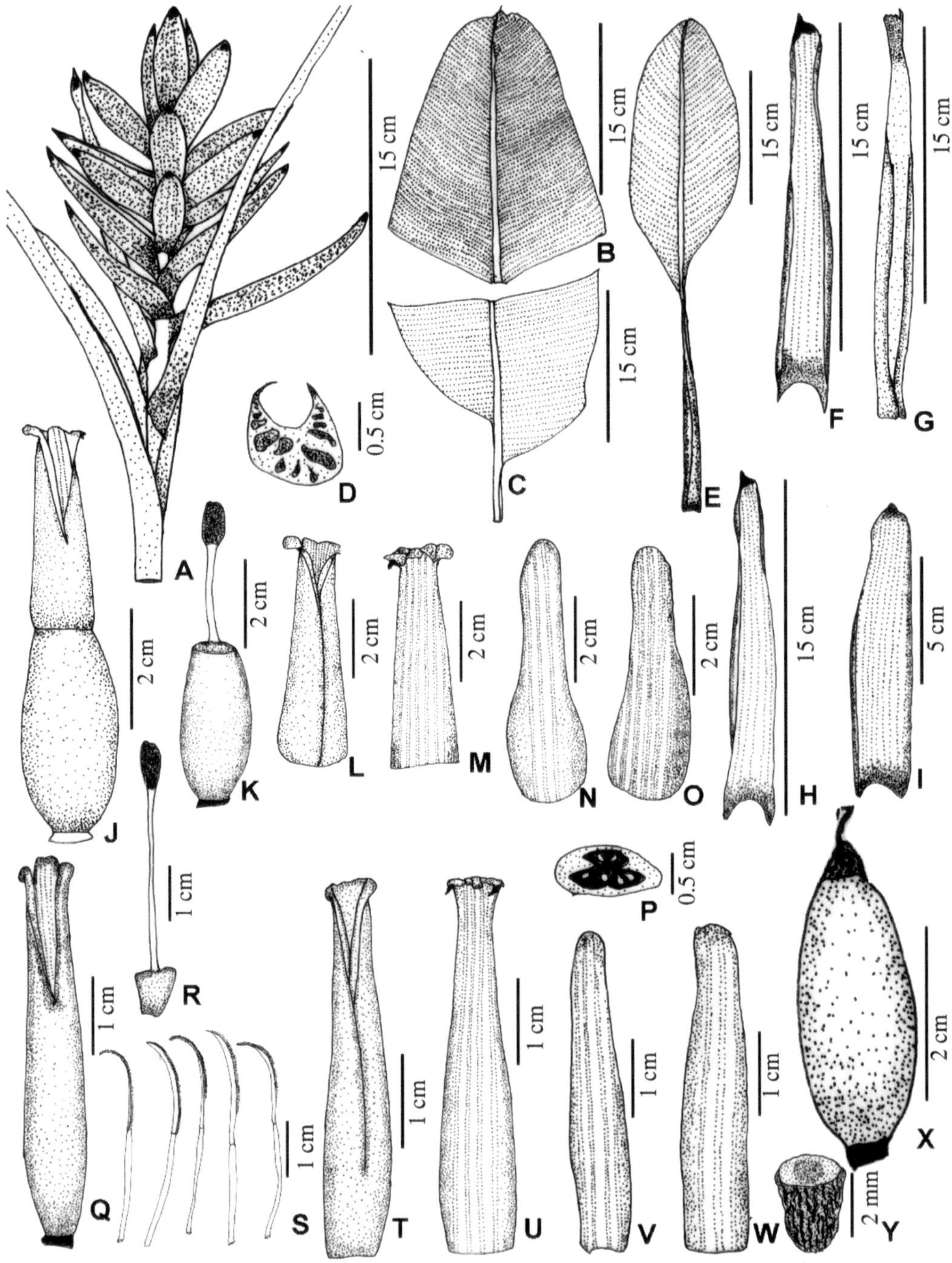

Fig. 1. Illustration of *Musa* × *parahaekkinenii* **hybrid nov.** A. Inflorescence; B. Leaf apex; C. Leaf base; D. Cross-section of petiole; E. Flag leaf; F&G. Sterile bracts; H. Female bract; I. Male bract; J. Female flower; K. Ovary with style and stigma; L&M. Compound tepals; N&O. Free tepals; P. Cross-section of ovary; Q. Male flower; R. Rudimentary ovary with style and stigma; S. Stamens; T&U. Compound tepals; V&W. Free tepals; X. Fruit; Y. Seed.

Fig. 2. *Musa × parahaekkinenii* **hybrid nov.** A. Habit; B. Inflorescence; C. Corm; D. Pseudostem; E. Cross-section of petiole; F. Leaf apex; G. Leaf base; H. Flag leaf; I. Sterile bracts; J. Female bract; K. Male bract; L. Female flower; M. Ovary with style and stigma; N. Compound tepals; O. Free tepals; P. Staminodes; Q. Cross-section of ovary; R. Male flower; S. Compound tepals; T. Free tepals; U. Stamens; V. Fruit; W. Seed.

1.6–2.5×0.1–0.2 cm, light yellow. Style rudimentary, straight, 3.5–4.4×0.1–0.2 cm, pale yellow with olive green tinge at apex. Stigma rudimentary, terete, yellowish orange, narrowly oblong, 0.6–0.7×0.3–0.4 cm. Ovary rudimentary, oblong, 0.5–0.7×0.4–0.8 cm, pale yellow, waxy. Fruit berry, 1 per bract, mature fruit narrowly oblong, 5.0–6.5 cm long, yellow, not waxy, straight, bear persistent floral relicts; 3.5–4.0 cm long. Seeds numerous, cylindrical, brown, warty, 1.5–2.0 cm long.

Etymology: The epithet has been named for the bract colour that resemblance with the male parent (*M. haekkinenii*) at the first appearance.

Key to the new hybrid *Musa* × *parahaekkinenii* and its parents

1.	Bracts obliquely upward to axis.	2
–	Bracts curving downwards to axis.	*M. haekkinenii*
2.	Horn-like appendages only on lateral lobes of compound tepal; bracts orange red.	*M.* × *parahaekkinenii*
–	Horn-like appendages on all lobes of compound tepal; bracts scarlet red.	*M. coccinea*

Table 1. Comparison of characters of *Musa* × *parahaekkinenii* hybrid nov. with its parents.

Characters	*M. haekkinenii*	*M.* × *parahaekkinenii*	*M. coccinea*
Plant height (cm)	250–300	80–110	150–200
Sap	Watery	Milky	Watery
Sucker number	8–13	9–12	8–10
Petiole canal margins	Wide, erect	Narrow, erect	Narrow, erect
Leaf base	One side rounded, one pointed	Both side pointed	Rounded
Female/Male bract behaviour	Persistent, curving downward to axis	Persistent, obliquely upward to axis	Persistent, obliquely upward to axis
Female/Male bract	Bright orange red on both sides	Bright orange red on both sides	Deep scarlet red on adaxial and abaxial side
Female/Male flowers per bract	1–4	1–2	1–2
Stigma shape	Flat	Narrowly flat	Narrowly flat
Appendage	Absent	Present only on lateral lobes of compound tepal	Present on all lobes of compound tepal
No. of fruit/s per bract and shape	3–4, straight	1, straight	1–2, straight

The petiole canal margins are clasping, ovary deep yellow and the seeds barrel-shaped and warty in the parents and the artificial hybrid.

A new hybrid *Musa* × *parahaekkinenii*, developed here by manual cross pollination is highly relevant for breeding application and to explore the potential ornamental value. The reciprocal cross between *M. haekkinenii* (female parent) and *M. coccinea* (male parent) also resulted in fruit set and seed production. Hybridization experiments of *M. coccinea* (female parent) with male parents of *M. beccarii*, *M. ornata*, *M. siamensis*, *M. laterita*, *M. velutina* and *M. markkuana* did not result in any fruit set. But the cross between *M. haekkinenii* (female parent) and *M. beccarii* (male parent) resulted in both fruit and seed set.

Musa coccinea is distributed in Indochina and China (Leong-Škorničková and Gallick, 2012). The species is remarkable for its highly attractive scarlet red bracts. The bract is oriented obliquely erect to axis and persistent that improves the potential ornamental value of the species. On the other hand, *Musa haekkinenii* is a native wild banana of Vietnam and known only under cultivation (in Calicut University Botanical Garden, Kerala, India and Suriana Botanic Garden, Penang, Malaysia) today. The centre of origin of the species is northern Vietnam and no wild population of it so far reported. Its existence in the wild is still uncertain. It was recognized as a Data Deficient species according to IUCN Red List categories and criteria (Lý *et al.*, 2012). The plant has a potential ornamental value because of the presence of highly attractive orange red persistent bract.

The new interspecific hybrid *M.* × *parahaekkinenii* which express intermediate characters of *M. coccinea* (♀) and *M. haekkinenii* (♂) and exhibits new characters also. The basal unisexual female flowers are fertile with receptive stigma and unisexuality is recognized as a unique adaptation for cross pollination. The male flowers are fertile with pollen grain production and recognized as a self-compatible hybrid. However, the new hybrid significantly adds the ornamental value because of the presence of highly attractive persistent bracts and flowers of inflorescence which lasts up to 3–4 months. Currently, many newly explored wild *Musa* species are used as staple food, medicine and ornamentals and also got socio-economic relevance. So, the interspecific hybridization between wild species has an immense value for breeding purposes and to explore the aspects of genetic variability which form the basis of genetic diversity.

Acknowledgements

The authors thank Kerala State Council for Science, Technology and Environment (KSCSTE), providing necessary funds for research work (Order No. 402/2015/KSCSTE dated 18.08.2015).

References

Aziz, N.A.A., Ho, L.-H., Azanari, A., Bhat, R., Cheng, L.-H. and Ibrahim, M.N.M. 2011. Chemical and functional properties of the native banana (*Musa acuminata* × *balbisiana* Colla cv. Awak) pseudo-stem and pseudo-stem tender core flours. Food Chem. **128**: 748–753.

Chiu, H.-L., Shii, C.-T. and Yang, T.Y.A. 2017. *Musa* × *formobisiana* (Musaceae), a new interspecific hybrid banana. Taiwania **62**: 147–150.

Cordeiro, N., Belgacem, M.N., Torres, L.C. and Moura, J.C.V.P. 2004. Chemical composition and pulping of banana pseudo-stems. Ind. Crops Prod. **19**: 147–154.

Dafni, A., Kevan, P.G. and Husband, B.C. 2005. Practical Pollination Biology. Enviroquest Ltd., Canada, pp. 1–590.

IPGRI-INIBAP/CIRAD. 1996. Description for Bananas (*Musa* spp.). International Plant Genetic Resources Institute, Rome, Italy/ International Network for the Improvement of Banana and Plantain, Montpellier, France/ Centre de Cooperation Internationale en Recherche Agronomique pour le Development, Montpellier, France, pp. 1–58.

Joe, A. and Sabu, M. 2016. Wild ornamental bananas in India: an overview. Sourth Ind. J. Biol. Sci. **2**: 213–221.

Kornerup, A. and Wanscher, J.H. 1978. Methuen Handbook of Colour, 3rd edn. Methuen, London, pp. 1–252.

Leong-Škorničková, J. and Gallick, D. 2012. The Ginger Garden. National Parks Board Singapore Botanic Garden, Singapore, 114 pp.

Lý, N.S., Lê, C.-K., Triệu, T.-D., Haevermans, A., Lowry II, P.P. and Haevermans, T. 2012. A distinctive new species of wild banana (*Musa*, Musaceae) from northern Vietnam. Phytotaxa **75**: 33–42.

Novák, P., Hřibová, E., Neumann, P., Koblížková, A., Doležel, J. and Macas, J. 2014. Genome wide analysis of repeat diversity across the family Musaceae. PLoS ONE **9**: e98918.

Shivanna, K.R. and Rangaswamy, N.S. 1992. Pollen Biology: A Laboratory Manual. Narosa Publishing House, New Delhi, pp.1–199.

Wallace, R. and Häkkinen, M. 2009. *Musa* × *georgiana,* a new intersectional hybrid banana with edible banana breeding relevance and ornamental potential. Nordic J. Bot. **27**: 182–185.

21

MERICARP MORPHOLOGY OF THE TRIBE SELINEAE (APIACEAE, APIOIDEAE) AND ITS TAXONOMIC IMPLICATIONS IN KOREA

CHANGYOUNG LEE[1], JINKI KIM, ASHWINI M. DARSHETKAR[2], RITESH KUMAR CHOUDHARY[2], SANG-HONG PARK[3], JOONGKU LEE[4] AND SANGHO CHOI[5]

International Biological Material Research Center, Korea Research Institute of Bioscience & Biotechnology, 125 Gwahak-ro, Yuseong-gu, Daejeon 34141, South Korea

Keywords: Mericarp surface characters; SEM; NMDS; UPGMA.

Abstract

Mericarp morphology of 24 taxa belonging to nine genera of the tribe Selineae (Family: Apiaceae) in Korea was studied by Scanning Electron Microscopy. UPGMA and NMDS analyses were performed based on 12 morphological characters. The mericarp surface characters like mericarp shape, rib number and shape, surface pattern, surface appendages and mericarp symmetry proved useful in distinguishing the genera of the tribe Selineae.

Introduction

The family Apiaceae comprises about 455 genera and is widely distributed across temperate regions of the world (Pimenov and Leonov, 1993). Many members of Apiaceae can be easily distinguished by umbellate inflorescence, fruits consisting of two one-seeded mericarps suspended from a split central column or carpophore and numerous minute epigynous flowers (Downie *et al.*, 1998). Fruits of Apiaceae are known as cremocarps which in the dry state split into two mericarps. Each mericarp has a flat commissural and convex dorsal surface. Drude (1898) proposed a sound system of classification of Apiaceae with three subfamilies Hydrocotiloideae, Saniculoideae and Apioideae and 12 tribes. Apioideae is the largest subfamily consisting of 404 genera and about 2,935 species (Pimenov and Leonov, 1993) and can be distinguished from the other two subfamilies by the synapomorphies like the presence of compound umbels, well-developed vittae (secretory canals) and free carpophores. Selineae is one of the tribes of Apioideae and is represented by genera *Angelica* L., *Carlesia* Dunn., *Cnidium* Cusson *ex* Juss., *Cymopterus* Raf., *Dystaenia* Kitag., *Glehnia* F., *Libanotis* Haller *ex* Zinn, *Ligusticum* L., *Peucedanum* L. and *Ostericum* Hoffm. Species of the tribe Selineae are mostly distributed in China, Japan and Korea (Spalik *et al.*, 2004).

Drude (1898) and many other workers like Calestani (1905), Koso-Poljansky (1916) considered fruit morphology and anatomy as important characters for classification of subfamily Apioideae. While Heywood (1971), Davis (1972), Cronquist (1982), doubted the validity of characters considered by Drude (1898) to diagnose evolutionary relationships. Later, various

[1]VNU University of Science, 334 Nguyen Trai, ThanhXuan, Hanoi, 100000, Vietnam.
[2]Biodiversity & Palaeobiology Group (Plants), Agharkar Research Institute, G.G. Agarkar Road, Pune 411004, India.
[3]National Institute of Ecology, 1210, Geumgang-ro, Maseo-Myeon, Seocheon-gun, Chungnam, 325-813, Republic of Korea.
[4]Department of Environment and Forest Resources, Chungnam National University, Yuseong-gu, Daejeon, 34134, Republic of Korea.
[5]Corresponding author. E-mail: decoy0@kribb.re.kr

molecular phylogenetic analyses using nrITS, and plastid markers like *rbc*L, *mat*K, *rpo*C1 provided little supports to classifications based on anatomy and morphology of fruits (Kondo *et al.*, 1996; Plunkett *et al.*, 1996; Downie *et al.*, 1998, 2001). In some analyses, the *Arracacia* clade arose within the *Angelica* clade (Plunkett *et al.*, 1996; Downie *et al.*, 1998) and Downie *et al.* (2001) suggested that the *Arracacia* clade may eventually be subsumed within the same. Both groups comprise many genera that were traditionally placed in tribe Peucedaneae subtribes Angelicinae and Ferulinae (Drude, 1898) or tribes Peucedaneae and Angeliceae (Pimenov and Leonov, 1993). Later, based on nrITS data Spalik *et al.* (2004) suggested that the *Angelica* and *Arracacia* clades form a strong monophyletic group. They described it as tribe Selineae which consists of 62 genera. However, various attempts to identify structural characters useful for delimiting the species under Selineae failed, even those based on molecular data (Ajani *et al.* 2008; Feng *et al.*, 2009; Zhou *et al.*, 2009).

Despite the consistent picture of relationships that has emerged in Apiaceae by molecular data, it has always been difficult to identify structural characters that could be used to classify and define the clades in Apiaceae with a major exception of fruit micromorphology and anatomy (Feng *et al.*, 2009). Bagchi and Srivastava (1989) studied epicarp surfaces of some medicinally important Apiaceae members and concluded that the surface characters of epicarp of fruits are useful in distinguishing the species. Liu *et al.* (2006) studied the taxonomic value of fruit wing types in order Apiales suggesting that their structural data could complement DNA with observable features to recognize and circumscribe taxa. Similar studies were carried out by various workers (Lee *et al.*, 2001; Spalik *et al.,* 2001; Liu *et al.*, 2006; Liao *et al.*, 2013) which proved that the fruit micromorphology is useful in classifying and identifying taxa of Apiaceae. Lee *et al.* (1997), for the first time, studied seed morphology of eight medicinally important species of Apiaceae. Further, the importance of seed trait in the taxonomy of tribe Scandinae was also proved by molecular data (Lee *et al.,* 2001). Recently, Liao *et al.* (2013) provided new insights into the phylogeny of *Angelica* and its allies based on nrDNA, cpDNA and morphological characters which included fruit anatomy and micromorphology.

In Korea, research on Selineae have been performed by various workers, but considerably less number of taxa were taken into account (Choi *et al.*, 1998; Yoon, 2001; Koo and Kim, 2008). The present study was therefore conducted to analyze the usefulness of mericarp characters in distinguishing genera of the tribe Selineae.

Material and Methods

The present study includes 20 Korean taxa belonging to the tribe Selineae and four taxa representing the tribe Scandiceae. Seeds of collected specimens and herbarium specimens were used for the study and are deposited in KRIB and KBH. Details of voucher deposition are mentioned in Table 1. Fruits were carefully dissected without removing testa under dissecting microscope (Nikon, AZ100). Size and colour of the seeds were noted under dissecting microscope. For micromorphological observations, mature fruits were mounted on stubs using double sided adhesive tape. Each sample was coated with a thin layer (20–40 nm) of gold using Hitachi E-1010 sputter coater and examined at 20 KV using a Hitachi S3400-N Scanning Electron Microscope.

For phenetic analysis, 12 qualitative characters of 24 Korean taxa were considered as mentioned in Table 2. The data was analyzed with the help of PAUP*4.0 (Swofford, 2001) to obtain a phenogram using the unweighted pair-group method with arithmetic mean (UPGMA) from the data matrix. Quantitative values were log-transformed to reduce the effect of zero values. The data was then subjected to non-metric multidimensional scaling (NMDS) following Quinn and Keough (2002).

Table 1. Voucher information of the specimens examined in the study.

Species	Place of collection	Collectors and date of collection	Accession number
Angelica anomala Avé-Lall.	Namwon-eup, Seogwipo-si, Jeju-do, Korea	Joongku Lee *et al.*; 24.1.2008	KRIB 0015965
A. cartilaginomarginata (Makino *ex* Y.Yabe) Nakai	Deokchi-ri, Jucheon-myeon, Namwon-si, Jeollabuk-do, Korea; Sanbuk-ri, Unju-myeon, Wanju-gun, Jeollabuk-do, Korea	S.M. Lee & H.Y. Lee; 21.10.2009	KRIB 0029410
A. dahurica (Hoffm.) Benth. & Hook.f. *ex* Franch. & Sav.	Yeongpyeong-dong, Jeju-si, Jeju-do, Korea	C.S. Kim; 7.8.2001	KRIB 0001002
A. decursiva (Miq.) Franch. & Sav.	Sikjangsan, Sechon-dong, Dong-gu, Daejeon, Korea	Changyoung Lee; 29.10.2008	KRIB 0020712
A. genuflexa Nutt.	Oesam-ri, Bongseong-myeon, Bonghwa-gun, Gyeongsangbuk-do, Korea	Geonrae Kim & Jinki Kim; 17.0.2003	KRIBBSD1040
A. gigas Nakai	Daeamsan, Seohwa-myeon, Inje-gun, Gangwon-do, Korea	Joongku Lee *et al.*; 6.10.2004	KRIB 0004085
A. grosseserrata Maxim.	Unjangsan, Jeongcheon-myeon, Jinan-gun, Jeollabuk-do, Korea	Hyeong-Kyu Lee & Taejin Kim; 19.7.2001	KRIB 0001031
A. japonica A. Gray	Geomundo, Deokchon-ri, Samsan-myeon, Yeosu-si, Jeollanam-do, Korea	Jinki Kim & Changyoung Lee; 2.2.2009	KRIB 0029071
A. polymorpha Maxim.	Oebang-ri, Sudong-myeon, Namyangju-si, Gyeonggi-do, Korea	J.H. Kim *et al.*; 12.11.2007	KRIB 0013530
A. tenuissima Nakai	Geumdaebong, Gohan-eup, Jeongseon-gun, Gangwon-do, Korea	Hyeong-Kyu Lee & Taejin Kim; 12.9.2001	KRIB 0001020
Anthriscus sylvestris (L.) Hoffm.	Ulleungdo, Ulleung-gun, Gyeongsangbuk-do, Korea	Shinho Kang *et al.*; 14.6.2006	KRIB 0006619
Cnidium japonicum Miq.	Susan-ri, Seongsan-eup, Seogwipo-si, Jeju-do, Korea	C.S. Kim *et al*; 15.11.2006	KRIB 0007560
C. monnieri (L.) Cusson	Goyang-si, Gyeonggi-do, Korea	W.K. Paik; 10.11.2001	KRIB 0010062
Cymopterus melanotilingia (H. Boissieu) C.Y. Yoon	Gayasan, Gaya-myeon, Hapcheon-gun, Gyeongsangnam-do, Korea	S.M. Lee & H.Y. Lee; 1.11.2006	KRIB 0006906
Dystaenia takesimana (Nakai) Kitag.	Ulleungdo, Ulleung-gun, Gyeongsangbuk-do, Korea	Taejin Kim; 13.10.2000	KRIB 0003115
Glehnia littoralis (A. Gray) F. Schmidt*ex* Miq.	Saekdal-dong, Seogwipo-si, Jeju-do, Korea	J.H. Kim *et al.*; 19.6.2002	KRIB 0011773
Libanotis seseloides (Fisch. & C.A. Mey. *ex* Turcz.) Turcz.	Baekunsan, Mitan-myeon, Pyeongchang-gun, Gangwon-do, Korea	Jinki Kim & Sanghong Park; 20.10.2005	KBH1053185
Ligusticum tachiroei (Franch. & Sav.) M. Hiroe & Constance	Jeombongsan, Girin-myeon, Inje-gun, Gangwon-do, Korea	Jinki Kim & Sanghong Park; 14.10.2004	KBH1032523
Osmorhiza aristata (Thunb.) Rydb.	Juwangsan, Budong-myeon, Cheongsong-gun, Gyeongsangbuk-do, Korea	G.Y. Chung *et al.*; 22.8.2002	KRIB 0011782

Table 1 Contd.

Species	Place of collection	Collectors and date of collection	Accession number
Ostericum sieboldii (Miq.) Nakai	Yongsan-ri, Doam-myeon, Pyeonchang-gun, Gangwon-do, Korea	W.K. Paik; 19.10.2000	KRIB 0010040
Peucedanum japonicum Thunb.	Oeyeondo, Ocheon-myeon, Boryeong-si, Chungcheongnam-do, Korea	Changyoung Lee; 2.10.2009	KBH1259205
P. terebinthaceum (Fisch. *ex* Trevir.) Ledeb.	Daejin-ri, Yeonghae-myeon, Yeongdeok-gun, Gyeongsangbuk-do, Korea	J.H. Kim *et al.*; 12.10.2001	KRIB 0010050
Torilis japonica (Houtt.) DC.	Cheondeungsan, Sancheok-myeon, Chungju-si, Chungcheongbuk-do, Korea	G.Y. Chung *et al.*; 5.8.2002	KRIB 0011785
T. scabra (Thunb.) DC.	Sanbangsan, Sagye-ri, Andeok-myeon, Seogwipo-si, Jeju-do, Korea	Taejin Kim *et al.*: 28.5.2002	KRIB 0002834

Table 2. Codes used for for mericarp character analysis.

No.	Mericarp characters
1.	Mericarp beak: absent (0), present (1)
2.	Mericarp shape: broadly elliptic or elliptic (0), triangular (1) oblong (2) linear or linear-oblong (3)
3.	Mericarp surface: irregular reticulate (0), colliculate (1), irregular-winkled (2), granulate-aculeate (3)
4.	Mericarp surface appendage: absent (0), hair (1), papillate (2), spiny or bristly (3)
5.	Mericarp rib number: 5 (0), 3-4 (1), 9 (2), absent (3)
6.	Mericarp rib: not developed (0), dorsal and lateral ribs filiform/inflated (1), dorsal ribs filiform/inflated and lateral ribs winged (2), dorsal ribs and lateral ribs winged (3)
7.	Dorsal and lateral ribs comparison: rib not developed (0), dorsal and lateral ribs identical (1), dorsal and lateral ribs different (2)
8.	Secondary ribs: absent (0), present (1)
9.	Mericarp compressed: compressed laterally or not compressed (0), dorsally compressed (1)
10.	Mericarp symmetry: homomorphic (0), heteromorphic (1)
11.	Apex of mericarp: non-emarginate (0), deeply emarginate (1)
12.	Base of mericarp: non-emarginate (0), deeply emarginate (1)

Stress values resulting from NMDS reflect the match between the pairwise differences originally calculated between the species and the pairwise distances in the ordination. Lower stress values indicate a better match (Quinn and Keough, 2002). This statistical analysis was performed using PAST (Hammer *et al.*, 2001). Names and authorities of the species are as per The Plant List (2013). Terminology to describe the seed coat and seed surface sculpturing follows Harris and Harris (1994), Stearn (1992), and Webb and Simpson (2001).

Results and Discussion

The results obtained from the study are summarized in Table 3. Photomicrographs of mericarps and mericarp surfaces are presented in Figures 1 and 2, respectively. In the tribe Selineae, large variation can be seen in sizes of mericarps. The length of mericarp varies from

2.7±0.2 to 11.4±0.6 mm (Fig. 3), while the width of mericarp varies from 1.4±0.2 to 9.3±0.4 mm (Fig. 4). Seed size of *Glehnia littoralis* is highly divergent than other species included in the study (Figs 3 & 4). Three different shapes of mericarp were observed during our study. The genera *Ligusticum* and *Libanotis* are characterized by oblong mericarps, the genus *Glehnia* is characterized by triangular mericarp, while broadly elliptic-elliptic mericarp was observed in the genera *Cnidium*, *Angelica*, *Ostericum*, *Cymopterus* and *Peucedanum*. In all species of Selineae, mericarps are symmetrical with five ribs except in *Cymopterus melanotilingia*, where asymmetric mericarps with three or four ribs were observed.

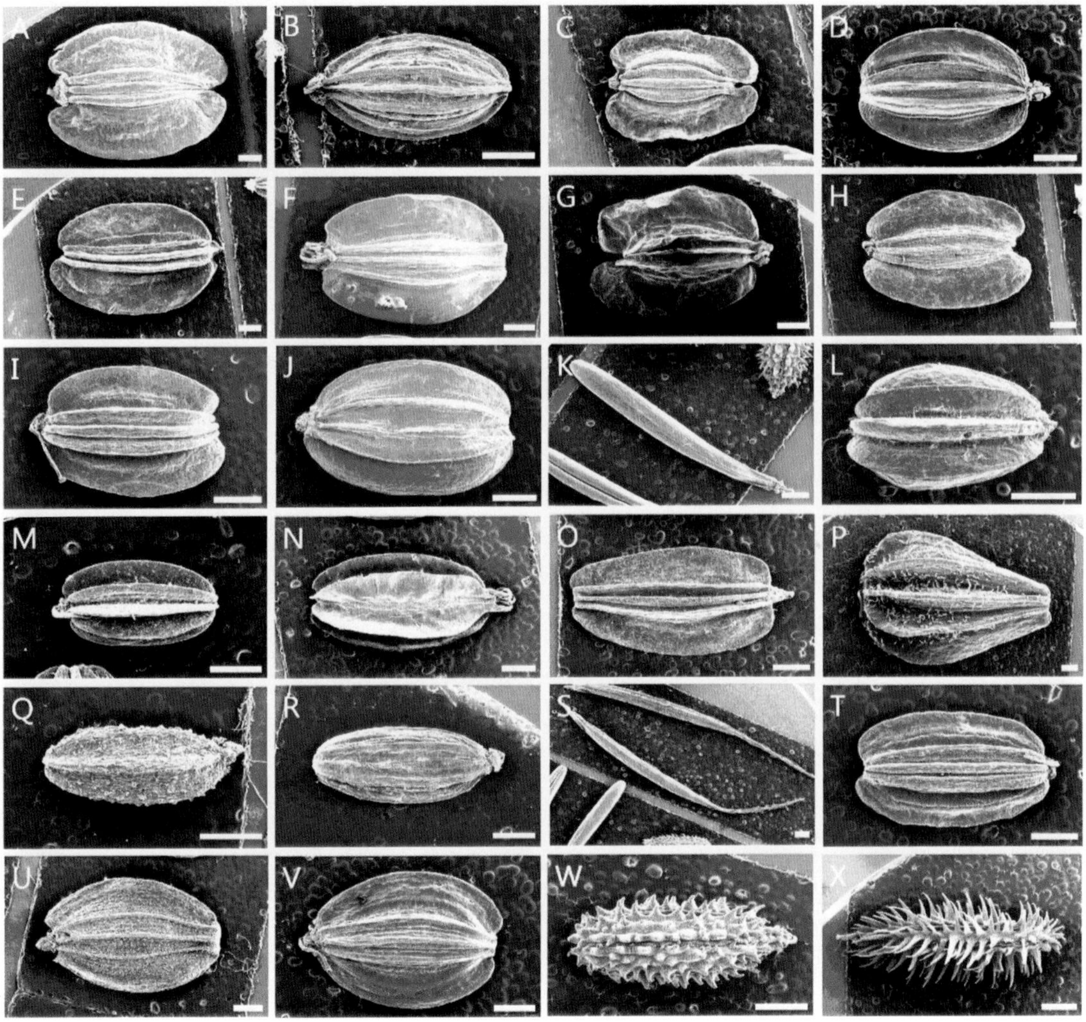

Fig. 1. SEM images of mericarps of 24 taxa of Selineae: A. *Angelica anomala*; B. *A. cartilaginomarginata*; C. *A. dahurica*; D. *A. decursiva*; E. *A. genuflexa*; F. *A. gigas*; G. *A. grosseserrata*; H. *A. japonica*; I. *A. polymorpha*; J. *A. tenuissima*; K. *Anthriscus sylvestris*; L. *Cnidium japonicum*; M. *C. monnieri*; N. *Cymopterus melanotilingia*; O. *Dystaenia takesimana*; P. *Glehnia littoralis*; Q. *Libanotis seseloides*; R. *Ligusticum tachiroei*; S. *Osmorhiza aristata*; T. *Ostericum sieboldii*; U. *Peucedanum japonicum*; V. *P. terebinthaceum*; W. *Torilis japonica*; X. *T. scabra*. Scale bar: 1 mm.

Three different surface patterns characterize Selineae mericarps. The irregular reticulate surface pattern was observed in *Cnidium*, *Dystaenia*, *Glehnia* and *Angelica cartilaginomarginata*, *A. genuflexa*, *A. japonica*, *A. anomala*, *A. gigas*, and *Peucedanum terebinthaceum*. While

Ligusticum, *Ostericum*, *Cymopterus*, *Angelica tenuissima*, *A. polymorpha*, *A. dahurica*, and *A. decursiva* showed colliculate surface. The irregular-wrinkled surface pattern was observed in *Libanotis seseloides* and *Peucedanum japonicum.* Papillate surface pattern characterizes the genus *Libanotis*. Hairs were observed on mericarps of *Glehnia littoralis* and *Peucedanum japonicum* (Fig. 2). In all species of Selineae employed in this study, mericarps are symmetrical with five ribs except in *Cymopterus melanotilingia*, where asymmetric mericarps with three or four ribs were observed.

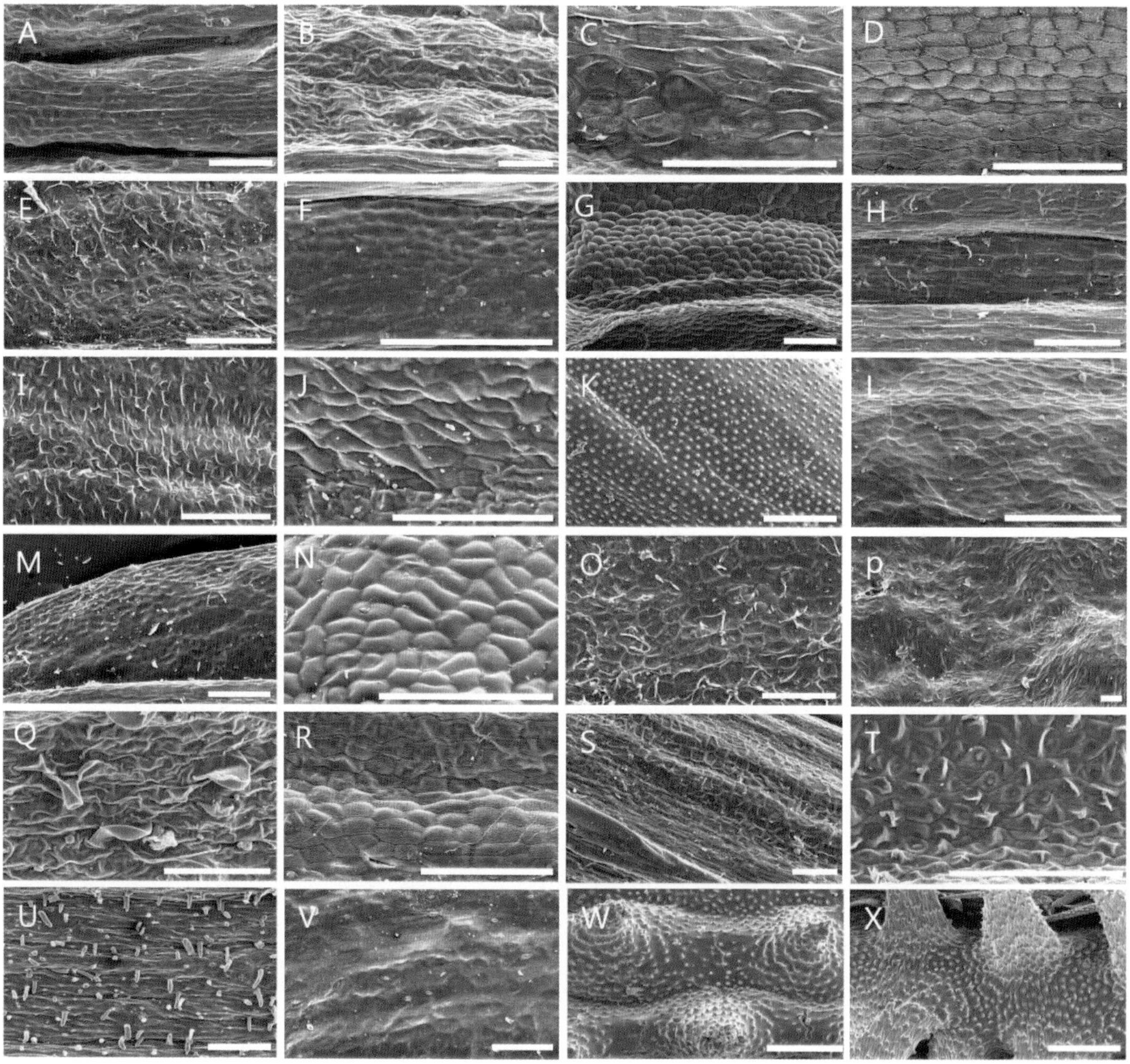

Fig. 2. Mericarp surface characters of 24 taxa of Selineae: A. *Angelica anomala*; B. *A. cartilaginomarginata*; C. *A. dahurica*; D. *A. decursiva*; E. *A. genuflexa*; F. *A. gigas*; G. *A. grosseserrata*; H. *A. japonica*; I. *A. polymorpha*; J. *A. tenuissima*; K. *Anthriscus sylvestris*; L. *Cnidium japonicum*; M. *C. monnieri*; N. *Cymopterus melanotilingia*; O. *Dystaenia takesimana*; P. *Glehnia littoralis*; Q. *Libanotis seseloides*; R. *Ligusticum tachiroei*; S. *Osmorhiza aristata*; T. *Ostericum sieboldii*; U. *Peucedanum japonicum*; V. *P. terebinthaceum*; W. *Torilis japonica*; X. *T. scabra*. Scale bar: 200 μm.

Table 3. Morphological characters of mericarp of 24 species in the tribe Selineae.

Species	Length (mm)	Width (mm)	Mericarp shape	Mericarp surface	Appendage	Ribs	
						Lateral	Dorsal
Anthriscus sylvestris	8.2±0.5	1.0±0.1	linear-oblong	granulate-aculeate	Absent	Absent	Absent
A. anomala	5.9±0.5	5.0±0.4	broadly elliptic	reticulate	Absent	winged	inflated
A. cartilagino-marginata	3.3±0.3	1.9±0.3	broadly elliptic or elliptic	irregular-reticulate	Absent	winged	inflated
A. dahurica	5.8±0.5	5.2±0.5	broadly elliptic	colliculate	Absent	winged	inflated
A. decursiva	4.8±0.6	3.1±0.4	broadly elliptic or elliptic	colliculate	Absent	winged	inflated
A. genuflexa	6.2±0.6	4.6±0.4	broadly elliptic	irregular-reticulate	Absent	winged	inflated
A. gigas	6.4±0.5	4.8±0.4	broadly elliptic	irregular-reticulate	Absent	winged	inflated
A. japonica	10.1±0.6	6.1±0.3	broadly elliptic or elliptic	reticulate	Absent	winged	inflated
A. polymorpha	4.2±0.3	3.1±0.2	broadly elliptic	colliculate	Absent	winged	inflated
Angelica grosseserrata	5.6±0.7	4.6±0.8	broadly elliptic	colliculate	Absent	winged	inflated
A. tenuissima	5.4±0.5	3.6±0.6	broadly elliptic or elliptic	colliculate	Absent	winged	inflated
Cnidium japonicum	3.0±0.2	2.0±0.2	Broadly elliptic or elliptic	reticulate	Absent	winged	winged
C. monnieri	2.9±0.2	1.8±0.2	broadly elliptic or elliptic	irregular-reticulate	Absent	winged	winged
Cymopterus melanotilingia	6.0±0.3	3.1±0.2	elliptic	colliculate	Absent	winged	winged
Dystaenia takesimana	5.7±0.6	3.2±0.4	broadly elliptic or elliptic	reticulate	Absent	winged	winged
Glehnia littoralis	11.4±0.6	9.3±0.4	triangular	irregular-reticulate	hairy	winged	winged
Libanotis seseloides	2.7±0.2	1.4±0.2	oblong	irregular-winkled	papillate	inflated	inflated
Ligusticum tachiroei	4.0±0.2	2.0±0.2	oblong	colliculate	Absent	inflated	inflated
Osmorhiza aristata	17.6±1.0	1.0±0.1	linear	reticulate	bristly	filiform	filiform
Ostericum sieboldii	3.8±0.3	2.6±0.2	broadly elliptic or elliptic	colliculate	Absent	winged	inflated
Peucedanum japonicum	5.1±0.5	2.1±0.2	elliptic	irregular-winkled	hairy	winged	filiform
P. terebinthaceum	4.6±0.2	3.3±0.2	broadly elliptic	irregular-reticulate	Absent	winged	inflated
Torilis japonica	3.8±0.3	1.4±0.1	oblong	granulate-aculeate	spiny	filiform	filiform
T. scabra	5.9±0.5	1.2±0.2	oblong	granulate-aculeate	spiny	filiform	filiform

We found three different mericarp rib shapes in Selineae. *Ligusticum tachiroei* and *Libanotis seseloides* have dorsal and lateral ribs which are filiform or inflated. The genera *Angelica*, *Ostericum* and *Peucedanum* are characterized by filliform or inflated dorsal ribs and winged lateral ribs, while the genera *Cnidium, Cymopterus, Glehnia* and *Dystaenia* showed winged dorsal and lateral ribs.

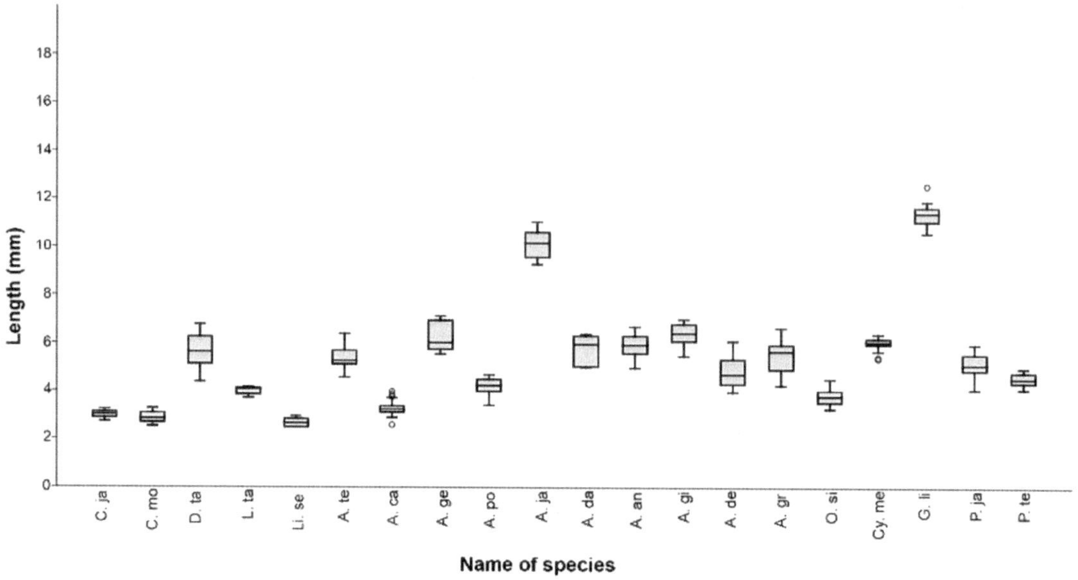

Fig. 3. Length of seeds summarized in Boxplot. The boxes represent the second and third quartiles, and the vertical line within each of the boxes represents the median. Outliers are identified with asterisks (C. ja: *Cnidium japonicum*; C. mo: *C. monnieri;* D. ta: *Dystaenia takesimana*; L. ta: *Ligusticum tachiroei;* Li. se: *Libanotis seseloides;* A. te: *Angelica tenuissima;* A. ca: *A. cartilaginomarginata;* A. ge: *A. genuflexa;* A. po: *A. polymorpha;* A. ja: *A. japonica;* A. da: *A. dahurica;* A. an: *A. anomala;* A. gi: *A. gigas;* A. de: *A. decursiva;* A. gr: *A. grosseserrata;* O. si: *Ostericum sieboldii;* Cy. me: *Cymopterus melanotilingia;* G. li: *Glehnia littoralis;* P. ja: *Peucedanum japonicum;* P. te: *Peucedanum terebinthaceum*).

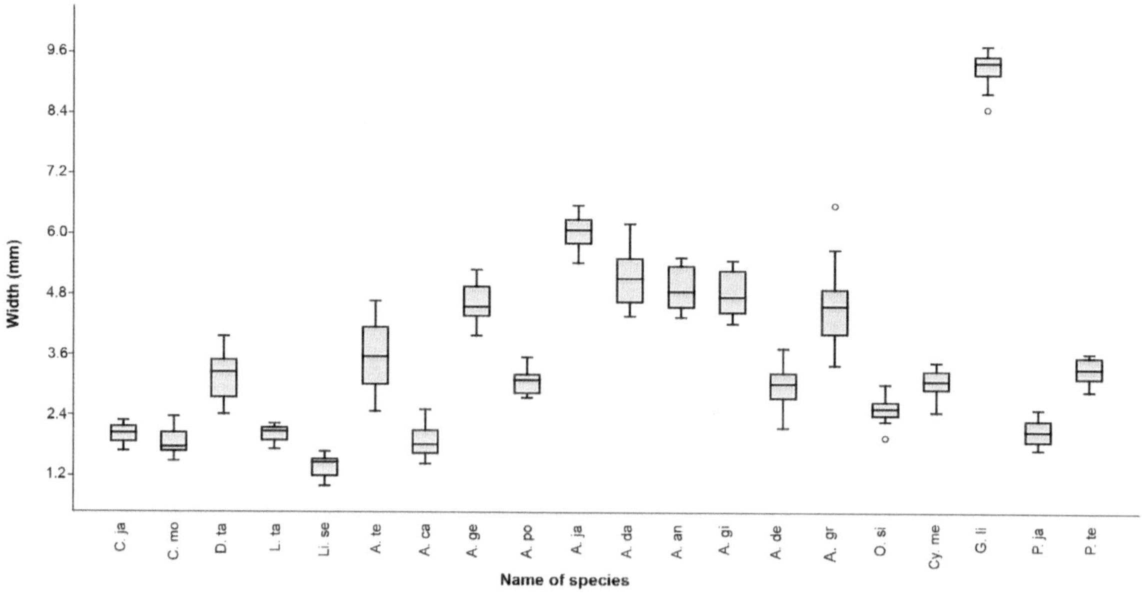

Fig. 4. Width of seeds summarized in Boxplot. The boxes represent the second and third quartiles, and the vertical line within each of the boxes represents the median. Outliers are identified with asterisks. (C. ja: *Cnidium japonicum*; C. mo: *C. monnieri;* D. ta: *Dystaenia takesimana*; L. ta: *Ligusticum tachiroei;* Li. se: *Libanotis seseloides;* A. te: *Angelica tenuissima;* A. ca: *A. cartilaginomarginata;* A. ge: *A. genuflexa;* A. po: *A. polymorpha;* A. ja: *A. japonica;* A. da: *A. dahurica;* A. an: *A. anomala;* A. gi: *A. gigas;* A. de: *A. decursiva;* A. gr: *A. grosseserrata;* O. si: *Ostericums ieboldii;* Cy. me: *Cymopterus melanotilingia;* G. li: *Glehnia littoralis;* P. ja: *Peucedanum japonicum;* P. te: *Peucedanum terebinthaceum*).

Emarginated mericarp apex was observed in *Angelica japonica, A. dahurica,* and *A. anomala*, while in other studied species mericarp apex was non-emarginate. Deeply emarginated mericarp base was observed in *Angelica genuflexa*, *A. polymorpha*, *A. japonica*, *A. dahurica*, *A. anomala*, *A. grosseserrata*, *Ostericum sieboldii*, *Cymopterus melanotilingia* and *Glehnia littoralis,* while in other species non-emarginated mericarp base was noticed.

Phenetic analysis

UPGMA phenogram was evaluated to identify characters that distinguish the genera of Apioideae. The analysis revealed seven major groups. The genus *Glehnia* can be distinguished from other genera of Selineae by triangular mericarp and hairy mericarp surface (Pairwise distance 2.474). *Ligusticum tachiroei* and *Libanotis seseloides* appear to be sister groups as oblong mericarps characterize them. Both the species can be distinguished by mericarp surface pattern and appendages. *Ligusticum tachiroei* has colliculate surface pattern and lacks appendages (pairwise distance 1.000), whereas *Libanotis seseloides* has irregularly wrinkled and papillate surface. *Cymopterus melanotilingia* can be distinguished from other genera like *Peucedanum, Angelica, Ostericum, Cnidium* and *Dystaenia* as it has asymmetric mericarp and rib number 3-4 (pairwise distance 2.188). *Cnidium* and *Dystaenia* appear to be sister groups (pairwise distance 1.064) and are characterized by winged dorsal and lateral ribs. Based on UPGMA analysis of mericarp surface characters, the genera *Glehnia, Cymopterus, Ligusticum, Libanotis, Cnidium* and *Dystaenia* can be distinguished, however, *Angelica*, *Peucedanum* and *Ostericum* cannot be distinguished. NMDS analysis of Bray-Curtis similarity coefficients (Fig. 5) calculated from 12 morphological characters showed similar separation of species as in UPGMA (Fig. 6).

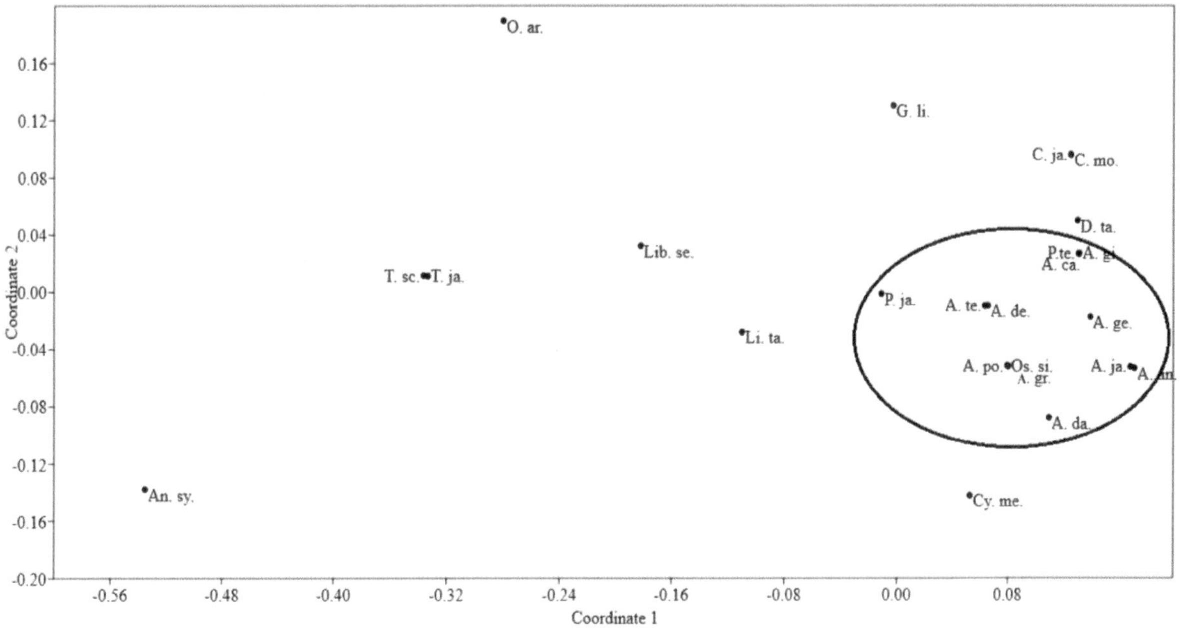

Fig. 5. Non-metric multidimensional scaling (NMDS) using Bray-Curtis similarity index of 12 morphological characters. (C. ja: *Cnidium japonicum*; C. mo: *C. monnieri;* D. ta: *Dystaenia takesimana*; L. ta: *Ligusticum tachiroei;* Li. se: *Libanotis seseloides;* A. te: *Angelica tenuissima;* A. ca: *A. cartilaginomarginata;* A. ge: *A. genuflexa;* A. po: *A. polymorpha;* A. ja: *A. japonica;* A. da: *A. dahurica;* A. an: *A. anomala;* A. gi: *A. gigas;* A. de: *A. decursiva;* A. gr: *A. grosseserrata;* O. si: *Ostericum sieboldii;* Cy. me: *Cymopterus melanotilingia;* G. li: *Glehnia littoralis;* P. ja: *Peucedanum japonicum;* P. te: *Peucedanum terebinthaceum*).

To conclude, characters like mericarp shape, rib number, and shape, surface pattern, surface appendages and mericarp symmetry can be considered as useful traits to distinguish the genera of tribe Selineae. These results are in congruence with the molecular systematic study of *Angelica* and allied genera based on nuclear DNA sequences (Feng *et al.*, 2009).

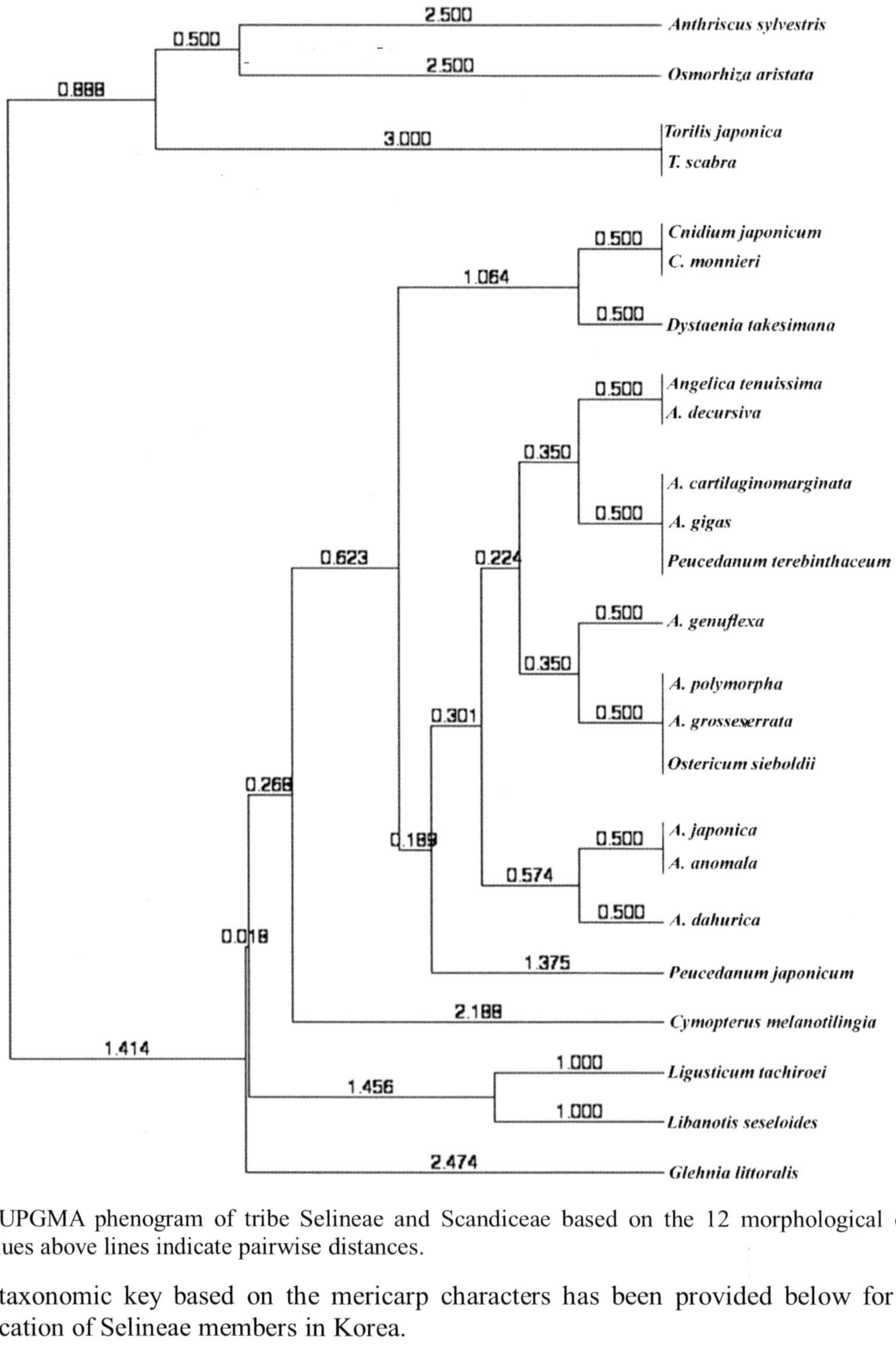

Fig. 6. UPGMA phenogram of tribe Selineae and Scandiceae based on the 12 morphological characters. Values above lines indicate pairwise distances.

A taxonomic key based on the mericarp characters has been provided below for the easy identification of Selineae members in Korea.

Key to the Korean taxa of Selineae based on mericarp characters:

1.	Mericarp heteromorphic and rib number 3-4.	*Cymopterus*
-	Mericarp homomorphic and rib number 5	2
2.	Mericarp triangular, surface appendage hairy	*Glehnia*
-	Mericarp broadly elliptic or elliptic or oblong, surface appendage absent or hairy	3
3.	Mericarp oblong; dorsal and lateral ribs filiform/inflated	4
-	Mericarp broadly elliptic or elliptic; dorsal ribs filiform/winged and lateral ribs winged	5
4	Mericarp surface colliculate	*Ligusticum*
-	Mericarp surface irregularly-wrinkled	*Libanotis*
5.	Dorsal ribs filiform and lateral ribs winged	*Angelica*, *Ostericum*, *Peucedanum*
-	Dorsal and lateral ribs winged	6
6.	Dorsal and lateral ribs identical	*Cnidium*
-	Dorsal and lateral ribs different	*Dystaenia*

Acknowledgements

We are grateful to the curator of the herbarium at Korea National Arboretum (KNA), Korea for permitting us to study the voucher specimens. This study was supported by Korea Research Institute of Bioscience and Biotechnology Research Initiative Program of the Republic of Korea. The 3rd (AMD) and 4th (RKC) authors acknowledge to the Memorandum of Understanding between IBMRC, KRIBB and Agharkar Research Institute, India, that facilitated this research.

References

Ajani, Y., Ajani, A., Cordes, J.M., Watson, M.F. and Downie, S.R. 2008. Phylogenetic analysis of nrDNA ITS sequences reveals relationships within five groups of Iranian Apiaceae subfamily Apioideae. Taxon **57**(2): 383–401.

Bagchi, G.D. and Srivastava, G.N. 1989. SEM of epicarp surfaces of some medicinally important Apiaceae. International J. Crude Drug Res. **27**(3): 171–177.

Calestani, V. 1905. Contributo alla sistematica delle ombrellifere d'Europa. Webbia **1**(1): 89–280.

Choi, H.K., Kim, Y., Sun, B.Y. and Shin, H. 1998. Phylogeny of *Dystaenia* in subfamily Apioideae (Family Apiaceae) based on ITS sequences. Korean J. Plant Taxon. **28**(2): 139–149.

Cronquist, A. 1982. Reduction of *Pseudotaenidia* to *Taenidia* (Apiaceae). Brittonia **34**: 365–367.

Davis, P.H. 1972. Umbelliferae. *In*: Davis, P.H. (Ed.), Flora of Turkey and the East Aegean Islands, Vol. **4**. Edinburgh University Press, Edinburgh, pp. 265–538.

Downie, S., Ramanath, S., Katz-Downie, D. and Llanas, E. 1998. Molecular systematics of Apiaceae subfamily Apioideae: phylogenetic analyses of nuclear ribosomal DNA internal transcribed spacer and plastid rpo C1 intron sequences. Am. J. Bot. **85**: 563–591.

Downie, S.R., Plunkett, G.M., Watson, M.F., Spalik, K., Katz, D., Valiejo, C., Terentieva, E.I., Troitsky, A.V., Lahham, J. and El, A. 2001. Tribes and clades within Apiaceae subfamily Apioideae: the contribution of molecular data. Edinburgh J. Bot. **58**: 301–330.

Drude, C.G.O. 1898. Umbelliferae. *In*: Engelmann, W. (Ed.), Die Natürlichen Pflanzenfamilien, Vol. **3**. Leipzig, Germany, pp. 63–250.

Feng, T., Downie, S.R., Yu, Y., Zhang, X., Chen, W., He, X. and Liu, S. 2009. Molecular systematics of *Angelica* and allied genera (Apiaceae) from the Hengduan Mountains of China based on nrDNA ITS sequences: phylogenetic affinities and biogeographic implications. J. Plant Res. **122**: 403–414.

Hammer, Ø., Harper, D.A.T. and Ryan, P.D. 2001. PAST: Paleontological statistics software package for education and data analysis. Palaeontol. Electronica **4**: 1–9.

Harris, J.G. and Harris, M.W. 1994. Plant identification terminology: An illustrated glossary (No. QK9 H37 2001). Spring Lake Publishing, Spring Lake, Utah, pp. 1–216.

Heywood, V.H. 1971. Systematic survery of Old World Umbelliferae. *In*: Heywood, V.H. (Ed.), The Biology and Chemistry of the Umbelliferae. Academic Press, London, pp. 31–41.

Kondo, K., Terabayashi, S., Okada, M., Yuan, C. and He, S. 1996. Phylogenetic relationship of medicinally important *Cnidium offcinale* and Japanese Apiaceae based on *rbc*L sequences. J. Plant Res. **109**: 21–27.

Koo, J.C. and Kim, M.Y. 2008. Taxonomic review for the *Cyrnopterus melanotilingia* (H. Boissieu) C.Y. Yoon in Apiaceae. Korean J. Pl. Taxon. **38**: 345–358.

Koso-Poljansky, B.M. 1916. Sciado phytorum systematis lineamenta. Bull. Soc. Imp. Nat. Moscou **29**: 93–222.

Lee, E.I, Ko, Y.O., Choi, C.H., Lee, J.K. and Kim, S.H. 1997. Morphological characteristics of seeds in medicinal plants of Umbelliferae. Korean J. Crop Sci. **42**(4): 429–434.

Lee, B.Y., Levin, G.A. and Downie, S.R. 2001. Relationships within the spiny-fruited umbellifers (Scandiceae: subtribes Daucinae and Torilidinae) as assessed by phylogenetic analysis of morphological characters. Syst. Bot. **26**(3): 622–642.

Liao, C., Downie, S.R., Li, Q., Yu, Y., He, X. and Zhou, B. 2013. New insights into the phylogeny of *Angelica* and its allies (Apiaceae) with emphasis on East Asian species, inferred from nrDNA, cpDNA, and morphological evidence. Syst. Bot. **38**: 266–281.

Liu, M.R., Plunkett, G.M., Lowry, P.P., Van Wyk, B.E. and Tilney, P.M. 2006. The taxonomic value of fruit wing types in the order Apiales. Am. J. Bot. **93**: 1357–1368.

Pimenov, M.G. and Leonov, M.V. 1993. The Genera of the Umbelliferae: A Nomenclator. Royal Botanic Gardens, Kew, London, pp. 1–156.

Plunkett, G.M., Soltis, D.E. and Soltis, P.S. 1996. Higher level relationships of Apiales (Apiaceae and Araliaceae) based on phylogenetic analysis of *rbc*L sequences. Am. J. Bot. **83**: 499–515.

Quinn, G.P. and Keough, M.J. 2002. Experimental Design and Data Analysis for Biologists. Cambridge University Press, Cambridge, UK.

Spalik, K., Reduron, J.P. andDownie, S.R. 2004. The phylogenetic position of *Peucedanum sensu lato* and allied genera and their placement in tribe Selineae (Apiaceae, subfamily Apioideae). Plant Syst. Evol. **243**: 189–210.

Spalik, K., Wojewódzka, A. and Downie, S.R. 2001. The evolution of fruit in Scandiceae subtribe Scandicinae (Apiaceae). Can. J. Bot. **79**: 1358–1374.

Stearn, W.T. 1992. Botanical Latin. David and Charles Publishers, London, pp. 489–491.

Swofford, D.L. 2001. Paup*: Phylogenetic analysis using parsimony (and other methods) 4.0. B5.

The Plant List 2013. Version 1.1. Published on the Internet; http://www.theplantlist.org/ (accessed 1st January 2018).

Webb, C.J. and Simpson, M.J. 2001. Seeds of New Zealand gymnosperms and dicotyledons. Manuka Press, pp. 1–428.

Yoon, C.Y. 2001. Taxonomic position of *Cymopterus melanotiligia* (H. Boissieu) C.Y. Yoon based on morphological and anatomical characters. Korean J. Pl. Taxon. **31**: 253–265.

Zhou, J., Gong, X., Downie, S.R. and Peng, H. 2009. Towards a more robust molecular phylogeny of Chinese Apiaceae subfamily Apioideae: Additional evidence from nrDNA ITS and cpDNA intron (rpl16 and rps16) sequences. Mol. Biol. Evol. **53**: 56–68.

22

FLORAL POLYMORPHISM IN *POLYGONUM BISTORTA* L.

MING-LIN CHEN[1], XIAO-YU WANG AND AUBREY L. FUNKE[2]

Provincial Key Laboratory of Biotic Environment and Ecological Safety in Anhui, Anhui Normal University, Wuhu 241000, Anhui Province, China

Keywords: Polygonum bistorta; Tristylous like flowers; SEM; Evolution; Style morph ratios.

Abstract

In the present study, new type of floral polymorphism (tristylous-like flowers) in *Polygonum bistorta* L., i.e., L-, M-, and S-styles (long, medium and short, respectively) is reported. This tristylous was not only present in different populations but also in a single inflorescence. Although the heights of styles and anthers in the three morphs of flowers differed, their ancillary tepal size, pollen polymorphism, and stigma polymorphism were found similar. The ratio of M-flowers was far lower than those of L- and S-flowers. Unlike the solid nectar of other *Polygonum* species, fluid nectar was present in *Polygonum bistorta*, which reflects its specialized state. Different from most of other hetereostylous species with tubular flowers and *Polygonum jucundum* with open distylous flowers, *Polygonum bistorta* contains semi-open flowers. Our findings suggest that *P. bistorta* represents special floral polymorphism in *Polygonum* and is an ideal plant species for studying the evolution of floral breeding system in *Polygonum*.

Introduction

Heterostyly is a floral polymorphism in which the various morphs differ in terms of the sequence of heights at which the anthers and stigmas are presented within its flowers. In these flowers, the sex organs are spatially separated and reciprocally arranged in the morphs; two levels of reproductive organs are present in distylous species, and three levels are present in tristylous species (Lloyd *et al.*, 1990). Tristyly, which occurs in only seven angiosperm families, is thought to be one of the most complex breeding systems in plants (Barrett, 1993; Naiki, 2012). Traditionally, tristyly was thought to only occur in different individuals or in different populations, and three morphs of flowers (with long, medium and short styles) were never found in the same inflorescence. Tristyly inheritance involves two diallelic loci (*S* and *M*), with the S-locus epistatic to the M-locus (Nowak *et al.*, 2015).

Heterostyly in Polygonaceae was first documented in *Fagopyrum esculentum* (Hildebrand, 1867). Subsequently, many distylous species have been reported for the genera *Oxygonum* and *Aconogonum* (Hong, 1999). The genus *Polygonum*, which includes over 300 species worldwide (Li and Bao, 2003), is seldom reported to be heterostylous. Reddy *et al.* (1977) reported the first distylous *P. chinense* in *Polygonum*. Recently, some distylous species in *Polygonum* were described (Chen and Zhang, 2010; Chen, 2012), and style-stamen dimorphism in this genus was also reported (Hassan and Khan, 1987), but to date, other new floral polymorphisms have never been reported in the genus.

Polygonum bistorta L. is a perennial herb that reaches up to 90 cm tall. This plant blooms from late spring to autumn, producing tall stems ending in single terminal racemes comprising club-like spikes, 5–7 cm long, which contain rose-pink flowers. *P. bistorta* is distributed in the Americas, western and central Asia, and northern and central Europe, including Britain and the mountains of

[1]Corresponding author. Email: jesschen28@126.com
[2]College of Engineering, Forestry and Natural Sciences, Northern Arizona University, USA

southern Europe. Most studies of this species have focused on its physiology (Kumar *et al.*, 2012), ecology (Egory Starr *et al.*, 2000), and medical applications (Demiray *et al.*, 2009). To date, floral polymorphisms have never been described in detail in this species. Therefore, this species provides unsurpassed opportunities for performing integrated evolutionary studies. The objectives of the current study include: (i) to examine the characteristics of floral polymorphisms and to determine whether *P. bistorta* is tristylous or exhibits anomalous heteromorphisms; (ii) to investigate the relationship between the reproductive system of *P. bistorta* and its insect visitors; and iii) to explore the systematic and evolutionary significance of floral polymorphism in *Polygonum*.

Material and Methods

Studied species and population sampling

This study was conducted in 2013-2016 in China and in 2014 in Flagstaff, Arizona, USA. Herbarium specimens or living materials from thirteen populations of *P. bistorta* in Anhui (China) and Arizona (USA) were used in this study (Table 1). All vouchers were deposited at the Herbarium of Anhui Normal University, China and Northern Arizona University (NAU), USA. Some voucher specimens for surveying L-, M-, and S-flower ratios of the species (deposited at NAU herbarium) were also collected.

Morphometrics

Floral morphological characters were observed in the field and in the laboratory. To characterize floral polymorphism in *P. bistorta*, tepal height, stigma and anther heights, and the size and number of pollen grains were measured. Since the stamens were present at different positions and angles, their heights were measured in the natural state instead of measuring the length of single dissected stigmas, stamens, and tepals (measure after fully matured so as to rule out the possibility of developmental changes) (Fig. 1). The floral traits recorded included the following: (i) corolla height; (ii) style length (from the basal ovary to the stigmatic surface); (iii) stamen height (from the basal ovary to the midpoint of each anther); (iv) style-nearer anther separation (SNAS); and v) Long-Short anther separation (LSAS) (Fig. 1). The flowers were slit longitudinally and measurements were made from digital photos using the image *Analysis 5.0* software. Forty flowers were randomly sampled from three separate populations (FS1: 14 flowers; FS2: 13 flowers; FS3: 13 flowers) in Flagstaff. All photographs were taken under a Zeiss Discovery V8 stereomicroscope.

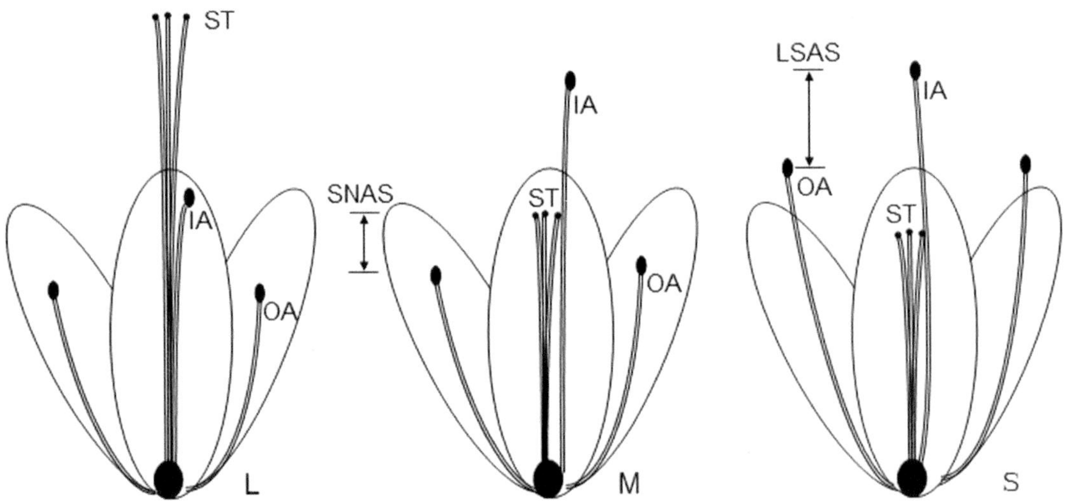

Fig. 1. L-, M-, and S- flowers of *P. bistorta*. ST=Style; IA=Inner anther; OA=Outer anther; SNAS=Style-nearer anther separation; LSAS=Long-Short anther separation.

Size and number of pollen grains

To measure pollen size, samples were collected from fresh flowers. One long (L)-, medium (M)-, and short (S)-level anther was chosen per bud from 25 flowers. Pollen was mounted on each slide in a drop of glycerol and photographed under a Zeiss Axio Scope A1 optical microscope (200X). The polar and equatorial diameters of 75 pollen grains (three moderately sized pollen grains per flower) were measured based on photographs taken with *Analysis 5.0* software. To determine the number of pollen grains produced, anthers from 75 flowers per type (L, M, and S) from the Flagstaff populations were placed on individual microscope slide. Using standard methods (Chen and Zhang, 2010), the number of pollen grains per flower was counted using a light-emitting microscope (LEM), the number of pollen grains in 75 flowers from three populations (25 flowers per population) was counted, and the number of pollen grains from outer and inner anthers was compared. Pollen counts for flowers with different style sizes were compared using Student's *t*-test.

Community ecology

1 m × 1 m sample was selected in 13 study populations, respectively. Plant species compositions, altitude and habitats were recorded for each sample. Some species occurred in populations were recorded as 1, or 0.

SEM observation

To characterize pollen shape, samples from L-, M-, and S- flowers stored in 70% ethanol were prepared according to the acetolysis method (Erdtman, 1960). Pollen grains were mounted in glycerine jelly and sealed with paraffin. The pollen was then observed under a Zeiss Supra 40 VP scanning electron microscope (SEM) operated at 20 kV. The shape and exine ornamentation of pollen grains from each sample were observed. Fresh stigmas and seeds were mounted on double-sided carbon tape and observed under SEM.

Table 1. Populations of *Polygonum bistorta* L. employed in the present study.

Codes	Sites	Habitat	Altitude (m)	Experiments
QLF1	Qingliangfeng, China	Grassy slopes	1650	SR, CE
QLF2	Qingliangfeng, China	Broadleaved deciduous forest	1450	CE
QLF3	Qingliangfeng, China	Broadleaved deciduous forest	1250	CE
QLF4	Qingliangfeng, China	Mixed forest	1050	CE
QLF5	Qingliangfeng, China	Mixed forest	850	CE
BSH1	Bashang 1, China	Uplands	1700	SR, CE
BSH2	Bashang 2, China	Uplands	1500	SR, CE
BSH3	Bashang 3, China	Mixed forest	1300	CE
BSH4	Bashang 4, China	Mixed forest	1100	CE
BSH5	Bashang 5, China	Mixed forest	900	CE
FS1	Flagstaff, Arizona, USA	Slope wastelands	2100	SR, MA, PS, PN, SEM, SSR, CE
FS2	Flagstaff, Arizona, USA	Mountain forest	2100	SR, CE
FS3	Flagstaff, Arizona, USA	Mountain forest	2100	SR, CE

MA: Morphometric analysis; PS: Pollen size; PN: Pollen number; SR: Style morph ratios; SSR: Seed setting rate; CE: Community ecology.

Surveys of style morph ratios and seed set

The relative frequencies of the three morphs of flowers were determined in six populations of *P. bistorta*, as shown in Table 1. At the same time, 24 mature inflorescences were harvested from each individual in the FS1, FS2 and FS3 populations (eight per population) to survey the style morph ratios. Seed set was recorded under natural or bagged conditions in the FS1 and FS2 populations (ten inflorescences were selected in each study population).

Results

Flower morphometrics

The present study reveals that the mature *P. bistorta* flowers are semi-open, and this plant exhibits an unusual type of floral polymorphism, i.e., L-, M-, and S- flowers are present not only in different populations, but also in a single inflorescence. Anthers and stigma are spatially separated in a flower and reciprocally arranged in the morphs, with three levels observed in this species (Fig. 2A, B, C). Normally, eight stamens are positioned at the base of the tepals in two whorls, with five stamens present in the outer whorl and three in the inner whorl. The inner whorl filaments are stronger than the outer whorl filaments, and the basal filaments of the inner whorl are red and inflated compared with those of the outer whorl (Fig. 2D). Fluid nectar was found at the base of the stamen and tepal (Fig. 2E). The stigma and anther heights of the three morphs of flowers are shown in Table 2 and Figure 3. There was no obvious difference in tepal size between the long-, medium-, and short-styled flowers (Table 2; LM: $t_{78} = 1.81$, $P > 0.05$; LS: $t_{78} = 1.66$, $P > 0.05$; MS: $t_{78} = 0.25$, $P > 0.05$) (Note: Here after, flowers with different style-morphs are referred to as L-, M-, and S-flowers and the anther levels are referred to as l, m, and s.) The stigma height of L-flowers was longer than the l anther height of M-flowers ($t_{78} = 14.14$, $P < 0.05$) and the l anther height of S-flowers ($t_{78} = 10.27$, $P < 0.05$). The stigma height of M-flowers was shorter than the m anther height of L-flowers ($t_{78} = 3.44$, $P < 0.05$) and the m anther height of S-flowers ($t_{78} = 13.89$, $P < 0.05$). The stigma height of S-flowers was longer than the s anther height of L-flowers ($t_{78} = 3.70$, $P < 0.05$) and the s anther height of M-flowers ($t_{78} = 2.09$, $P < 0.05$). The LSAS of M-flowers was the largest among the three morphs of flowers (ML: $t_{78} = 71.28$, $P < 0.05$; MS: $t_{78} = 19.51$, $P < 0.05$), while the SNAS of M-flowers was smaller than those of L-flowers ($t_{78} = 29.02$, $P < 0.05$) and S-flower ($t_{78} = 14.27$, $P < 0.05$). There were no obvious differences in pollen number and size among the three morphs of flowers.

The stigma height differed among L-, M-, and S-flowers of *P. bistorta* (Fig. 4A, B, C). The stigmas of the three morphs of flowers were present in a spherical, cerebriform pattern. There were no obvious differences in stigma surface ornamentation among the three morphs of flowers (Fig. 4D, E, F). Anthers of *P. bistorta* are of the adnate type, pink to purple in colour, and elliptical to oblong in shape. The pollen type of *P. bistorta* is 3-zonocolporate, and is elliptical to slightly rectangular in the equatorial view and triangular, angles obtuse, apertures in convex sides in the polar view, and the exine is reticulate. However, many perforations of unequal size and smaller columellae were observed in two polar views and near the colporate pollen. Both the shape and size of pollen were similar in L-flower and S-flower (Fig. 4G–L).

Surveys of style morph ratios and seed set

The style morph ratios of L-, M-, and S-flowers were approximately 237:18:33 among the studied populations (Table 3). No M-flowers in the FS1 population were found. However, in a single inflorescence, some inflorescences composed entirely of long-styled flowers were often found, and some inflorescences which mainly composed of medium-, and short-styled flowers were rarely found (Fig. 5A, B). The most common flowers were L- and S-styled morphs. L-flowers were always situated in the upper part of the inflorescence, while S-flowers were always found in the lower part of the inflorescence (Fig. 5C). The ratio of L-, M-, and S-type flowers was approximately 103:4:49 among the inflorescences investigated (Table 3). The ratio of M-styled flowers was far lower than those of L-, and S-styled flowers. Under natural conditions, the seed sets were 36.6 ± 8.45%, but under bagged conditions, the seed sets were 0 indicating cross pollination.

Fig. 2. L-, M-, and S-flowers of *P. bistorta.* A=L-styled flower; B=M-styled flower; C=S-styled flower; D=Inner anther (IA) and outer anther (OA); E=Fluid nectar at the base of the tepal (l: long anther; m: medium anther; s: short anther; FN: fluid nectar).

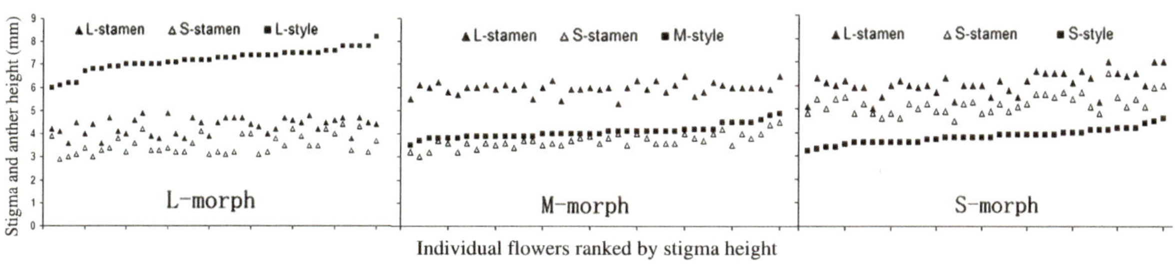

Fig. 3. Characteristics of flowers of *P. bistorta* ranked by style height to illustrate the reciprocal correspondence of stigma and anther positions in the L-, M-, and S-flowers. Positions of stigmas are indicated by solid rectangles (■); Positions of long and short anther are indicated by solid triangles (▲) and open triangles (Δ), respectively.

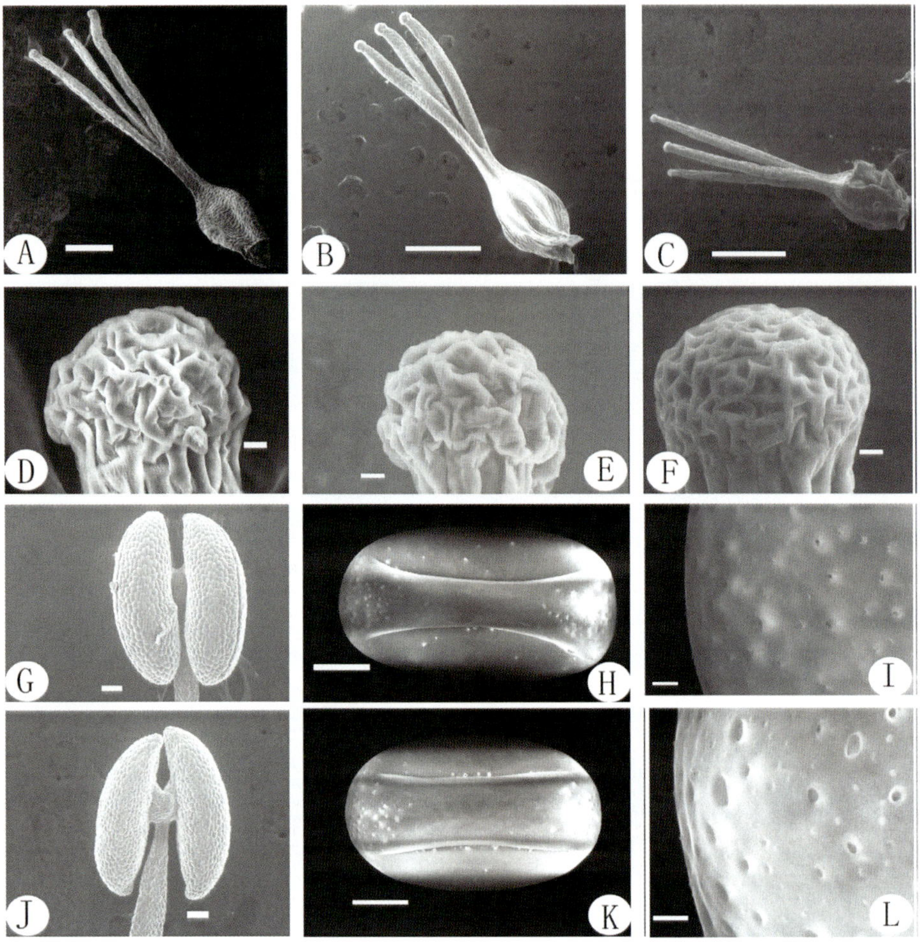

Fig. 4. Micromorphological characteristics of *P. bistorta* (SEM). A=Pistil of L-styled flower; B=Pistil of M-styled flower; C=Pistil of S-styled flower; D=Stigma of L-styled flower; E=Stigma of M-styled flower; F=Stigma of S-styled flower; G=Inner anther; H=Pollen of inner anther; I=Exine ornamentation of pollen of inner anther; J=Outer anther; K=Pollen of outer anther; L=Exine ornamentation of pollen of outer anther. Bars: A,B,C=1 mm; D,E,F,H,K=10 μm; G,J=100 μm; I,L=1 μm.

Fig. 5. Inflorescences of *P. bistorta* showing different flowers. A: L-styled; B: S-styled; C: L- and S-styled (Note: L: Long; S: Short).

Table 2. Measurement of floral characters of L-, M-, and S-flowers of *P. bistorta*. Differences between the means were analyzed by Student's *t*-test. [Mean ± standard error (SE)].

Floral characters	L-flower	M-flower	S-flower	N	*P*<
Height of tepal (mm)	5.02±0.286[a]	4.91±0.249[a]	4.93±0.228[a]	40	-
Height of L-anther (mm)	4.34±0.366[a]	5.97±0.261[b]	6.09±0.48[c]	40	0.01
Height of S-anther (mm)	3.53±0.419[a]	3.69±0.312[b]	5.20±0.415[c]	40	0.01
Height of stigma (mm)	7.19±0.482[a]	4.09±0.292[b]	3.84±0.309[c]	40	0.05
SNAS (mm)	2.85±0.291[a]	0.395±0.196[b]	1.37±0.281[c]	40	0.01
LSAS (mm)	0.81±0.212[a]	2.27±0.347[b]	0.88±0.28[c]	40	0.05
Pollen long-diameter (μm)	49.5±5.01[a]	48.6±4.94[a]	50.0 ±4.74[a]	75	-
P/E Ratio	c. 1.42	c. 1.41	c. 1.43	75	
Pollen number per flower	402±48[a]	395±58[a]	387±49[a]	90	-

Note: Different letters in the same row indicate significant differences between values.SNAS: Style-nearer anther separation; LSAS: Long-short anther separation.

Community ecology

Thirteen populations of *P. bistorta* were surveyed. Based on data matrix, cluster group was constructed using Canoco for windows 4.5 software. The populations 11, 12 and 13 were clustered together, and the populations 1, 2, 6 and 7 were clustered together, while the other populations were clustered together (Fig. 6). We also found that there was no *P. bistorta* in population 3, 4, 5, 8, 9, 10, whereas *P. bistorta* could be found in the former two cluster groups.

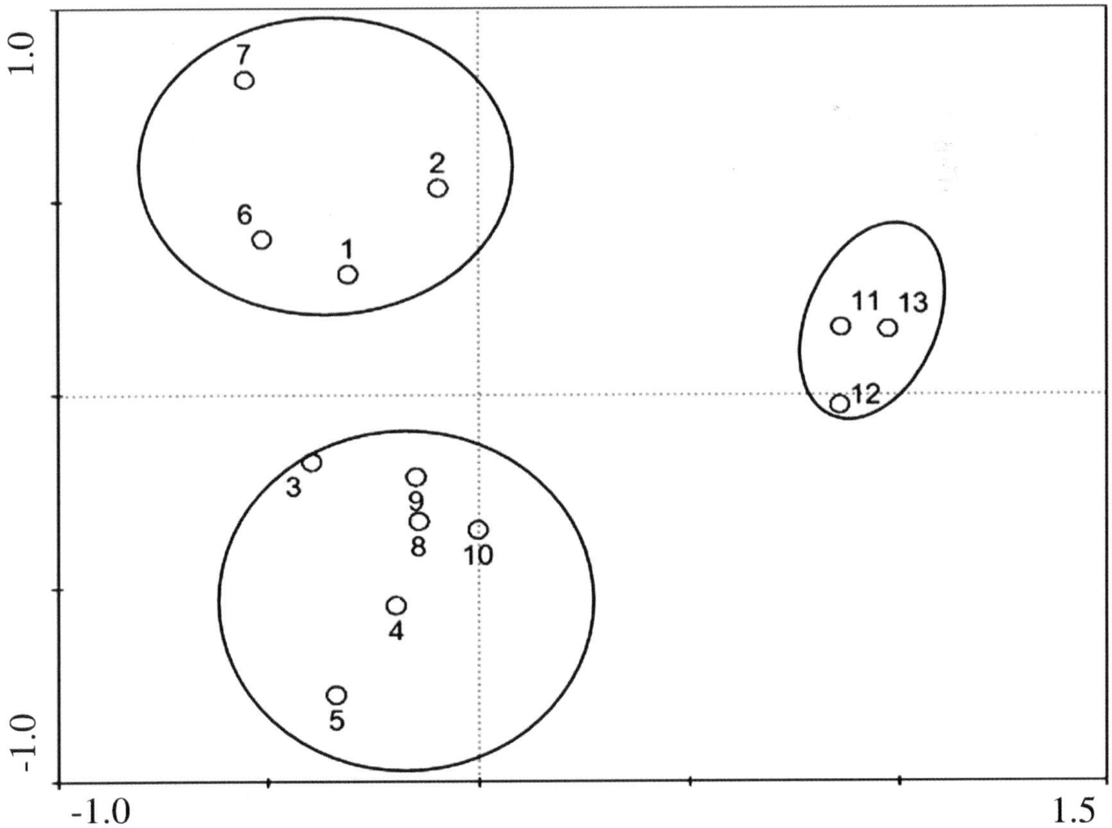

Fig. 6. PCA analysis for species compositions from 13 studied populations of *P. bistorta*.

Table 3. Style morph ratios of *P. bistorta* among populations and among inflorescences.

Scopes	L-styled	M-styled	S-styled	n
Among populations	237.17±47.34	18.17±6.18	33±5.77	6
Among inflorescences	103.38±7.60[a]	4.21±0.58[b]	49.21±3.87[c]	24

Note: Different letters in the same row indicate significant differences between values ($P<0.05$)

Discussion

Various stochastic and deterministic factors can cause biased morph ratios in tristylous populations. Founder events and genetic drift in small populations are a common cause of anisoplethy and morph loss (Barrett, 1993). Theoretical studies have indicated that the S-morph should be lost most often and the L-morph least often (Costa *et al.*, 2016). However, in the present study, we found that L-, M-, and S-flowers of *P. bistorta* not only could be found in different populations but also in a single inflorescence. This phenomenon is difficult to explain based on the traditional supergene locus control theory (Naiki, 2012). Clearly, *P. bistorta* is an intriguing plant material for studying the floral polymorphism. In trimorphic species, a 1:1:1 style morph ratio (i.e., isoplethy) is expected in populations at equilibrium (Costa *et al.*, 2016), but in field populations, many species present altered morph ratios (Lloyd *et al.*, 1990). For example, no L-morphs are present in tristylous *Oxalis corymbosa* in China (Chen *et al.*, 2007), while in the self-compatible, perennial *Decodon verticillatus* (Lythraceae), populations commonly lack style morphs, particularly M-styled flowers (Eckert and Mavraganis, 1996), which may provide this transitional variation. We found that the L-styled ratio of *P. bistorta* in populations was the largest in a field population, and inflorescences composed primarily of M-styled morphs were seldom found in field populations.

Darwin (1877) hypothesized that the function of the stamen-style polymorphism in heterostylous plants is to increase the likelihood of pollen transfer between sexual organs at the same level. In typical tristylous species, the style height of L-morph is usually approximately the same as that of the l stamen of M-morph and S-morph, and the style height of M-morph is approximately same as that of the m stamens of L-morph and S-morph. The same scenario occurs in *Eichhornia azurea* (Alves Dos Santos and Wittmann, 2000). However, this scenario was not detected in *P. bistorta.* We found that the style height of L-morph was higher than that of the l stamen of M-morph and S-morph. Therefore, the increased reciprocal placement between sexual organs of S- and L-styled floral morphs is expected to have occurred during the evolution of *P. bistorta.* In addition, two whorls of stamens are often the same height in distylous taxa of *Polygonum* (Chen and Zhang, 2010; Chen, 2012), while in *P. bistorta*, the inner whorl stamens are higher than those of the outer whorl, which suggests that the change in stamen height may have occurred in an evolutionary event.

There are no heterostylous taxa in the basal angiosperm groups and basal 'eudicots' (Naiki, 2012), and heterostyly has always been found in medium evolutionary-state tubular flowers (Darwin, 1877; Lloyd and Webb, 1992), with the exception in Polygonaceae and *Turnera* (Barrett, 1992). The Polygonaceae frequently have dish-shaped corolla, which is true for the heterostylous genera, *Fagopyrum* and *Polygonum*. For example, the distylous flowers of *Polygonum jucundum* are broadly open (Chen and Zhang, 2010), but based on the present study, *P. bistorta* flowers are semi-open, which represents a specialized form during petal evolution in *Polygonum.* Along with the generalistic pollination phenomenon of this species, the formation of floral tubes in heterostylous flowers might have coevolved with specialistic pollinators during later evolution.

Typical heterostylous species are always dimorphic or trimorphic for the intrinsic features of style height, anther height, and a series of ancillary characteristics, such as pollen size, exine sculpturing, pollen number and production, and stigma exine ornamentation (Gettys and Wofford,

2008). However, in *P. bistorta*, except for the intrinsic feature of style and anther heights, we observed no difference in these ancillary characteristics. Moreover, eight yellow nectaries are present at the base of the perianth in distylous species such as *P. jucundum* (Chen and Zhang, 2010) and *P. hastato-sagittatum* (Chen, 2012), but fluid nectar rather than solid nectaries are present in *P. bistorta*. Based on the results of community ecology, we found all populations of *P. bistorta* were occurred in high altitude. Both of the findings reflected the specialization in this species and its adaptation to the high altitude, which is consistent with the molecular evidence (Sanchez *et al.*, 2011).

The rate of seed set, which is closely correlated with ovule development, can reflect the degree of successful reproduction in a species (Zhou *et al.*, 1996). However, because we occasionally found two or three morphs in a single inflorescence, it was difficult to survey the seed sets of each morph, and we only obtained two sets of data (0 and 36.5% under bagged and natural conditions, respectively). This result demonstrates that this species can reproduce under natural conditions.

As mentioned above, the results demonstrate that *P. bistorta* is not a typical tristylous species, but it has tristylous-like flowers, which have not previously been found in angiosperms. Therefore, it would be important to investigate the origin and evolution of floral polymorphism in this species. Heterostylous polymorphisms provide opportunities for ecological and genetic investigations into the evolution of plant sexual systems (Barrett, 1992). Although elucidating the evolutionary origins of heterostyly continues to be difficult (Barrett, 1993), significant progress has been made towards understanding these evolutionary pathways (Weller, 1992). At present, several hypotheses about the evolutionary mechanisms favouring heterostyly have been proposed. According to Darwin (1877), heterostyly can be considered as a strategy for reducing the likelihood of self- and intramorph pollination and for promoting legitimate intermorph pollination. Barrett (2002) stated that reciprocal sex-organ positions in style-morphs help increase male fertility by actively promoting more precise pollen dispersal among plants than would occur in populations with uniform sexual organs and herkogamy. However, in the present study, the heights of stigmas and anthers of L-, M-, and S-morphs in *P. bistorta* did not exactly coincide. Lloyd and Webb (1992) proposed a model describing how heterostyly is established, from approach herkogamy to style polymorphism, followed by reciprocal herkogamy (heterostyly). Their model demonstrates that heterostyly could be established from a population that exhibits continuous phenotypic variation, which is not the case in *P. bistorta*. The evolution of distyly from tristyly was investigated in populations of *Oxalis alpina* (Weller *et al.*, 2007), and the evolutionary shift from tristyly to distyly appears to have been particularly frequent in the Lythraceae (Weller, 1992). Two of the largest genera (*Nesaea* and *Lythrum*) contain both tristylous and distylous taxa, and some of these distylous taxa have retained features that suggest a tristylous ancestry (Ornduff, 1979). Elsewhere in this family, distylous *Pemphis acidula* appears to have evolved distyly through loss of the mid-styled morph from a tristylous ancestor (Lewis, 1975). In the evolutionary model from tristyly to distyly in *Pemphis*, the m anthers of the L-morph are closer to the s level, whereas the m anthers of S-morph are closer to the l level (Lewis and Rao, 1971). Similar patterns were also evident in *P. bistorta*, as we found that the m anthers of L-morph were shorter than the tepals of L-morph, while the m anthers of S-morph were longer than the tepals of S-morph. The hypothesis of the evolution of floral polymorphism toward distyly predicts divergence in m anther position between the L- and S-morphs (mS > mL) in dimorphic populations, which is also found in *P. jucundum* and *P. hastato-sagittatum* (Chen and zhang, 2010; Chen, 2012). Thus, we argue that heterostyly may arise from floral polymorphism in a single population, followed by heterostyly in different populations, and that *P. bistorta* appears to represent a specialized state from floral polymorphism to heterostyly in *Polygonum*.

Acknowledgements

The work was supported by grants from Natural Science Foundation of Anhui Province (1808085MC76), the Restoration of Degeneration Ecosystem in the City Belt Along the Yangtze River in Anhui project and the Research Platform Project Fund (Comparative study on breeding system of five species of *Polygonum*) at AHNU, China. We also thank Dr. Russell Benford and Nashelly Menese of Northern Arizona University, USA for all of their help with fieldwork.

References

Alves Dos Santos, I. and Wittmann, D. 2000. Legitimate pollination of the tristylous flowers of *Eichhornia azurea* (Pontederiaceae). Plant Syst. Evol. **223**: 127–137.

Barrett, S.C.H. 1992. Heterostylous genetic polymorphisms: Model systems for evolutionary analysis. *In*: Barrett, S.C.H.(Ed.), Evolution and Function of Heterostyly. Berlin: Springer-Verlag, pp. 1–29.

Barrett, S.C.H. 1993. The evolutionary biology of tristyly. *In*: Futuyma, D. and Antonovics, J. (Eds), Oxford Survey in Evolutionary Biology. Oxford University Press, Oxford, pp. 283–326.

Barrett, S.C.H. 2002. The evolution of plant sexual diversity. Nat. Rev. Genet. **3**: 274–284.

Chen, M.L. 2012. Floral morphology and breeding system in *Polygonum hastato-sagittatum* Mak. (Polygonaceae). Flora **207**: 365–371.

Chen, M.L., Liu, D.Y. and Li, S.S. 2007. Observation of micromorphological characters of five species in *Oxalis*. J. Plant Resour. & Environ. **16**(3): 7–18.

Chen, M.L. and Zhang, X.P. 2010. Distyly in *Polygonum jucundum* Meisn. Plant Syst. Evol. **288**: 139–148.

Costa, J., Castro, S., Loureiro, J. and Barrett, S.C.H. 2016. Variation in style morph frequencies in tristylous *Lythrum salicaria* in the Iberian Peninsula: the role of geographical and demographic factors. Ann. Bot. **117**: 331–340.

Darwin, C. 1877. The different forms of flowers on plants of the same species. Appleton and Co, New York.

Demiray, S., Pintado, M.E. and Castro, P.M.L. 2009. Evaluation of phenolic profiles and antioxidant activities of Turkish medicinal plants: *Tilia argentea*, *Crataegi folium* leaves and *Polygonum bistorta* roots. World Acad. Sci. Eng. & Tech. **3**: 251–256.

Eckert, C.G. and Mavraganis, K. 1996. Evolutionary consequences of extensive morph loss in tristylous *Decodon verticillatus* (Lythraceae): A shift from tristyly to distyly? Am. J. Bot. **83**(8): 1024–1032.

Egory Starr, G.R., Oberbauer, S.F. and Pop, E.R.I.C.W. 2000. Effects of lengthened growing season and soil warming on the phenology and physiology of *Polygonum bistorta*. Glob. Change Biol. **6**(3): 357–369.

Erdtman, G. 1960. The acetolysis technique: a revised description. Svensk Botanisk Tisdkr. **54**:561–564.

Gettys, L.A. and Wofford, D.S. 2008. Genetic control of floral morph in tristylous Pickerel weed (*Pontederia cordata* L.). J. Hered. **99**(5): 558–563.

Hassan, M.A. and Khan, M.S. 1987. Style-stamen dimorphism in *Polygonum* L. (Polygonaceae). Bangladesh J. Bot. **16**: 93–95.

Hildebrand, F. 1867. Die Geschlechter-Vertheilungbei den Pflanzen und das Gesetz der vermiedenen und unvortheilhaftenstetigen Selbstbefruchtung. Leipzig: W. Engelmann, 32.

Hong, S.P. 1999. Pollen dimorphism in heterostylous species of *Oxygonum*. Syst. & Geogr. Plants **68**: 245–252.

Kumar, M.D., Deepmala, J. and Sangeeta, S. 2012. Antioxidant, antipyretic and choleretic activities of crude extract and active compound of *Polygonum bistorta* Linn. in albino rats. International J. Pharm. & Biol. Sci. **2**(1): 25–31.

Lewis, D. and Rao, A.N. 1971. Evolution of dimorphism and population polymorphism in *Pemphis acidula* Forst. Proc. Royal Soc. London. Ser. B. **178**: 79–94.

Lewis, D. 1975. Heteromorphic incompatibility systems under disruptive selection. Proc. Royal Soc. London. Ser. B. **188**: 247–256.

Li, A.J. and Bao, B.J. 2003. Polygonaceae. *In*: Wu, Z.Y., Raven, P.H. and Hong, D.Y. (Eds), Flora of China. Beijing, Science Press and St. Louis: Missouri Botanical Garden Press **5**: 277–350.

Lloyd, D.G. and Webb, C.J. 1992. The selection of heterostyly. *In*: Barrett, S.C.H. (Ed.), Evolution and Function of Heterostyly. Springer-Verlag, Berlin, pp. 151–178.

Lloyd, D.G., Webb, C.J. and Dulberger, R. 1990. Heterostyly in species of *Narcissus* (Amaryllidaceae) and *Hugonia* (Linaceae) and other disputed cases. Plant Syst. Evol. **172**: 215–227.

Naiki, A. 2012. Heterostyly and the possibility of its breakdown by polyploidization. Plant Species Biol. **27**: 3–29.

Nowak, M.D., Russo, G., Schlapbach, R., Huu, C.N., Lenhard, M. and Conti, E. 2015. The draft genome of *Primula veris* yields insights into the molecular basis of heterostyly. Genome Biol. **16**(1): 1–17.

Ornduff, R. 1979. The morphological nature of distyly in *Lythrum* section *Euhyssopifolia*. Bull. Torrey Bot. Club **106**: 4–8.

Reddy, N.P., Bahadur, B. and Kumar, P.V. 1977. Heterostyly in *Polygonum chinense*. J. Genet. **63**(2): 79–82.

Sanchez, A., Schuster, T.M., Burke, J.M. and Kron, K.A. 2011. Taxonomy of Polygonoideae (Polygonaceae): A new tribal classification. Taxon **60**(1): 151–160.

Weller, S.G. 1992. Evolutionary modifications to tristylous breeding systems. *In*: Barrett, S.C.H. (Ed.), Evolution and Function of Heterostyly. Springer-Verlag, Berlin, pp. 247–272.

Weller, S.G., Dominguez, C.A., Molina-Freaner, F.E., Fornoni, J. and Lebuhn, G. 2007. The evolution of distyly from tristyly in populations of *Oxalis alpina* (Oxalidaceae) in the Sky islands of the Sonoran desert. Am. J. Bot. **94**(6): 972–983.

Zhou, S.L., Pan, K.Y. and Hong, D.Y. 1996. Comparative studies on pollination biology of *Mosla hangchouensis* and *M. chinensis* (Labiatae). Acta Bot. Sin. **38**(7): 530–540.

23

A COMPARATIVE STUDY OF ACHENE MORPHOLOGY IN KOREAN POLYGONACEAE

MIN-JUNG KONG, JUN-HO SONG[1], BEOM-CHEOL AN, SUNG-WON SON[2], GANG-UK SUH[2], MI-JIN CHUNG[2] AND SUK-PYO HONG[3]

Laboratory of Plant Systematics, Department of Biology, Kyung Hee University, Seoul 02447, South Korea

Keywords: Achene surface; Fruit morphology; Micromorphology; Polygonoideae; SEM.

Abstract

A comparative macro- and micromorphological study was conducted on achenes of Korean Polygonaceae comprising 45 taxa under 10 genera using a Stereo-microscope (SM) and Scanning Electron Microscope (SEM) to evaluate the taxonomical relevance of achene morphological characteristics. The achene shape is lenticular, biconvex, trigonous, or spheroidal. The largest achenes are found in *Fagopyrum esculentum* Moench (5.80–6.70 × 3.90–4.40 mm) and *Rheum rhabarbarum* L. (5.00–6.50 × 4.00–5.20 mm), and the smallest is found in *Rumex maritimus* L. (1.17–1.41 × 0.60–0.84 mm). Three types of embryo shape (curved, straight, and folded) and two types of embryo position (plane and parietal) can be distinguished. Most taxa have a curved embryo shape with plane position. Four types of surface patterns (smooth, papillae, tuberculate, or small pits) are observed. The surface sculpture or embryo type is rather consistent at the generic or tribal level, and the comprehensive consideration of achene characteristics is also useful at the species level. The achene morphology is described in detail, compared, and illustrated. The taxonomic significance of the achene morphology is discussed.

Introduction

Polygonaceae Juss. consists of ca. 1,200 species in 48 genera and has a worldwide distribution (Freeman and Reveal, 2005). The family Polygonaceae is usually divided into subfamilies Eriogonoideae Arnott and Polygonoideae Eaton (Freeman and Reveal, 2005), and the subfamily Symmerioideae Meisn. was suggested based on recent molecular phylogeny with number of stamens and pyramidal achene (Brandbyge, 1993; Sanchez *et al.*, 2011). The Polygonoideae has been characterized by herb or sub-shrubs, the presence of ochreae and swollen nodes, flowers subtended by bracteoles but not enclosed in involucres, and cosmopolitan distribution (Haraldson, 1978; Freeman and Reveal, 2005; Sanchez *et al.*, 2011). The subfamily is also consistently confirmed by molecular sequence data (Sanchez *et al.*, 2009, 2011).

However, the taxonomic circumstance of intra-subfamily in Polygonoideae (e.g., tribal, generic, or sectional level) have been controversial, and new tribal compositions were suggested according to recent molecular phylogenetic studies (Freeman and Reveal, 2005; Sanchez *et al.*, 2009, 2011; Schuster *et al.*, 2015). For example, the proposed taxonomic system by these studies consist of five to seven tribes, such as Calligoneae C.A. Mey, Fagopyreae Yoneak., Oxygoneae T.M. Schust. & Reveal, Persicarieae Dumort., Polygoneae Rchb., Pteroxygoneae T.M. Schust. & Reveal and Rumiceae Dumort. (*mat*K, *ndh*F and *rbc*L, Sanchez *et al.*, 2011; ITS, *mat*K and *trn*L-F,

---

[1]Present address: Herbal Medicine Resources Research Center, Korea Institute of Oriental Medicine, Naju 58245, Korea.
[2]Plant Conservation Division, Korea National Arboretum, Pocheon 11186, Korea.
[3]Corresponding author. Email: sphong@khu.ac.kr

Schuster *et al.*, 2015). The taxonomic composition in the Persicarieae or Polygoneae is also most controversial as large group within Polygonoideae, namely the generic or sectional delimitation of these tribes are constantly changing (Hedberg, 1946; Haraldson, 1978; Brandbyge, 1993; Lamb Frye and Kron, 2003; Li *et al.*, 2003; Park and Hong, 2007; Sanchez *et al.*, 2011). For example, *Persicaria* was treated into *Polygonum s.l.* (Li *et al.*, 2003; Zhu *et al.*, 2007), however recent studies suggested that *Persicaria* is separated as independent genus (Haraldson, 1978; Ronse Decraene *et al.*, 2000; Lamb Frye and Kron, 2003). *Aconogonon* and *Bistorta* were merged into *Polygonum* in some studies (Li *et al.*, 2003; Park and Hong, 2007), while *Aconogonon* and *Bistorta* were usually recognized as independent genera (Hedberg, 1946; Haraldson, 1978; Freedman and Reveal, 2005). However, a recent molecular study treated *Aconogonon* as a section of *Koenigia* (Schuster *et al.*, 2015). Various research approaches have been attempted to clarify the taxonomic positions within the Polygonoideae; thus, a comparative study of the morphological characteristics or re-evaluation of recently suggested taxonomic system is necessary to determine taxonomic delimitation.

The importance of fruit and seed morphology in flowering plants has been recognized for systematic information at various taxonomic levels; furthermore, several recent studies have shown the phylogenetic value of fruit and seed characteristics (Donoghue *et al.*, 2004; Jacobs *et al.*, 2008, 2010; Choi *et al.*, 2012; Song *et al.*, 2015). The description of anatomical features or micromorphology of the achenes of Polygonaceae were often performed, but only these were performed on represented taxa or regional taxa (e.g., Ronse Decraene *et al.*, 2000; Yurteseva, 2001; Hou *et al.*, 2007; Kantachot and Chantaranothai, 2011), or were simply described without any taxonomic implications (Martin, 1946, 1954). The Korean Polygonaceae is known to that consist of 85 taxa (10 genera) which belong to the tribes Fagopyreae, Persicarieae, Polygoneae and Rumiceae (except Calligoneae, Oxygoneae and Pteroxygoneae) within the subfamily Polygonoideae (Chang *et al.*, 2014; Schuster *et al.*, 2015). However, the study of achene morphology in Korean Polygoanceae has been mainly in the form of brief notes or only on sectional or generic level (Kim *et al.*, 2001; Nakayama *et al.*, 2004; Lee *et al.*, 2010), and is still lacking comprehensive discussion. Thus, the present study aims to investigate and provide detailed descriptions of achene macro- and micromorphology in the selected Korean Polygonaceae, and to evaluate the taxonomic or diagnostic importance of fruit morphological features.

Materials and Methods

The achene morphology of 45 taxa from 10 genera in the Korean Polygonaceae (Table 1) was investigated. The achenes were collected from herbarium specimens of KH and KHUS, and donated from the seed bank of the Korean National Arboretum, Pocheon, South Korea [see Table 1; abbreviations according to Thiers (2018) and continuously updated].

Fully matured fruits were selected and observed under a stereomicroscope (SM; SteREO Discovery.V8; Carl Zeiss Microscopy GmbH, Germany) to confirm their normality and taxon identification. Cross-sections of the achenes were also taken, and examined under a SM. To observe detailed sculpturing patterns of the achenes, dry fruits were rehydrated overnight in the Agepon wetting agent (Agepon®:DW = 1:200), and fresh fruits were fixed in FAA (formaldehide: acetic acid: alcohol) for 48 h, and stored in 70 % ethanol. Prepared achenes were dehydrated through an ethanol series (in 50 %, 70 %, 90 %, and 95 % ethanol for 10 min each and in absolute ethanol for about 20 min) and then replaced ethanol with carbon dioxide for critical point drying

Table 1. Voucher specimens of the Korean Polygonaceae that are examined in the present study (Some taxa have only seed bank management number).

Taxon	Voucher specimens
Aconogonon (Meisn.) Rchb.	
A. alpinum (All.) Schur	Korea, Chungcheongnam-do, Oct.2002, G.W. Seo and J.Y.Kim (KH-154)
A. divaricatum (L.) Nakai *ex* T. Mori	Korea, Gyunggi-do, Oct.2004, J.Y. Kim (KH-L1455)
Bistorta (L.) Scop.	
B. manshuriensis (Petrov *ex* Kom.) Kom.	Korea, Gyunggi-do, Aug.2006, J.Y. Kim (KH-L2769)
Fagopyrum Mill.	
F. esculentum Moench	Korea, Is. Jeju, Oct. 2011, S.H. Kang 111024004 (KH)
Persicaria (L.) Mill.	
P. chinensis (L.) H. Gross var. *chinensis*	Korea, ls, Jeju, Nov. 2012, M.J. Kongs, n. (KHUS)
P. dissitiflora (Hemsl.) H. Gross *ex* T. Mori	Korea, Gyungsangbuk-do, Sep.2012, S.G. Son and H.W.Choi (KH-L10091)
P. filiformis (Thunb.) Nakai	Korea, Gyunggi-do, Sep.2005, J.Y. Kim (KH-L2208)
P. hydropiper (L.) Delarbre	Korea, Gyunggi-do, Oct.2002, G.W. Seo and J.Y. Kim (KH-L153)
P. japonica (Meisn.) Nakai	Korea, Gyunggi-do, Nov.2004, J.Y. Kim (KH-L1835)
P. lapathifolia (L.) Delarbre var. *lapathifolia*	Korea, Gyunggi-do, Sep.2012, J.Y. Kim (KH-L10035)
P. lapathifolia var. *salicifolia* (Sibthorp) Miyabe	Korea, Incheon, Nov.2014, S.H. Park *et al.* (KH-L14267)
P. longiseta (Bruijn) Kitag.	Korea, Gyungsangbuk-do, Sep.2004, G.W. Seo *et al.* (KH-L1329)
P. maackiana (Regel) Nakai *ex* T. Mori	Korea, Gyungsangnam-do, Oct.2012, S.D. Lee (KH-L10050)
P. muricata (Meisn.) Nemoto	Korea, Gyunggi-do, Oct.2007, J.Y. Kim (KH-L3238)
P. nepalensis (Meisn.) H. Gross	Korea, Gangwon-do, Sep.2011, M.S. Kim *et al.* (KH-L8969)
P. nodosa (Pers.) Opiz	Korea, Gyunggi-do, Nov.2004, J.Y. Kim (KH-L1821)
P. orientalis (L.) Spach	Korea, Gyunggi-do, Oct.2003, J.Y. Kim (KH-L798)
P. perfoliata (L.) H. Gross	Korea, Gyungsangbuk-do, Aug.2014, G.H. Gang (KH-L13798)
P. posumbu (Buch.-Ham. *ex* D. Don) H.Gross	Korea, Gyunggi-do, Oct.2007, J.Y. Kim (KH-L3309)
P. pubescens (Blume) H. Hara	Korea, Gyungsangbuk-do, Oct.2014, G.H. Gang *et al.* (KH-13817)
P. sagittata (L.) H. Gross	Korea, Gyunggi-do, Oct.2006, J.Y. Kim (KH-L2914)
P. senticosa (Meisn.) H. Gross	Korea, Gyungsangbuk-do, Sep.2012, S.G. Son and H.W.Choi (KH-L10094)
P. thunbergii (Siebold & Zucc.) H. Gross	Korea, Gyunggi-do, Oct.2007, J.Y. Kim (KH-L3380)
P. tinctoria (Aiton) H. Gross	Korea, Gyunggi-do, Oct.2007, J.Y. Kim (KH-L3236)
P. viscofera (Makino) H. Gross	Korea, Gyungsangbuk-do, Aug. 2012, S. G. Son and H.Y. Choi (KH-L10095)
P. viscosa (Buch.-Ham. *ex* D. Don) H. Gross *ex* T. Mori	Korea, Gyungsangbuk-do, Sep.2012, S.G. Son and H.S.Kwon (KH-L10096)
P. vulgaris Webb & Moq.	Korea, Gyungsangnam-do, Oct.2012, J.S. Kim (KH-L10059)
Fallopia Adanson	
F. convolvulus (L.) A. Löve	Korea, Gyunggi-do, Nov.2005, J.Y. Kim (KH-L2618)
F. dentatoalata (F. Schmidt) Holub	Korea, Jeollabuk-do, Sep.2013, G.S. Jung and S.J. Kwon (KH-L13170)
F. dumetorum (L.) Holub	Korea, Gyunggi-do, Nov.2005, J.Y. Kim (KH-L2602)
F. japonica (Houtt.) Ronse Decr.	Korea, Gyunggi-do, Nov.2005, J.Y. Kim (KH-L2625)
F. koreana B.U. Oh & J.G. Kim	Korea, Chungcheongbuk-do, Oct.2011, G.Y. Lee (KH-L9470)
F. sachalinensis (F. Schmidt) Ronse Decr.	Korea, Is. Ulleung, Gyungsangbuk-do, Oct.2011, M.S.Kim *et al.* (KH-L9025)

Table 1 Contd.

Taxon	Voucher specimens
Knorringia (Czukav.) Tzvelev	
K. sibirica (Laxm.) Tzvelev subsp. *sibirica*	Korea, Incheon, Sep.2006, S.H.Park 61732 (KH)
Polygonum L.	
P. aviculare L.	Korea, Gyunggi-do, Oct.2004, J.Y.Kim (KH-L1772)
Oxyria Hill.	
O. digyna (L.) Hill	D.P.R.K., Mt. Baekdu, Aug.1997, K.W.Park L-61153 (KH)
Rheum L.	
R. rhabarbarum L.	Mongolia, Tariat, Jul.2002, S.H. Park ParkSH23435 (KH);
Rumex L.	
R. acetosa L.	Korea, Gyunggi-do, Jun.2007, J.Y. Kim *et al.* (KH-3324)
R. acetosella L.	Korea, Gyunggi-do, Jul.2005, J.Y. Kim (KH-L2135)
R. conglomeratus Murray	Korea, Gyungsangnam-do, Jul.2004, J.M. Jung and M.H.Park (KH-1073)
R. crispus L.	Korea, Gyungsangnam-do, Jul.2004, J.M. Jung and J.Y.Kim (KH-L1047)
R. japonicus Houtt.	Korea, Gyunggi-do, Jul.2005, J.Y. Kim (KH-L2126)
R. maritimus L.	Korea, Gangwon-do, Sep.2004, J.M. Jung *et al.* (KH-L1282)
R. obtusifolius L.	Korea, Gyunggi-do, Jul.2005, J.Y. Kim (KH-L2127)
R. patientia L.	Korea, Chungcheongbuk-do, Jul.2011, E.S. Jeon (KH-L8856)

Herbarium acronyms are in accordance with Thiers (2018) [continuously updated]

(CPD, SPI-13200J-AB). All samples for SEM analysis were coated with platinum using an ion-sputtering device (E-1045; Hitachi, Tokyo, Japan). Samples were examined under a field emission SEM (FE-SEM; S-4700; Hitachi, Tokyo, Japan) operating at 10 kV with a working distance of 10–13 mm. A total of 10 achenes for CPD were measured using the Magnification 2.0 version software (Orbicule, Leuven, Belgium).

We followed the currently accepted taxonomic treatment for the taxa studied to avoid any taxonomic confusion (Freeman and Reveal, 2005; Sanchez *et al.*, 2011). The terminology for fruits and embryos was mainly adopted from Ball *et al.* (1962), Brandbyge (1993), and Ronse Decraene *et al.* (2000).

Results and Discussion

Achene shape, size, colour, embryo type, and micromorphology

Shape: The various achene shapes in Korean Polygonaceae were lenticular, biconvex, trigonous, and spheroidal (Table 2; Fig. 1). Most of the taxa studied had trigonous achenes, and the lenticular shape was observed in five taxa within *Persicaria* (*P. hydropiper*, *P. japonica*, *P. lapathifolia* var. *salicifolia*, *P. nodosa*, and *P. orientalis*; Table 2; Fig. 1F). Biconvex achenes were found in four taxa (*Persicaria filiformis*, *P. nepalensis*, *P. tinctoria*, and *Oxyria digyna*; Table 2; Fig. 1C, D & M), and spheroidal achenes were only observed in *Persicaria perfoliata*.

The longitudinal-section (L.S.) shapes were found to be elliptic (Fig. 2C, O & S), elliptic rhomboid (Fig. 2G & W), ovoid (Fig. 2E, I & K), broadly ovoid (Fig. 2A, Q & U), and circular (Fig. 2M). The cross-section (C.S.) shapes were mostly triangular (Fig. 2B, D, F, H, P, T, V & X), but also sometimes narrowly elliptic (Fig. 2N), elliptic (Fig. 2J), rhombic (Fig. 2R), and triangular- ovoid (Fig. 2L). Most of the taxa studied had a beak (Figs. 1A, B, D, E, F, G, I, J, K, L, N, O, & 2I). Some had a prominent beak, such as a caudate or cuspidate (Fig. 1B, D, E, F, G, I, & J), but five taxa had achenes without a beak (e.g., *Persicaria filiformis, P. senticosa*, *P. thunbergii*,

Fallopia dumetorum, *Rheum rhababarum*, *Rumex acetosa*, and *R. acetosella*; Fig. 1C & H). The absence of beak could help to identify some taxa.

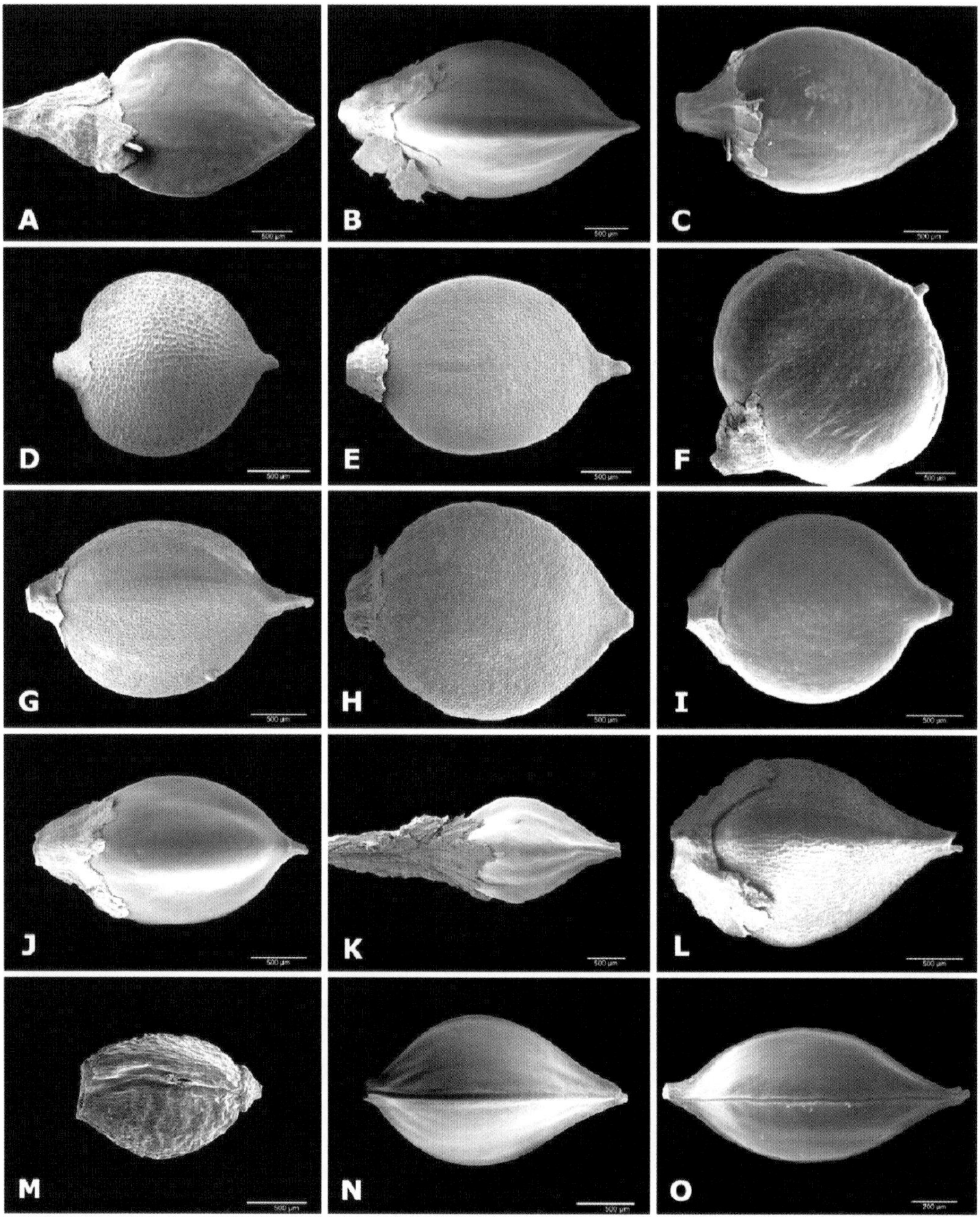

Fig. 1. SEM micrographs of achenes of Polygonaceae in Korea. A. *Aconogonon alpinum*; B. *Bistorta manshuriensis*; C. *Persicaria filiformis* (sect. *Tovara*); D. *Persicaria nepalensis* (sect. *Cephalophilon*); E. *Persicaria maackiana* (sect. *Echinocaluon*); F. *Persicaria orientalis* (sect. *Persciaria*); G. *Persicaria sagittata* (sect. *Echinocaluon*); H. *Persicaria thunbergii* (sect. *Echinocaluon*); I. *Persicaria nodosa* (sect. *Persicaria*); J. *Persicaria pubescens* (sect. *Persicaria*); K. *Fallopia sachalinensis*; L. *Polygonum aviculare*; M. *Oxyria digyna* (beak was broken; see Fig. 2I); N. *Rumex obtusifolius*; O. *Rumex maritimus*.

Table 2. Details of achene characteristics in the Korean Polygonaceae taxa. Infrageneric affiliation (e.g., subgenera or sectional) was obtained from the most accepted taxonomic treatment for the investigated taxa according to Freeman and Reveal (2005) and Sanchez *et al.* (2011).

Taxon	Size range (mm)		Gloss	Colour	Shape	L.S.	C.S.	Beak	Embryo		Sculpture
	length	width							shape	position	
Tribe Fagopyreae Yonek.											
Fagopyrum Mill.											
F. esculentum Moench	5.80–6.70	3.90–4.40	-	d.b	trg	ovo	tria	+	folded	~	smooth
Tribe Persicarieae Dumort.											
Aconogonon (Meisn.) Rchb.											
A. alpinum (All.) Schur	3.79–5.61	2.07–2.77	++	o.b	trg	bov	tria	+	curved	parietal	smooth
A. divaricatum (L.) Nakai *ex* T.Mori	4.24–4.84	2.16–2.98	++	o.b	trg	ovo	tria	+	curved	parietal	smooth
Bistorta (L.) Scop.											
B. manshuriensis (Petrov *ex* Kom.) Kom.	3.17–3.67	2.01–2.23	++	bl	trg	ell	tria	+	curved	parietal	smooth
Persicaria (L.) Mill.											
Persicaria* sect. *Persicaria											
P. hydropiper (L.) Delarbre	1.81–2.46	1.35–1.80	+	d.b	lent	bov	nell	+	curved	parietal	smooth
P. japonica (Meisn.) Nakai	1.92–2.39	1.35–1.68	+	bl	lent	bov	nell	+	curved	parietal	smooth
P. lapathifolia (L.) Delarbre var. *lapathifolia*	1.77–2.34	1.22–1.47	++	bl	trg	ell	tria	+	curved	parietal	shallow papillae
P. lapathifolia var. *salicifolia* (Sibthorp) Miyabe	2.03–2.85	1.41–1.99	+	d.b	lent	cc	nell	+	curved	parietal	smooth
P. longiseta (Bruijn) Kitag.	1.76–2.25	1.17–1.46	++	bl	trg	ell	tria	+	curved	parietal	smooth
P. nodosa (Pers.) Opiz	2.09–2.61	1.50–1.88	+	d.p.b	lent	bov	nell	+	curved	parietal	smooth
P. orientalis (L.) Spach	3.09–3.60	2.63–3.40	+	bl	lent	cc	nell	+	curved	parietal	smooth
P. posumbu (Buch.-Ham. ex D.Don) H. Gross	2.43–3.28	1.56–2.06	+	bl	trg	ovo	ell	+	curved	parietal	tubercle wall
P. pubescens (Blume) H. Hara	2.16–2.64	1.40–1.71	++	bl	trg	ell	tria	+	curved	parietal	shallow papillae
P. tinctoria (Aiton) H. Gross	2.13–2.51	1.10–1.54	++	bl	bicv	ell	rhom	+	curved	parietal	smooth
P. viscofera (Makino) H. Gross	1.51–1.86	1.09–1.33	++	bl	trg	ovo	tria	+	curved	parietal	smooth
P. viscosa (Buch.-Ham. *ex* D.Don) H.Gross *ex* T.Mori	2.54–2.98	1.58–1.97	++	bl	trg	ovo	tria	+	curved	parietal	smooth
P. vulgaris Webb & Moq.	1.95–2.54	1.22–1.73	+	d.p.b	trg	ell	tria	+	curved	parietal	shallow papillae
Persicaria* sect. *Cephalophilon (Meisn.) H. Gross											
P. chinensis (L.) H. Gross var. *chinensis*	2.57–3.41	1.99–2.31	-	bl	trg	bov	tria	+	curved	parietal	tubercle wall
P. nepalensis (Meisn.) H. Gross	1.37–2.46	1.25–1.74	-	d.b	bicv	bov	rhom	+	curved	parietal	tubercle wall
Persicaria* sect. *Echinocaulon (Meisn.) H. Gross											
P. dissitiflora (Hemsl.) H. Gross *ex* T. Mori	3.02–4.00	2.02-2.38	++	d.b	trg	ovo	trov	+	curved	parietal	smooth
P. maackiana (Regel) Nakai *ex* T. Mori	2.52–3.11	1.47–1.94	+	bl	trg	ovo	tria	+	curved	parietal	tuberculate
P. muricata (Meisn.) Nemoto	1.96–2.45	1.20–1.47	++	p.b	trg	ell	tria	+	curved	parietal	smooth
P. perfoliata (L.) H. Gross	3.33–4.12	2.83–3.18	++	bl	sph	bov/cc	cc	+	curved	parietal	smooth
P. sagittata (L.) H. Gross	2.23–2.73	1.48–1.77	+	bl	trg	ovo	tria	+	curved	parietal	tuberculate
P. senticosa (Meisn.) H. Gross	3.09–3.81	2.38–2.83	-	bl	trg	bov	trov	-	curved	parietal	tuberculate
P. thunbergii (Siebold & Zucc.) H. Gross	3.37–4.25	1.91–2.80	+	br	trg	ovo	trov	-	curved	parietal	smooth

Table 2 Contd.

Taxon	Size range (mm)		Gloss	Colour	Shape	L.S.	C.S.	Beak	Embryo		Sculpture
	length	width							shape	position	
Persicaria* sect. *Tavara (Adanson) H. Gross											
P. filiformis (Thunb.) Nakai	2.78–3.49	1.73–2.01	+	r.b	bicv	bov	rhom	-	curved	parietal	smooth
Tribe Polygoneae Rchb.											
Fallopia Adanson											
Fallopia* sect. *Fallopia											
F. convolvulus (L.) A. Löve	2.73–3.50	1.65–2.03	++	bl	trg	elrh	tria	+	curved	parietal	papillae
F. dentatoalata (F.Schmidt) Holub	3.91–5.67	2.21–2.64	++	pu	trg	elrh	tria	+	curved	parietal	smooth
F. dumetorum (L.) Holub	1.37–1.62	0.87–1.09	++	bl	trg	elrh	tria	-	curved	parietal	smooth
Fallopia sect. *Reynoutria* (Houtt.) Ronse Decr.											
F. japonica (Houtt.) Ronse Decr.	3.64–5.08	2.34–3.08	++	d.p.b	trg	elrh	tria	+	curved	parietal	shallow papillae
F. koreana B.U.Oh & J.G.Kim	5.13–7.45	2.15–2.64	++	d.p.b	trg	nell	tria	+	curved	parietal	smooth
F. sachalinensis (F.Schmidt) Ronse Decr.	2.44–2.98	1.22–1.60	++	bl	trg	elrh	tria	+	curved	parietal	smooth
Knorringia (Czukav.) Tzvelev											
K. sibirica (Laxm.) Tzvelev subsp. *sibirica*	2.90–3.50	1.90–2.20	++	br	trg	elrh	tria	+	curved	parietal	smooth
Polygonum L.											
Polygonum* sect. *Polygonum											
P. aviculare L	2.65–3.13	1.25–1.76	-	br	trg	ovo	trov	+	curved	parietal	tubercle wall
Tribe Rumiceae Dumort.											
Oxyria Hill											
O. digyna (L.) Hill	1.30–2.10	0.80–1.80	-	o.b	bicv	ovo	ell	+	straight	plane	smooth
Rheum L.											
R. rhabarbarum L.	5.00–6.50	4.00–5.20	-	br	trg	bov	tria	-	straight	plane	tubercle wall
Rumex L.											
Rumex* subg. *Rumex											
R. conglomeratus Murray	1.18–1.58	0.88–1.22	++	d.b	trg	elrh	tria	+	straight	plane	smooth
R. crispus L.	1.88–2.36	1.15–1.53	++	d.b	trg	elrh	tria	+	straight	plane	small pits
R. japonicus Houtt.	1.88–2.23	1.03–1.27	++	o.b	trg	elrh	tria	+	straight	plane	smooth
R. maritimus L.	1.17–1.41	0.60–0.84	++	o.b	trg	ell	tria	+	straight	plane	small pits
R. obtusifolius L.	1.68–2.38	1.03–1.41	++	d.b	trg	elrh	tria	+	straight	plane	small pits
R. patientia L.	2.04–2.78	1.32–1.72	++	d.b	trg	elrh	tria	+	straight	plane	small pits
Rumex* subg. *Acetosa											
R. acetosa L.	1.51–2.01	0.88–1.15	++	bl	trg	ell	tria	-	straight	plane	smooth
Rumex* subg. *Acetosella											
R. acetosella L.	1.10–1.41	0.90–1.00	+	d.b	trg	bov	tria	-	curved	plane	small pits

Gloss: +, glossy; ++, strong glossy; -, dull; **Colour**: d.b, dark brown; o.b, orange brown; r.b, red brown; p.b, purple brown; d.p.b, dark purple brown; br, brown; bl, black; pu, purple; **Shape**: C.S., cross-sectioned shape; L.S., longitudinal-sectioned shape: bov, broadly ovoid; bicv, biconvex; cc, circular; elrh, elliptic rhomboid; ell, elliptic; nell, narrowly-elliptic; ovo, ovoid; rhom, rhombic; sph, sphrodial; trg, trigonous; tria, triagular; trov, trigonous-ovoid; /, or; **Beak**: +, present; -, absent; ~, not applicable.

Fig. 2. Cross and longitudinal section of achenes of Polygonaceae in Korea. A, B. *Aconogonon alpinum* (Type III); C, D. *Bistorta manshuriensis* (Type III); E, F. *Fagopyrum esculentum* (Type I); G, H. *Fallopia dentatoalata* (Type III); I, J. *Oxyria digyna* (Type II); K, L. *Persicaria dissitiflora* (sect. *Echinocaulon*; Type III); M, N. *Persicaria lapathifolia* var. *salicifolia* (sect. *Persicaria*; Type III); O, P. *Persicaria pubescens* (sect. *Persicaria*; Type III); Q, R. *Persicaria tinctoria* (sect. *Persicaria*; Type III); S, T. *Polygonum aviculare* (Type III); U, V. *Rumex acetosella* (Type IV); W, X. *Rumex obtusifolius* (Type II). A, C, E, G, I, K, M, O, Q, S, U, and W, longitudinal section of achenes; B, D, F, H, J, L, N, P, R, T, V, and X, cross-section of achenes in investigated taxa.

The achene shapes of the taxa studied agreed with the descriptions in earlier studies (Ronse Decraene *et al.*, 2000; Li *et al.*, 2003; Freedman and Reveal, 2005; Lee *et al.*, 2010; Kantachot and Chantaranothai, 2011). According to previous studies, some taxa showed heteromorphy in their achene shape; moreover, collecting season has implications on the achene shape (Yurtseva, 2001; Li *et al.*, 2003; Freedman and Reveal, 2005; Kantachot and Chantaranothai, 2011). For example, *Persicaria hydropiper* has either convex or trigonous achenes (Lee, 2003; Freedman and Reveal, 2005). The achene shape was consistent for each taxon in present study, however, taxonomic application of achene shape could be needed careful application.

Size: The size of achenes ranged from 1.10–7.45 mm in length and 0.60–5.20 mm in width (Table 2). The large achenes were found on *Fagopyrum esculentum* (5.80–6.70 × 3.90–4.40 mm) and *Rheum rhabarbarum* (5.00–6.50 × 4.00–5.20 mm), whereas the smallest achenes were found in *Rumex maritimus* (1.17–1.41 × 0.60–0.84 mm). The taxa in *Oxyria* and *Rumex* had the small achenes compared to other genera (1.10–2.78 × 0.60–1.80 mm). Most taxa, except *Fagopyrum esculentum*, *Rheum rhabarbarum*, *Rumex acetosella* and *R. maritimus*, showed to have similar size ranges of achenes (1.37–5.67 × 0.87–3.40 mm). The achene sizes in the taxa studied were also similar to those recorded in earlier studies (Li *et al.*, 2003; Kantachot and Chantaranothai, 2011). The largest or smallest achene sizes could be used to classify to the genus level.

Colour: Fully matured achenes of most of the taxa were brownish to black, and dull or shiny (Table 2); this colour is commonly found in Polygonaceae (Kantachot and Chantaranothai, 2011). The colour of fully matured fruit is usually not an important character to identify genus or species, except for in a few taxa that have odd colours such as grey or greyish green (Kantachot and Chantaranothai, 2011; Sadeghian *et al.*, 2014). Most of the taxa studied had glossy achenes except six examined taxa (Table 2), which had dull achenes (e.g. *Fagopyrum esculentum*, *Oxyria digyna*, *Persicaria nepalensis*, *P. senticosa*, *Polygonum aviculare,* and *Rheum rhabarbarum*). The existence of dull achenes is useful for the identification of the species.

Embryo shape and position: Three embryo shapes were observed in this study, *viz.* curved, straight, and folded (Table 2; Fig. 2). Most taxa studied had curved embryos (e.g. *Aconogonon*, *Bistorta*, *Fagopyrum*, *Fallopia*, *Knorringia*, *Persicaria*, and *Polygonum*; Fig. 2A, C, G, K, M, O, Q, S & U). While *Oxyria*, *Rheum,* and *Rumex* (except *R. acetosella*) had straight embryos (Fig. 2I & W), and folded embryos were only observed in *Fagopyrum* (Fig. 2E & F). The embryo shapes of the taxa studied were similar to those in previously published data (Martin, 1946; Brandbyge, 1993; Freedman and Reveal, 2005; Sanchez *et al.*, 2011). In the present study, all taxa studied had one shape of embryo; however, embryos of Rumiceae were observed to be curved or straight (Freedman and Reveal, 2005). In this study, *Rumex acetocella* of the *Rumex* subg. *Acetosella* had a different shape from other Rumiceae taxa. Additionally, two types of embryo position (plane and parietal) were observed (Table 2; Fig. 2). Most of the taxa had aparietal position (Fig. 2B, D, H, J, L, N, P, R & T), while *Oxyria*, *Rheum,* and *Rumex*, which belong to the tribe Rumiceae, had a plane-positioned embryo (Fig. 2V & X).

The embryos can be divided into four types based on shape and position: Type I – folded embryo (*Fagopyrum* - Fagopyreae); Type II – straight embryo with plane position (*Oxyria*, *Rheum*, and *Rumex* - Rumiceae); Type III – curved embryo with parietal position (*Aconogonon*, *Bistorta*, *Persicaria*, *Fallopia*, *Knorringia*, and *Polygonum* - Persicarieae and Polygoneae); Type IV–curved embryo with plane position (*Rumex acetosella*). The types are congruent with the delimitation of suggested tribes (Sanchez *et al.*, 2011; Fig. 4). For example, the Type I embryo, which is an unusual type in Polygonoideae, was found only in *Fagopyrum*. This genus which was separated into the tribe Fagopyreae according to the recent taxonomic system (Sanchez *et al.*, 2011), and its embryo type could be considered to be an apomorphy (Fig. 4). The Persicarieae and Polygoneae both show Type III embryos, which could be assumed to be a parallelism of characteristics when compared with current phylogenetic studies (Fig. 4). The tribe Rumiceae is strongly supported by embryo position; however, embryo shape was varied at generic level.

Achene micromorphology: Four types of achene surfaces were recognized: smooth, papillae, tuberculate, and small pits (Table 2; Fig. 3). Most taxa had smooth to rugose surfaces without appendages (Fig. 3A, B, H & M). These types had remarkable anticlinal cell walls, generally undulated, and of these, *Fagopyrum esculentum* was characterized by longitudinal grooves on the achene surface (Fig. 3C).

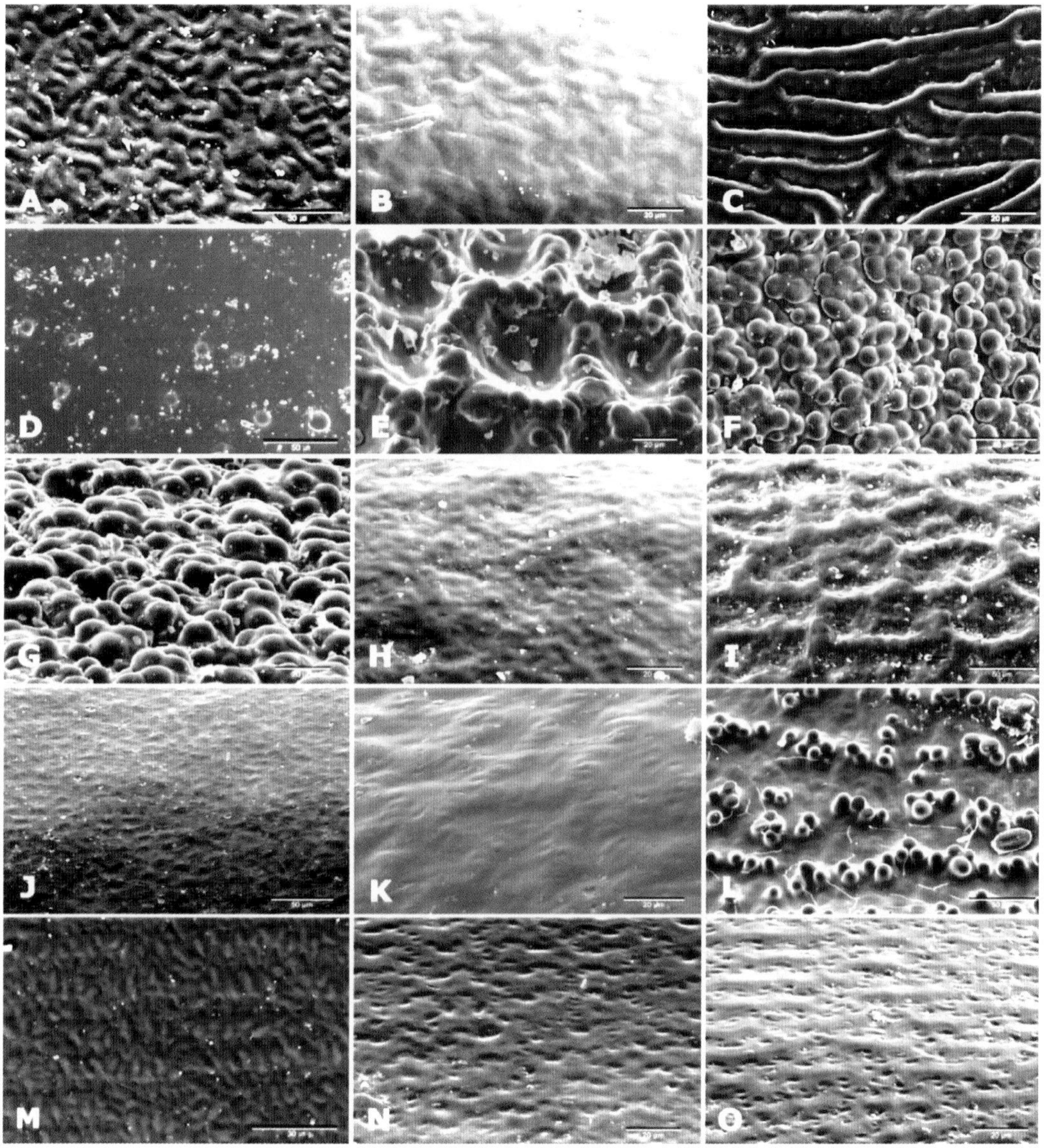

Fig. 3. The surface of achenes of Polygonaceae in Korea. A. *Aconogonon alpinum* (smooth); B. *Bistorta manshuriensis* (smooth); C. *Fagopyrum esculentum* (smooth); D. *Fallopia convolvulus* (papillae distributed irregularly); E. *Persicaria nepalensis* (sect. *Cephalophilon*; tubercles along the anticlinal cell walls); F. *Persicaria maackiana* (sect. *Echinocaulon*; tubercles distributed throughout entire pericarp); G. *Persicaria sagittata* (sect. *Echinocaulon*; tubercles distributed throughout entire pericarp); H. *Persicaria nodosa* (sect. *Persicaria*; smooth); I. *Persicaria posumbu* (sect. *Persicaria*; tubercles along the anticlinal cell walls); J. *Persicaria pubescens* (sect. *Persicaria*; shallow papillae); K. *Fallopia japonica* (shallow papillae); L. *Polygonum aviculare* (discontinued tubercles along the anticlinal cell walls); M. *Rumex acetosa* (smooth); N. *Rumex crispus* (small pits); O. *Rumex obtusifolius* (small pits).

Tubercles or papillae were divided into two types based on their distribution patterns: Type (i), covered the entire achene (Fig. 3E & F), and Type (ii), distributed along the anticlinal cell walls (Fig. 3E, I, J, K & L). Type (i) tubercles were observed in three taxa of *Persicaria*, and they were densely or regularly distributed (*P. maackiana*, *P. sagittata*, and *P. senticosa*; Fig. 3F & G). Type (i) papillae were only observed in *Fallopia convolvulus* (Fig. 3D; Table 2), and were loosely and irregularly distributed compared to the tubercles. Type (ii) tubercles were observed in four taxa (*Persicaria nepalensis*, *P. posumbu*, *Polygonum aviculare,* and *Rheum rhabarbarum*; Fig. 3E, I, J & L). Type (ii) tubercles of three taxa showed continuous distribution and were arranged along the anticlinal cell wall (Fig. 3E, I & J), while those of *Polygonum aviculare* showed a discontinuous arrangement (Fig. 3L). Similarly, Type (ii) papillae were observed in three taxa of *Persicaria* (*P. lapathifolia* var. *lapathifolia*, *P. pubescens*, and *P. vulgaris*; Fig. 3K), but these structures were represented by collapsed anticlinal cell walls (Ronse Decrane *et al.*, 2000). Yurtseva (2001) suggested that the xerophytic taxa of *Polygonum* subsect. *Polygonum* usually had verrucae on the achene surface, while the taxa that inhabit wet environments did not. Although the functions of these verrucae are not clear, smooth and glossy surface increase water repellency and prevent fungal or pathogenic infection (Barthlott, 1981; Yurtseva, 2001). Thus, the verrucae could be related to water absorption. In the present study, some taxa without tubercles or papillae within other genera were distributed through wet valleys (e.g., *Persicaria muricata*); thus this is seemed to support the early suggestion. While there are also taxa which inhabit both environments (e.g., *Persicaria senticosa* also inhabit wet valleys; Li *et al.,* 2003; Freedman and Reveal, 2005).

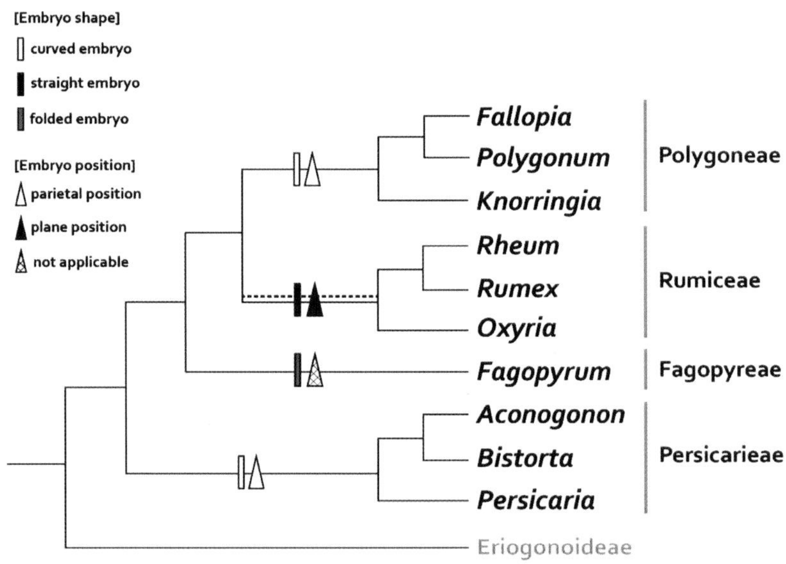

Fig. 4. Diagrams of embryo characters on most recent cladograms. Simplified phylogenetic tree is adopted from Schuster *et al.* (2015). Embryo shape was described as rectangular, and embryo position was described as triangular. The dotted line means that various type of embryo shape were observed in this group compared to those found in early studies (e.g., Freedman and Reveal, 2005; Sanchez *et al.*, 2011).

Small pits were observed to be scattered on the surface of most *Rumex* taxa (*R. acetosella*, *R. crispus*, *R. maritimus*, *R. obtusifolius*, and *R. patientia*; Table 2; Fig. 3N & O). According to a previous study, the small pits were also found in the *Persicaria* sect. *Tovara* (investigated taxa: *P. neofiliformis* and *P. virginiana*; Ronse Decrane *et al.*, 2000); however, the *Persicaria* sect. *Tovara* (*P. filiformis*) used in this study had a smooth surface (Table 2). This difference in surface sculpture could be considered to be interspecies difference.

Taxonomic implication of achene morphology in Korean Polygonaceae

The taxonomic position of *Aconogonon* has been controversial. The *Aconogonon* was usually acknowledged to be an independent genus although it was belonged to *Polygonum s.l.* in early studies (Hedberg, 1946; Haraldson, 1978; Freedman and Reveal, 2005). The sister group of the *Koenigia* with *Aconogonon* is *Bistorta*. In comparison with achene morphology, almost all characteristics, such as the shape, size, existence of beak, glossy, embryo and smooth surface, are similar between *Aconogonon* and *Bistorta*. The genus *Koenigia* in previous study had also same embryo type and smooth surface as in *Aconogonon* and *Bistorta*; however, the achene is distinguished from *Aconogonon–Bistorta* based on the lack of beak, glossiness and smaller size (cf., 1.10–1.84 ×1.79–1.02 mm; Ronse Decrane *et al.*, 2000; Kong and Hong, in prep.).

Persicaria lapathifolia var. *lapathifolia* is complicated taxonomically, thus many taxa were treated as synonyms of this taxon (e.g., Freedman and Reveal, 2005; Chang *et al.*, 2014). For example, *P. lapathifolia* var. *salicifolia* and *P. nodosa,* were treated as synonyms of *P. lapathifolia* var. *lapathifolia* (Freedman and Reveal, 2005; Chang *et al.*, 2014), or only one of them was treated as synonym (Timson, 1963). Li *et al.* (2003) suggested that *P. lapathifolia* var. *salicifolia* was admitted independent variety of *P. lapatifolia* var. *salicifolia.* The achene morphological characteristics between *P. lapathifolia* var. *salicifolia* and *P. nodosa* appeared to be more similar rather than those between *P. lapathifolia* var. *salicifolia* and *P. lapathifolia* var. *lapathifolia* (Table 2). Thus, *P. lapathifolia* var. *lapathifolia* and *P. lapathifolia* var. *salicifolia* could be separated as independent variety from each other, although the taxonomic position of *P. nodosa* is still unclear based on its achene characteristics.

In conclusion, achene morphological characteristics are useful to recognize certain taxa to the species, generic, or tribal level. Of the studied characteristics, the surface micromorphology of achene or embryo characteristics are rather consistent at the generic or tribal level; for example, small pits were found in most *Rumex* taxa, and papillae were only found in *Fallopia convolvulus.* In addition, the comprehensive consideration of achene characteristics are more useful to identify at the species level; for example, dull, broadly ovoid, biconvex and achenes with tubercle walls are characteristics of *P. nepalensis*, and regularly tuberculate on the whole achene surface and straight embryo in the plane position (Type IV) are characteristics of *Rumex acetosella.* This achene morphological study could improve our understanding of Korean Polygonaceae. A comparative study with molecular phylogeny will further improve our understanding of systematic tendencies in these groups.

Acknowledgements

We are grateful to our colleagues (Dr. Moon, H.-K. and Oak, M.-K.) in the Laboratory of Plant Systematics at Kyung Hee University, and to our anonymous reviewers and editor. We are grateful to the herbarium keepers of KH for their permission to examine specimens and for the loan of materials. This research was supported by the project ‘Studies on the Establishment of Seed Bank base for the Asian network’ through the Korea National Arboretum [grant numbers 20150437 and 20170334], and partially supported by the Basic Science Research Program through the National Research Foundation of Korea (NRF) funded by the Ministry of Education, Sciences and Technology [grant number NRF-2018R1D1A1A09083715] to S.P. Hong.

References

Ball, H.W., Exell, A.W., Harding, J.P., Léonard, J., Lewis, J., Melderis, A. and Vander Veken, P. 1962. Systematics association committee for descriptive biological terminology. II. Terminology of simple symmetrical plane shapes (Chart 1). Taxon **11**(5): 145–156.

Barthlott, W. 1981. Epidermal and seed surface characters of plants: systematic applicability and some evolutionary aspects. Nordic J. Bot. **1**(3): 345–355.

Brandbyge, J. 1993. Polygonaceae. *In*: Kubitzki, K. and Bittich, V. (Eds), The Families and Genera of Vascular Plants. Springer, Berlin, pp. 531–544.

Chang, C.S., Kim, H. and Chang, K.S. 2014. Provisional Checklist of Vascular Plants for the Korea (KPF) (Version 1.0). Designpost. Seoul, 561 pp.

Choi, B.K., Duretto, M.F. and Hong, S.P. 2012. Comparative seed morphology of *Boronia* and related genera (Boroniinae: Rutaceae) and its systematic implications. Nordic J. Bot. **30**(2): 241–256.

Donoghue, M.J., Baldwin, B.G., Li, J. and Winkworth, R.C. 2004. *Viburnum* phylogeny based on chloroplast *trn*K intron and nuclear ribosomal ITS DNA sequences. Syst. Bot. **29**(1): 188–198.

Freeman, C.C. and Reveal, J.L. 2005. Polygonaceae. *In:* Flora of North America Editorial Committee (Eds), Flora of North America, Vol. **5**. New York: Oxford Univ. Press, pp. 479–601.

Haraldson, K. 1978. Anatomy and taxonomy in Polygonaceae subfam. Polygonoideae Meissn. emend. Jaretzky. Symb. Bot. Ups. **22**(2): 1–95.

Hedberg, O. 1946. Pollen morphology in the genus *Polygonum* L. *s. lat.* and its taxonomical significance. Svensk Bot. Tidsk. **40**: 371–404.

Hou, Y.T., Xu, C.M., Qu, C.Y., Ba, X.G., Lu, F.J. and Li, F.Z. 2007. A study on fruit morphology of *Polygonum* sect. *Polygonum* (Polygnaceae) from China. Acta Phytotaxon. Sin. **45**(4): 523–537. [in Chinese with English summary]

Jacobs, B., Donoghue, M.J., Bouman, F., Huysmans, S., and Smets, E. 2008. Evolution and phylogenetic importance of endocarp and seed characters in *Viburnum* (Adoxaceae). Int. J. Plant Sci. **169**(3): 409–431.

Jacobs, B., Bell, C. and Smets, E. 2010. Fruits and seeds of the *Valeriana* clade (Dipsacales): Diversity and evolution. Int. J. Plant Sci. **171**(4): 421–434.

Kantachot, C. and Chantaranothai, P. 2011. Achene morphology of *Polygonum s.l.* (Polygonaceae) in Thailand. Tropic. Nat. Hist. **11**(1): 21–28.

Kim, M.H., Kwak, M., Song, J., Lee, S.J., Yoo, M.J. and Park, C.W. 2001. Morphology of hairs, flowers, and achenes of *Polygonum* section *Echinocaulon* Meisn. (Polygonaceae). Korean J. Plant Taxon. **31**(2): 143–160. [in Korean with English summary]

Lamb Frye, A.S. and Kron, K.A. 2003. Phylogeny and character evolution in Polygonaceae. Syst. Bot. **28**(2): 326–332.

Lee, T. 2003. Polygonaceae. *In*: Coloured Flora of Korea, Vol. **I**. Hyangmunsa, Seoul, pp. 245–279. [in Korean].

Lee, J.K., Kim, J.H., Lee, S.M., Park, S.H., Ali, M.A., Kim, J.K., Lee, C.Y. and Kim, G.R. 2010. Seeds of wild plants of Korea. <http://www.seedbank.re.kr/>. Korea Research Institute of Bioscience & Biotechnology, Dajeon, Korea.

Li, A.J., Bao, B., Grabovskava-Borodina, A.E., Hong, S.P., McNeill, J., Mosyakin, S.L., Ohba, H. and Park, C.W. 2003. Polygonaceae. *In:* The Flora of China Editorial Committee (Eds), Flora of China, Vol. **5**. Missouri Botanical Garden Press: Saint Louis, pp. 277–350.

Martin, A.C. 1946. The comparative internal morphology of seeds. Am. Midl. Nat. **36**(3): 513–660.

Martin, A.C. 1954. Identifying *Polygonum* seeds. J. Wildlife Manag. **18**(4): 514–520.

Nakayama, S., Inokuchi, M. and Minamitani, T. 2004. Seeds of wild plants in Japan. Tohoku University Press, Sendai. [in Japanese].

Navajas-Pérez, R., de la Herrán, R., López González, G., Jamilena, M., Lozano, R., Ruiz Rejón, C., Ruiz Rejón, M. and Garrido-Ramos, M.A. 2005. The evolution of reproductive systems and sex-determining mechanisms within *Rumex* (Polygonaceae) inferred from nuclear and chloroplastidial sequence data. Mol. Biol. Evol. **22**: 1929–1939.

Park, C.W. and Hong, S.P. 2007. Polygonaceae Juss. *In:* Flora of Korea Editorial Committee (Eds), The Genera of Vascular Plants of Korea. Academy Publishing Co., Seoul, pp. 334–360.

Rechinger, K.H. 1937. The North American species of *Rumex*. (Vorarbeiten zu einer Monographie der Gattung *Rumex* 5.) Publ. Field Mus. Nat. Hist., Bot. Ser. **17**: 1–150.

Ronse Decraene, L.P. and Akeroyd, J.R. 1988. Generic limits in *Polygonum* and related genera (Polygonaceae) on the basis of floral characters. Bot. J. Linn. Soc. **98**(4): 321–371.

Ronse Decraene, L.P., Hong, S.P. and Smets, E. 2000. Systematic significance of fruit morphology and anatomy in tribes Persicarieae and Polygoneae (Polygonaceae). Bot. J. Linn. Soc. **134**(1-2): 301–337.

Sadeghian, S., Zarre, S. and Heubl, G. 2014. Systematic implication of seed micromorphology in *Arenaria* (Caryophyllaceae) and allied genera. Flora **209**(9): 513–529.

Sanchez, A., Schuster, T.M. and Kron, K.A. 2009. A large-scale phylogeny of Polygonaceae based on molecular data. Int. J. Plant Sci. **170**(8): 1044–1055.

Sanchez, A., Schuster, T.M., Burke, J.M. and Kron, K.A. 2011. Taxonomy of Polygonoideae (Polygonaceae): A new tribal classification. Taxon **60**(1): 151–160.

Schuster, T.M., Reveal, J.L., Bayly, M.J. and Kron, K.A. 2015. An updated molecular phylogeny of Polygonoideae (Polygonaceae): Relationships of *Oxygonum*, *Pteroxygonum*, and *Rumex*, and a new circumscription of *Koenigia*. Taxon **64**(6): 1188–1208.

Song, J.H., Kim, J.M., Oak, M.K. and Hong, S.P. 2015. Comparative seed morphology of the tribe Alsineae (Caryophyllaceae) in Korea and its taxonomic implications. Korean J. Pl. Taxon. **45**(4): 369–379.

Timson, J. 1963. The taxonomy of *Polygonum lapathifolium* L., *P. nodosum* Pers, & *P. tomentosum* Schrank. Watsonia **5**: 386–395.

Thiers, B. 2018. [continuously updated] Index Herbariorum: A global directory of public herbaria and associated staff. <http://sweetgum.nybg.org/science/ih/>. New York Botanical Garden's Virtual Herbarium, Bronx, U.S.A.

van der Pijl, L. 1982. Principles of Dispersal in Higher Plants (2nd ed.). Springer-Velag. Berlin. 162 pp.

Yurtseva, O.V. 2001. Ultrasculpture of achene surface in *Polygonum* section *Polygonum* (Polygonaceae) in Russia. Nord. J. Bot. **21**(5): 513–528.

Zhu, L.T., Lu, F.J., Hou, Y.T. and Li, F.Z. 2007 Micro-morphology of leaf epidermis and its taxonomic significance of *Polygonum* section *Cephalophilon* in China. J. Wuhan Bot. Res. **25**(2): 136–142. [in Chinese].

Permissions

All chapters in this book were first published in BJPT, by Bangladesh Association of Plant Taxonomists; hereby published with permission under the Creative Commons Attribution License or equivalent. Every chapter published in this book has been scrutinized by our experts. Their significance has been extensively debated. The topics covered herein carry significant findings which will fuel the growth of the discipline. They may even be implemented as practical applications or may be referred to as a beginning point for another development.

The contributors of this book come from diverse backgrounds, making this book a truly international effort. This book will bring forth new frontiers with its revolutionizing research information and detailed analysis of the nascent developments around the world.

We would like to thank all the contributing authors for lending their expertise to make the book truly unique. They have played a crucial role in the development of this book. Without their invaluable contributions this book wouldn't have been possible. They have made vital efforts to compile up to date information on the varied aspects of this subject to make this book a valuable addition to the collection of many professionals and students.

This book was conceptualized with the vision of imparting up-to-date information and advanced data in this field. To ensure the same, a matchless editorial board was set up. Every individual on the board went through rigorous rounds of assessment to prove their worth. After which they invested a large part of their time researching and compiling the most relevant data for our readers.

The editorial board has been involved in producing this book since its inception. They have spent rigorous hours researching and exploring the diverse topics which have resulted in the successful publishing of this book. They have passed on their knowledge of decades through this book. To expedite this challenging task, the publisher supported the team at every step. A small team of assistant editors was also appointed to further simplify the editing procedure and attain best results for the readers.

Apart from the editorial board, the designing team has also invested a significant amount of their time in understanding the subject and creating the most relevant covers. They scrutinized every image to scout for the most suitable representation of the subject and create an appropriate cover for the book.

The publishing team has been an ardent support to the editorial, designing and production team. Their endless efforts to recruit the best for this project, has resulted in the accomplishment of this book. They are a veteran in the field of academics and their pool of knowledge is as vast as their experience in printing. Their expertise and guidance has proved useful at every step. Their uncompromising quality standards have made this book an exceptional effort. Their encouragement from time to time has been an inspiration for everyone.

The publisher and the editorial board hope that this book will prove to be a valuable piece of knowledge for researchers, students, practitioners and scholars across the globe.

List of Contributors

Lin Meng and Peichun Mao
Beijing Research and Development Center for Grass and Environment, Beijing Academy of Agriculture and Forestry Sciences, Beijing 100097, China

Yilmaz Aykut
Department of Molecular Biology and Genetics, Faculty of Science and Arts, Uşak University, 64200 Uşak, Turkey

Uslu Emel and Babaç M. Tekin
Department of Biology, Faculty of Science and Arts, Abant Izzet Baysal University, Bolu, Turkey

Muazaz Al-Hadeethy, Athiya Al-Mashhadani and Talib Al-Khesraji
Department of Biology, College of Education, University of Tikrit, Tikrit, Iraq
Department of Biology, College of Education-Ibn Al-Haitham, University of Baghdad, Baghdad, Iraq

Sahapat Barusrux
Centre for Research and Development of Medical Diagnostic Laboratories (CMDL), Faculty of Associated Medical Sciences, Khon Kaen University, Khon Kaen, Thailand

Hazim Al-Jewari Piyada Theerakulpisut and Pimwadee Pornpongrungrueng
Applied Taxonomic Research Center, Department of Biology, Faculty of Science, Khon Kaen University, Khon Kaen, Thailand

Wafaa Amer and Rim Hamdy
Department of Botany & Microbiology, Faculty of Science, Cairo University, Giza 12613, Egypt

Riham Mahdy
Timber Trees and Forestry Department, Horticulture Research Institute, Agriculture Research Center, Giza, Egypt

Ebadi-Nahari Mostafa
Department of Biology, Faculty of Science, Azarbaijan Shahid Madani University, Tabriz, Iran

Nikzat-Siahkolaee Sedigheh
Faculty of Biological Sciences, Shahid Beheshti University, Tehran, Iran

Ayesa Begum Momtaz Begum and Md. Oliur Rahman
Department of Botany, University of Dhaka, Dhaka 1000, Bangladesh

Nilofer Sheikh and Yogendra Kumar
Botany Department, North Eastern Hill University, Shillong-22, India

Vu Tien Chinh and Ritesh Kumar Choudhary
Biodiversity & Palaeobiology Group, Agharkar Research Institute, G.G. Agarkar Road, Pune, India

Tran The Bach, Bui Hong Quang, Do Van Hai and Sangmi Eum
International Biological Material Research Center, Korea Research Institute of Bioscience and Biotechnology, Daejeon, Republic of Korea

Joongku Lee
Department of Environment and Forest Resources, Chungnam National University, Daejeon, Republic of Korea

Ahmed Elkordy
Botany & Microbiology Department, Fac. of Science, Sohag University, 82524, Egypt

Ahmed Faried
Botany & Microbiology Department, Faculty of Science, Assiut University, 71515, Egypt

Maryam Naderifar, Ali Sonboli and Abbas Gholipour
Department of Biology, Medicinal Plants and Drugs Research Institute, Shahid Beheshti University, Tehran, Iran
Department of Biology, Payame Noor University, 19395-4697, Tehran, Iran

Tulay Aytas Akcin
Department of Biology, Faculty of Arts and Science, Ondokuz Mayıs University, Samsun, Turkey

Adnan Akcin
Department of Biology, Faculty of Arts and Science, Amasya University, Amasya 05100, Turkey

Mehmet Cengiz Karaismailoglu
Department of Botany, Faculty of Science, Istanbul University, 034116 Istanbul, Turkey

Mehmet Bona
Department of Botany, Faculty of Science, Istanbul University, Istanbul, Turkey

S. A. Deshmukh and Vinod B. Shimpale
Department of Botany, The New College, Kolhapur- 416 012, (MS), India

Savithri Sarojani Udage and Deepthi Yakandawala
Department of Botany, University of Peradeniya, Sri Lanka

Hideki Takahashi
Laboratory of Systematic Botany, Graduate School of Agriculture, Hokkaido University, Japan

A. K. M. Golam Sarwar
Department of Crop Botany, Bangladesh Agricultural University, Mymensingh 2202, Bangladesh

Ream I. Marzouk and Salama M. El-Darier
University of Alexandria, Faculty of Science, Botany and Microbiology Department, 21511, MoharamBek; Alexandria, Egypt

Abdel Baset M. Askar
Omar EL-Mukhtar University; Faculty of Science; El-Beyda; Libya

Neslihan Taşar
Munzur University, Tunceli Vocational School, Department of Organic Agriculture, Tunceli, Turkey

Gülden Doğan and Yaşar Kiran
Fırat University, Science Faculty, Biology Department, Elazığ, Turkey

M. Oliur Rahman
Department of Botany, University of Dhaka, Dhaka 1000, Bangladesh

Uğur Çakilcioğlu
Munzur University, Pertek Sakine Genç Vocational School, Tunceli, Turkey

Özgür Eminagaoglu and Melahat Ozcan
Department of Forest Engineering, Faculty of Forestry, Artvin Coruh University, 08000 Artvin, Turkey

Komban Parameswaran Smisha and Mamiyil Sabu
Angiosperm Taxonomy and Floristics Division, Department of Botany, University of Calicut, Kerala 673 635, India

Changyoung Lee
International Biological Material Research Center, Korea Research Institute of Bioscience & Biotechnology, 125 Gwahak-ro, Yuseong-gu, Daejeon 34141, South Korea

Jinki Kim, Ashwini M. Darshetkar and Ritesh Kumar Choudhary
VNU University of Science, 334 Nguyen Trai, ThanhXuan, Hanoi, 100000, Vietnam

Sang-Hong Park
Biodiversity & Palaeobiology Group (Plants), Agharkar Research Institute, G.G. Agarkar Road, Pune 411004, India

Joongku Lee
National Institute of Ecology, 1210, Geumgang-ro, Maseo-Myeon, Seocheon-gun, Chungnam, 325-813, Republic of Korea

Sangho Choi
Department of Environment and Forest Resources, Chungnam National University, Yuseong-gu, Daejeon, 34134, Republic of Korea

Ming-Lin Chen and Xiao-Yu Wang
Provincial Key Laboratory of Biotic Environment and Ecological Safety in Anhui, Anhui Normal University, Wuhu 241000, Anhui Province, China

Aubrey L. Funke
College of Engineering, Forestry and Natural Sciences, Northern Arizona University, USA

Min-Jung Kong and Jun-Ho Song
Laboratory of Plant Systematics, Department of Biology, Kyung Hee University, Seoul 02447, South Korea

Beom-Cheol An, Sung-Won Son, Gang-Uk Suh and Mi-Jin Chung
Herbal Medicine Resources Research Center, Korea Institute of Oriental Medicine, Naju 58245, Korea

Suk-Pyo Hong
Plant Conservation Division, Korea National Arboretum, Pocheon 11186, Korea

Index

A

A. Arabicum, 108, 113

A. Biebersteinii, 100, 102, 104-105

A. Grandifolia, 101-102, 104-105

Achene, 100-102, 104-106, 115-116, 120, 123, 125-126, 171, 173, 212-213, 215, 219-225

Achene Micromorphology, 100-101, 105, 221

Achillea, 100-102, 104-106

Aethionema, 107-108, 110-114

Ajugoideae, 154, 161

Anomocytic Stomata, 48, 50, 55, 60

Apiaceae, 189-190, 199-200

Apioideae, 189, 197, 199-200

Artificial Hybrid, 182, 186

Asteraceae, 100, 104-106, 114-116, 125-126, 162, 170-173, 181

B

Brassicaceae, 100, 107, 113-114, 126

C

C. Boissieri, 27-29, 32-35, 37

C. Cylindristachya, 27-29, 32-37

C. Polypodiifolia, 162, 166, 168-171

C. Urvillei Subsp Armata, 167

Caprifoliaceae, 40

Celastraceae, 68, 72-73, 180

Centaurea, 115-116, 119, 121, 125-126, 162-173

Chromosome Number, 45, 55, 162-163, 166, 169-172, 174, 179-180

Cluster Analysis, 11, 14-16, 40, 42, 57-58, 63, 76, 80, 108, 113, 127, 129, 133, 136-138

Convolvulaceae, 127, 133-134

Cordia L., 27-28, 39

Cymopterus, 189, 191, 193-197, 199-200

Cytology, 57, 162

D

D. Aucheri, 88-90, 93-95, 97-98

D. Lindbergii, 88-90, 94-97

Dioscorea L., 57

Dioscoreaceae, 57, 66-67

Dracocephalum, 88-90, 93-99

E

E. Elongata, 1-5, 7-9

E. Intermedia, 1-5, 7-9

Elliottia Pyroliflora, 145, 147

Elytrigia Caespitosa, 1-2

Elytrigia Desv, 1

Epigaea Asiatica, 145-147, 151

F

Fagopyrum Esculentum, 201, 212, 219-221

Floral Polymorphism, 201-202, 204, 208-209

Fruit Micromorphology, 174, 179, 190

G

Glehnia, 189, 191, 193-197, 199

H

Hippocrateoideae, 68, 72-73

I

Interspecific Hybrid, 182, 187

L

Lamiaceae, 9, 26, 39, 56, 88-89, 97-100, 105-106, 154, 160-161

Libanotis Seseloides, 191, 193-197

M

M. Aegyptia, 127, 129, 133

Mericarp Surface Characters, 189, 194, 197

Merremieae, 127-129, 133

Micromorphology, 1-2, 7, 10, 27, 39, 49, 53, 55-56, 100-101, 105, 160, 174, 179, 190, 212-213, 215, 221, 223, 225

Monochoria Vaginalis, 135, 143-144

Morphometric Analysis, 81, 133-137, 182, 203

Musa Coccinea, 182, 187

Musa Haekkinenii, 182, 187

N

Numerical Taxonomy, 56, 74, 76, 127, 134, 144

O

Oligadenia, 74, 76, 80, 86

P

Phyllodoce Caerulea, 145, 147, 150, 152

Phyllodoceae, 98, 145-147, 151-153

Plagiopteron, 68-73

Plagiopteron Suaveolens, 68-73

Pollen Dimorphism, 154, 161, 210

Pollen Morphology, 20, 22, 25-26, 40-41, 44-47, 74-75, 87-89, 96-98, 106-107, 112, 114, 126, 145-146, 152-155, 160-161, 171-172, 224

Pollen Tetrads, 145, 147, 149-151

Polyadenia, 74, 76, 80, 86

Polygonoideae, 211-213, 220, 224-225

Polygonum Bistorta, 201, 203, 210

Pontederiaceae, 135, 140, 143-144, 210

Principal Component Analysis, 11, 14, 16-17, 42, 44, 57-58, 63, 85, 108, 127

Q

Q. Aucheri, 11-13, 15-17

Q. Calliprinos, 11-12, 18

Q. Coccifera, 11-13, 15-18

Quercus, 11-12, 17-19, 56

R

Rhaponticoides, 115-116, 123-125

Rheum Rhabarbarum, 212, 220, 222

Rhus Chinensis, 174-177, 179-181

Rumex Maritimus, 212, 216, 220

S

S. Auriculata, 48-50, 52-55

S. Columbaria, 40-43, 45

S. Micrantha, 41-43, 45-46

Scabiosa, 40-41, 44-47

Selineae, 189-190, 192-195, 197-200

Senna Mill., 48-50, 52-54, 56, 134

Style Morph Ratios, 201, 203-204, 208

Sub-palisade Cavities, 135, 140, 143

T

T. Fruticans, 154-155, 157-158

Tamarix Nilotica, 74, 76

Tetrahedral Tetrads, 145, 147

Teucrium, 154-158, 160-161, 172

V

V. Palmyrense, 20-21, 23-25

V. Thapsus, 20-21, 23-25

Verbascum, 20-26, 172